QUANTUM MECHANICS IN CHEMISTRY

Third Edition

PHYSICAL CHEMISTRY TEXTBOOK SERIES

Edited by Walter Kauzmann, Princeton University

MOLECULAR THERMODYNAMICS

RICHARD E. DICKERSON, *California Institute of Technology*

RATES AND MECHANISMS OF CHEMICAL REACTIONS

W. C. GARDINER, JR., *University of Texas*

QUANTUM MECHANICS IN CHEMISTRY, Third Edition

MELVIN W. HANNA, *University of Colorado*

THERMAL PROPERTIES OF MATTER

WALTER KAUZMANN, *Princeton University*

VOLUME I: KINETIC THEORY OF GASES

VOLUME II: THERMODYNAMICS AND STATISTICS: WITH APPLICATIONS TO GASES

QUANTUM MECHANICS IN CHEMISTRY

Third Edition

Melvin W. Hanna

University of Colorado

THE BENJAMIN/CUMMINGS PUBLISHING COMPANY
Menlo Park, California • Reading, Massachusetts
London • Amsterdam • Don Mills, Ontario • Sydney

Library of Congress Cataloging in Publication Data

Hanna, Melvin W
Quantum mechanics in chemistry..

Bibliography: p.
Includes index.
1. Quantum chemistry. I. Title.
QD462.H3 1981 541.2'8 80-23226

ISBN 0-8053-3705-9
HIJK-AL-898

The Benjamin/Cummings Publishing Company, Inc.
2727 Sand Hill Road
Menlo Park, California 94025

Editor's Foreword

THOSE RESPONSIBLE for introducing physical chemistry to students must make difficult decisions in selecting the topics to be presented. Molecular physics and quantum mechanics are part of the subject matter of physical chemistry in the past 35 years and are essential to the chemist's training. Yet many of the more classical areas of physical chemistry continue to be important, not only in chemistry, but also to an increasing extent in biology, geology, metallurgy, engineering, and medicine. Consequently there is pressure on teachers and textbook writers to cover more and more material, but the time available in the curriculum has not increased corrspondingly, and there is a limit to the size of any textbook. Furthermore, it is difficult for any one author to write with authority about all of physical chemistry.

This textbook series is an attempt to make it easier to deal with this problem. The important basic topics of physical chemistry are covered at an introductory level in relatively, brief interrelated volumes. The volumes are written in such a way that if a topic of special interest to a student may not have been included in the course to which he has been exposed, he can learn about it through self-study. Consequently intructors can feel less reluctant to omit or condense material in their courses, and flexibility will be possible in the course plan, both from year to year and from institution to institution, in accordance with changing demands. The introductory presentation of physical chemistry in this form has the additional advantage that it permits a more detailed explanation of difficult points. It also permits the occasional inclusion of more advanced material to which the instructor can refer the more highly motivated students.

WALTER KAUZMANN
Princeton, New Jersey

Series Foreword

PHYSICAL CHEMISTRY has been defined (by a practitioner) as "That part of chemistry which is fundamental, molecular, and interesting." Although this may be a useful guide in selecting a research problem, it is of very little help in planning an introductory course. Such a course should not be put together from a little bit of everything that has ever been called physical chemistry. Instead, it should concentrate on fundamentals so that wherever students turn later, they can build on a secure foundation.

Physical chemistry has the general task of explaining the causes of chemical behavior. The essential, irreducible fundamentals of the subject are four in number:

1. Quantum mechanics: the mechanics of atoms and of their combinations in molecules;
2. Statistical mechanics: the framework by which molecular properties can be related to the macroscopic behavior of chemical substances;
3. Thermodynamics: the study of energy and order–disorder, and their connections with chemical changes and chemical equilibrium;
4. Kinetics: the study of the rates of chemical reactions and of the molecular processes by which reactions occur.

Many additional topics are found in introductory physical chemistry textbooks. These include methods of molecular structure determination, the several branches of spectroscopy, electrochemistry, surface chemistry, macromolecules, photochemistry, nuclear and radiation chemistry, and theories of condensed phases. These are essentially applications of the fundamental concepts, and in our books they are taught as such. The relative emphasis given to these topics will vary with the nature and level of the course, the needs of the students, and the inclination of the instructor. Yet a secure grounding in the four fundamentals will not give chemists everything they need to know in physical chemistry, but the ability to recognize and learn what they need to know as circumstances arise.

These books are an outgrowth of our experience in teaching the basic physical chemistry course over the past several years at the Universities of Illinois, Colorado, and Texas, and the California Institute of Technology. Each of us has written the part of the course that he knows best. Hopefully this approach will avoid the arid style of a book-by-committee, and yet allow each topic to be covered by an author who is vitally concerned with it. Although the present order—quantum mechanics, thermodynamics, kinetics—appears to us to be the most desirable pedagogically and the most obviously unified to the student, the material has been written so thermodynamics can precede quantum mechanics as in the more traditional course.

The authors feel strongly that basic physical chemistry should be presented as a unified whole rather than as a collection of disparate and difficult topics. We hope that our books will serve that end.

MELVIN W. HANNA
RICHARD E. DICKERSON
W. C. GARDINER, JR.

Preface

THIS BOOK was developed in an attempt to introduce undergraduates at the University of Colorado to some aspects of quantum mechanics, spectroscopy, and the electronic structure of atoms and molecules. The author believes that students in other fields of chemistry have long had an advantage over physical chemists in that good students could read research material after a one-year undergraduate course. In physical chemistry, any adequate discussion of quantum phenomena was usually reserved for graduate school, and, as a result, many good students were not able to catch some of the excitement inherent in the areas of physical chemistry dealing with quantum phenomena as undergraduates. In addition, students desiring to do research in the fields of quantum mehcanics and molecular structure had to spend their first year (and sometimes second year) in graduate school, developing the necessary background knowledge. Putting a fairly complete introduction to quantum phenomena at the undergraduate level gives these graduate students a chance to begin research earlier. There is the additional advantage that good students can do an undergraduate research project in these areas in their senior year.

To put any kind of adequate discussion of quantum phenomena into the undergraduate physical chemistry course requires that certain areas of "classical" physical chemistry be left out. The argument about whether such a procedure is justified or not will, no doubt, go on for some time. The author feels, however, that there is one compelling argument for the inclusion of quantum phenomena. Any discussion of quantum mechanics requires an extensive new vocabulary and mode of symbolism. Since physical chemistry is now so broad a field that something must be left out of an undergraduate course, it seems most logical to leave out those subjects that the student with a normal background can study and learn on his own. Because of the new language and symbolism, quantum mechanics, spectroscopy, and electronic structure are *not* fields that fall into this category. The main purpose of this book is to allow instructors to develop quantum ideas in an undergraduate course in conjunction with some selection of topics from classical physical chemistry.

This book has been written for students with a wide variety of mathematical backgrounds. For students who have had only calculus, Chapter 1 provides an introduction to the mathematical fundamentals used throughout the text. (Students with very good mathematical backgrounds may wish to skip Chapter 1 entirely.) Additional mathematics are then introduced as needed in conjunction with specific problems. The author believes that quantum mechanics can most logically be introduced to undergraduates, who usually know very little about the physics of wave motion, by the postulational approach. This approach begins with the classical Hamiltonian and then transforms to the appropriate quantum mechanical operators as is done in the Schrödinger formalism. For this reason, the student is introduced to certain features of classical mechanics in Chapter 2. The main point of this chapter is to teach the student how to write the classical Hamiltonian for any problem of interest. In addition, the student is introduced to the idea of generalized coordinates, conservative and nonconservative systems, and the separation of the motion of the center of mass in a many-particle system in which the potential energy depends only on the internal coordinates of the system.

With this introduction, the postulates of quantum mechanics are presented in Chapter 3 after a discussion of the historical events showing the necessity for a new mechanics. Then the postulates are applied to the specific example of a particle in a one-dimensional box. A discussion follows of vibrational and rotational energy levels and vibration–rotation spectroscopy, atomic structure, molecular structure and spectra, the electronic structure of conjugated systems, and electron and nuclear magnetic resonance spectroscopy. These subjects are developed in a logical sequence.

Although the treatment of each chapter builds on the material that has preceded it, there is material at several different levels in this book, and the individual instructor is free to choose the extent to which he wants to treat many subjects. For example, Chapters 2, 3, and parts of 4, 5, 6, and 7 can be used for a short introduction to quantum chemistry. Conversely, for advanced students, the material in this book can be used as a foundation on which to build a more advanced treatment in the lectures. Thus, each instructor can build a course according to the needs of his particular situation.

This book is not meant to be a comprehensive discussion of all the topics of quantum mechanics, spectroscopy, and electronic structure. It is the author's prejudice that undergraduate students should become used to the idea that all the necessary material for understanding a particular subject will not be found in a single text. For this reason, many subjects that are well covered in other books have not been repeated in this one. Rather, references to many points about which students are likely to want more information are given in the body of the text. Students using this book are encouraged to spend considerable time in the library reading other works. To facilitate this process, a rather substantial annotated bibliography has been given at the end of this edition.

The third edition of this book differs from the second in a number of ways. Chapters 3, 6, and 7 have been extensively revised to put the material in a more logical order to present background material for modern quantum mechanical treatments of atoms and molecules more adequately. Thus, the theorems about Hermitian operator properties have been moved from Chapter 6 to Chapter 3 where they can be dealt with when Hermitian operators are first introduced. The electronic structure of the lithium atom is treated in enough detail in Chapter 6 so that students can be introduced to SCF calculations on atoms, and so that the usual orbital filling diagram, used in freshman chemistry presentations of the periodic table can be critically evaluated. Chapter 7, on molecular electronic structure, has been rewritten to take advantage of the changes in Chapter 6 and to attempt to introduce the student to some of the effects that are used to rationalize molecule formation. Many of these changes were initiated by conversations and correspondence with students and faculty who had used the second edition, and the author is grateful to all who took time to communicate their comments.

Much of the credit for this book is due to Professor William Lipscomb, who first interested the author in quantum chemistry, and to Professors Norman Davidson, Harden McConnell, and J. de Heer, who nurtured that interest. Our knowledge is built on the foundation laid by excellent teachers. I am fortunate to have worked closely with such people.

Melvin W. Hanna
Boulder, Colorado

Contents

Chapter 1

MATHEMATICAL PRELIMINARIES

PHYSICAL CHEMISTRY is a science requiring the application of mathematics and mathematical reasoning to chemical problems. Emphasis on the use of mathematics is especially pronounced in the branch of physical chemistry dealing with quantum mechanics and molecular structure, and the purpose of this chapter is to outline briefly some of the mathematical tools needed by the student to understand the material that follows. Other mathematical procedures are introduced later as needed. The student interested in advanced work in physical chemistry should make his or her mathematical background as extensive as possible, and it is recommended that such students supplement the material in this chapter with additional study [1–4].

1-1 COORDINATE SYSTEMS

The purpose of a coordinate system is to make it convenient to describe a point, a curve, or a surface in space. There are many different kinds of coordinate systems, and four will be used in this text: (1) rectangular or Cartesian coordinates, (2) spherical polar coordinates, (3) cylindrical coordinates, and (4) confocal ellipsoidal coordinates. The choice of the kind of coordinates to use depends on the problem one is trying to solve. The coordinate system is always chosen to make the mathematical equations that describe the problem as simple as possible. Of course, any numerical result which one calculates must be independent of the choice of a coordinate system.

Cartesian coordinates are the most familiar. A point P in "Cartesian space" is represented by distances along three mutually perpendicular axes called X, Y, and Z (Figure 1-1). A rectangular coordinate system should always make use of the "right-hand rule." This rule states that, when the fingers of the *right* hand are curled so that they point from the X to the Y axis, the thumb points along Z.

The other kinds of coordinates are most conveniently expressed in terms of Cartesian coordinates. In spherical polar coordinates (Figure 1-2), a point $P(r, \theta, \phi)$ is represented by one distance r and two angles θ and ϕ. The coordinate r is the length of the line OP drawn from the origin to point P. The angle θ is called the polar angle and is the angle between the Z axis and line OP. The angle ϕ is called the azimuthal angle and is the angle between the X axis and the projection of line OP in the XY plane. It is left for the student to show that the Cartesian coordinates of point P are related to the spherical polar coordinates by the relations

$$\begin{aligned} x &= r \sin\theta \cos\phi \\ y &= r \sin\theta \sin\phi \\ z &= r \cos\theta \end{aligned} \tag{1-1}$$

EXERCISE 1-1 Using Eq. 1-1 show that $(x^2 + y^2 + z^2) = r^2$.

Cylindrical coordinates are shown in Figure 1-3. The location of point P is given by two distances and one angle. The two distances are z and the

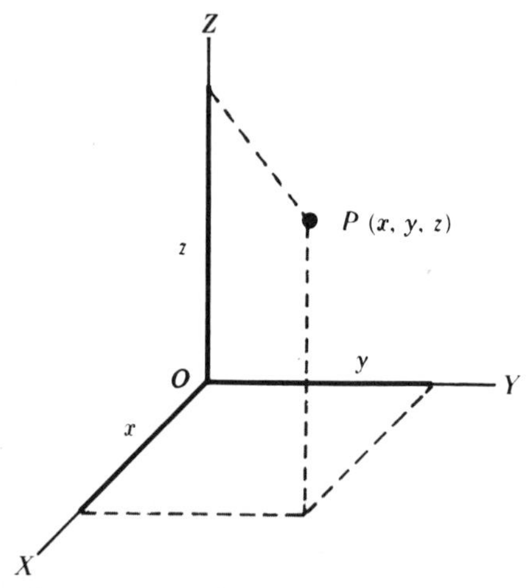

FIGURE 1-1 *Cartesian or rectangular coordinates. A point P(x, y, z) is defined by three distances along three mutually perpendicular axes.*

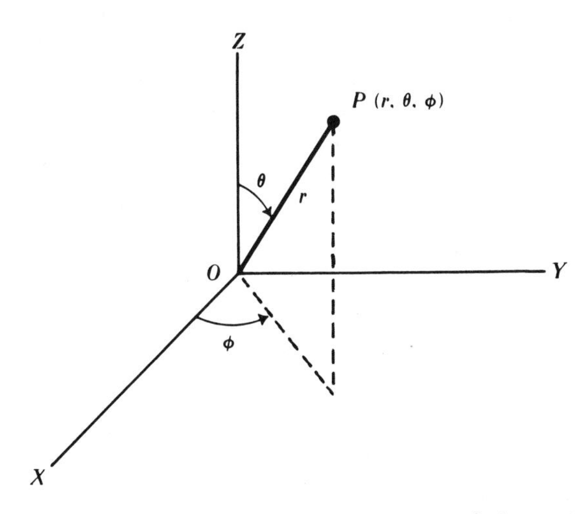

FIGURE 1-2 *Spherical polar coordinates. A point $P(r, \theta, \phi,)$ is defined by two angles and one distance.*

length of the projection of line OP in the XY plane, ρ. The angle ϕ is the same as in spherical polar coordinates. The student may easily verify the relations

$$\begin{aligned} x &= \rho \cos \phi \\ y &= \rho \sin \phi \\ z &= z \end{aligned} \tag{1-2}$$

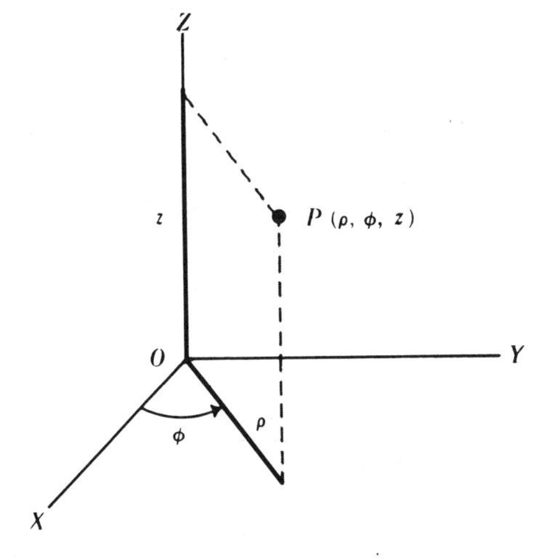

FIGURE 1-3 *Cylindrical coordinates. A point $P(\rho, \phi, z)$ is defined by two distances and one angle.*

Confocal ellipsoidal coordinates, sometimes just called elliptical coordinates, are shown in Figure 1-4. These coordinates are used for problems involving two centers, A and B, a fixed distance R apart. The lines AP and BP define a plane and the line formed by the intersection of this plane with the XY plane defines the angle ϕ. The point $P(\mu, \nu, \phi)$ can be defined by specifying the distances r_A and r_B along lines AP and BP, respectively, and the angle ϕ. Elliptical coordinates μ and ν are then defined as

$$\mu = \frac{r_A + r_B}{R} \qquad \nu = \frac{r_B - r_A}{R} \tag{1-3}$$

The third coordinate is the angle ϕ described above. Keeping μ constant defines an ellipsoid of revolution with the points A and B as foci. Surfaces of constant ν are paraboloids of revolution about the z axis. These surfaces are shown in Figure 1-4, and some of their properties are illustrated in Exercises 1-7 and 1-8. The equations that express x, y, and z in terms of μ, ν, and ϕ are

$$x = \frac{R}{2}(\mu^2 - 1)^{\frac{1}{2}}(1 - \nu^2)^{\frac{1}{2}} \cos \phi$$

$$y = \frac{R}{2}(\mu^2 - 1)^{\frac{1}{2}}(1 - \nu^2)^{\frac{1}{2}} \sin \phi \tag{1-4}$$

$$z = \frac{R}{2}\mu\nu$$

In problems of quantum mechanics, one will often be required to evaluate integrals over all space. To do this, the differential volume element, called $d\tau$, must be known for each kind of coordinate system. These volume elements for the various coordinate systems, and the limits of integration that include all space, are

Cartesian

$$d\tau = dx\, dy\, dz \qquad \begin{array}{l} -\infty \le x \le +\infty \\ -\infty \le y \le +\infty \\ -\infty \le z \le +\infty \end{array}$$

Spherical polar

$$d\tau = r^2 \sin\theta\, dr\, d\theta\, d\phi \qquad \begin{array}{l} 0 \le r \le +\infty \\ 0 \le \theta \le \pi \\ 0 \le \phi \le 2\pi \end{array}$$

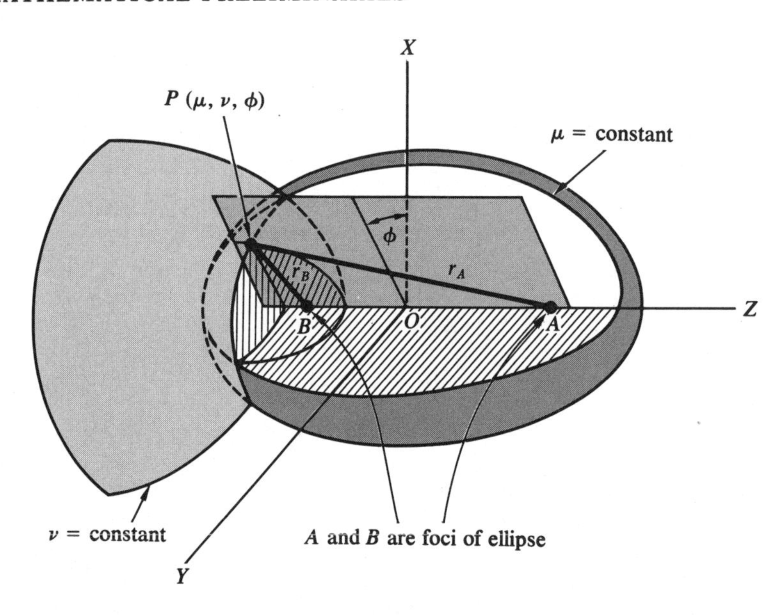

FIGURE 1-4 *Confocal ellipsoidal coordinates. Surfaces of constant μ are ellipsoids of revolution about the Z axis. Surfaces of constant ν are paraboloids of revolution. The intersection of these two surfaces defines a circle. The final coordinate φ specifies a point P(μ, ν, φ) on the circle.*

Cylindrical

$$d\tau = \rho \, d\rho \, d\phi \, dz \qquad \begin{array}{c} 0 \le \rho \le \infty \\ 0 \le \phi \le 2\pi \\ -\infty \le z \le +\infty \end{array}$$

Elliptical

$$d\tau = \frac{R^3}{8}(\mu^2 - \nu^2)\, d\mu \, d\nu \, d\phi \qquad \begin{array}{c} 1 \le \mu \le \infty \\ -1 \le \nu \le +1 \\ 0 \le \phi \le 2\pi \end{array}$$

The interested student may find a discussion of the derivation of these volume elements in [1, 5].

1-2 DETERMINANTS

There are many physical problems of interest that are most conveniently described by writing down an array of mathematical quantities. These arrays are then dealt with according to a set of predetermined rules. Two types of arrays commonly used in quantum chemistry are determinants and matrices. Matrices, although not difficult to handle, are not used in this book even though some problems are more conveniently formulated using matrix notation. It is necessary for us to use determinants, however, and some of their properties will be briefly discussed.

A determinant is an arrangement of N^2 quantities into a square array with N rows and N columns. The number N of rows or columns is called the order of the determinant. Thus the arrays

$$\begin{vmatrix} x - E & B \\ B & 2x - E \end{vmatrix} \tag{1-5a}$$

$$\begin{vmatrix} 8 & 5 & 3 \\ 3 & 5 & 8 \\ 5 & 3 & 8 \end{vmatrix} \tag{1-5b}$$

are determinants, the first of order 2 and the second of order 3. For a determinant, designated by a symbol enclosed between vertical lines, for example $|A|$, each element will have two subscripts. The first subscript defines the row; the second defines the column in which the element appears. Thus the quantity a_{ij} is the element from the ith row and the jth column of $|A|$.

Every determinant has a numerical value although in some cases the value of a determinant may be expressed in terms of unspecified quantities, such as in array 1-5a. The most convenient way to evaluate a determinant is to make use of the method of *signed minors* or *cofactors*. The minor of an element A_{ij} is the $(N - 1)$th order determinant remaining when the row i and column j of the original determinant are struck out. To form the cofactor, the minor is given a sign according to the position of the element A_{ij} in the original determinant. This sign is $(-1)^{i+j}$. A determinant is evaluated by picking either a row or a column, forming the product of each element in the row (or column) with its cofactor, and summing the products.

Example: Evaluate the determinant 1-5b by the method of cofactors.

$$\begin{vmatrix} 8 & 5 & 3 \\ 3 & 5 & 8 \\ 5 & 3 & 8 \end{vmatrix} = 8\begin{vmatrix} 5 & 8 \\ 3 & 8 \end{vmatrix} - 5\begin{vmatrix} 3 & 8 \\ 5 & 8 \end{vmatrix} + 3\begin{vmatrix} 3 & 5 \\ 5 & 3 \end{vmatrix}$$

$$= 8(40 - 24) - 5(24 - 40) + 3(9 - 25)$$

$$= 128 + 80 - 48 = 160$$

Two useful properties of determinants that will be used in this text are the following.

1. The value of a determinant changes sign when two rows or two columns are interchanged.

2. If two rows are identical, or if two columns are identical, the determinant is zero.

For proof of these properties and other characteristics of determinants, see [1–4].

EXERCISE 1-2 Evaluate the determinant

$$\begin{vmatrix} 4 & 1 & 2 & 3 \\ 1 & 2 & 3 & 4 \\ 2 & 3 & 4 & 1 \\ 3 & 4 & 1 & 2 \end{vmatrix}$$

by the method of cofactors.

1-3 *SUMMATION AND PRODUCT NOTATION*

In physical chemistry one often encounters equations of the form

$$y = a_1 + a_2 + \cdots a_i + \cdots a_n \tag{1-6a}$$

$$z = a_1 a_2 \cdots a_i \cdots a_m \tag{1-6b}$$

In order to simplify notation, sums of the type shown in Eq. 1-6a are designated by a capital sigma ($\sum$) with the limits of summation shown by index numbers placed below and above the $\sum$. A similar procedure is followed for products except that a capital pi ($\prod$) replaces the sigma. Using this notation, Eqs. 1-6 become

$$y = \sum_{i=1}^{n} a_i \tag{1-7a}$$

$$z = \prod_{i=1}^{m} a_i \tag{1-7b}$$

EXERCISE 1-3 Let the a_i be the series of even integers beginning with $a_1 = 2$. Evaluate

$$y = \sum_{i=1}^{4} a_i \qquad z = \prod_{i=1}^{4} a_i$$

1-4 *VECTORS*

Most numerical and algebraic quantities that the student is familiar with have been scalar quantities. These are quantities such as 106 and $x^2 + 3x + 2$, which have magnitude only. A vector is used to represent a physical quantity which has *both* magnitude and direction. Quantities such as force, an electric field, or an acceleration are all vector quantities. A vector will be represented in this book by **boldface type.** It can also be represented by a symbol with an arrow above or below it, that is, $\vec{r}$ or $\underset{\rightarrow}{E}$. The length of a vector is called its magnitude and is a scalar quantity. A vector that has a length of one unit is called a unit vector. A vector often used is the radius vector **r.** In Figure 1-2, **r** is the vector whose length or magnitude is r and whose direction is from the origin to the point P.

It is usually most convenient to work with vectors in terms of thier components. To do this, three mutually perpendicular unit vectors called **i, j,** and **k** are defined that point along the X, Y, and Z axes, respectively. Any vector can then be written in terms of its components (projections) along these three axes. The radius vector becomes simply

$$\mathbf{r} = x\mathbf{i} + y\mathbf{j} + z\mathbf{k} \tag{1-8}$$

Just as multiplication and division of numbers follow certain rules, so there are rules which define the combination of vectors. These rules are summarized as follows.

1. *Addition.* Addition of vectors can either be done graphically or analytically. Consider the vector sum

$$\mathbf{A} + \mathbf{B} = \mathbf{C} \tag{1-9}$$

In the graphical method, the tail of **B** is placed at the head of **A**. The sum **C** is the vector which starts at the tail of **A** and ends at the head of **B** (Figure 1-5a). The magnitude and direction of a vector are independent of the coordinate system used to describe the vector and, therefore, the origin of a vector can be taken at any point in space.

If the vectors **A** and **B** can be written in terms of their components, then the addition can be done analytically. Thus if

$$\mathbf{A} = A_x\mathbf{i} + A_y\mathbf{j} + A_z\mathbf{k} \tag{1-10a}$$

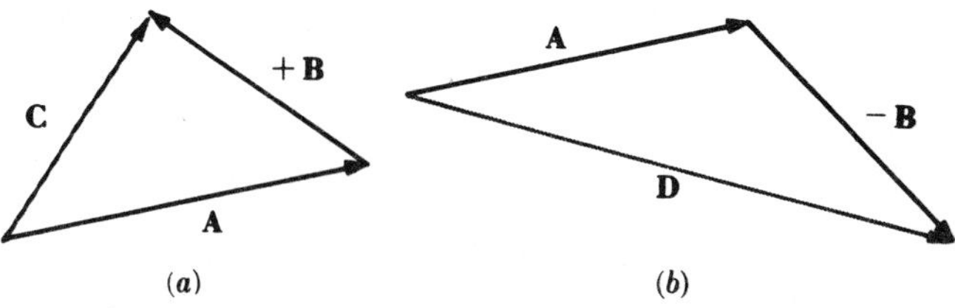

FIGURE 1-5 *Illustration of the graphical addition* (*a*) *and subtraction* (*b*) *of two vectors* **A** *and* **B**.

and

$$\mathbf{B} = B_x\mathbf{i} + B_y\mathbf{j} + B_z\mathbf{k} \tag{1-10b}$$

then

$$\mathbf{C} = (A_x + B_x)\mathbf{i} + (A_y + B_y)\mathbf{j} + (A_z + B_z)\mathbf{k} \tag{1-10c}$$

2. *Subtraction.* Vectors are subtracted by adding the negative of the appropriate vector. Thus $\mathbf{A} - \mathbf{B} = \mathbf{D}$ is illustrated in Figure 1-5b. Analytically,

$$\mathbf{D} = (A_x - B_x)\mathbf{i} + (A_y - B_y)\mathbf{j} + (A_z - B_z)\mathbf{k} \tag{1-11}$$

3. *Magnitude.* Frequently one needs to express the magnitude of a vector in terms of its components. By elementary trigonometry, the student may verify that the length of the radius vector **r** in Figure 1-3 is (see Exercise 1-1)

$$r = (x^2 + y^2 + z^2)^{\frac{1}{2}} \tag{1-12}$$

Similarly, the magnitude of any vector $\mathbf{A} = A_x\mathbf{i} + A_y\mathbf{j} + A_z\mathbf{k}$ is $|A| = (A_x^2 + A_y^2 + A_z^2)^{\frac{1}{2}}$.

4. *Multiplication.* Two different kinds of vector multiplication have been defined. The first kind, symbolized $\mathbf{A} \cdot \mathbf{B}$, is called the dot or scalar product and results in a scalar. The second kind, symbolized $\mathbf{A} \times \mathbf{B}$, is called the vector or cross product and results in a vector.

The scalar product $\mathbf{A} \cdot \mathbf{B}$ is defined as

$$\mathbf{A} \cdot \mathbf{B} \equiv AB \cos\theta \tag{1-13}$$

where A and B are the magnitudes of **A** and **B**, respectively, and where θ is the angle between **A** and **B**. The angle θ is always taken to be less than 180°. The student should note that there is a simple geometrical interpretation to the dot product. It is equal to the length **A** times the length of the projection of **B** on **A** *or vice versa.* If two vectors are perpendicular,

$$\cos\theta = \cos 90° = 0 \tag{1-14}$$

and $\mathbf{A} \cdot \mathbf{B} = 0$. Conversely, it is also true that if $\mathbf{A} \cdot \mathbf{B} = 0$, then **A** and **B** are perpendicular. When this is the case, the two vectors are said to be *orthogonal.* For two vectors written in terms of their components, we can show that

$$\mathbf{A} \cdot \mathbf{B} = A_x B_x + A_y B_y + A_z B_z \tag{1-15}$$

EXERCISE 1-4 Prove Eq. 1-15 making use of the definitions of $\mathbf{A} \cdot \mathbf{B}$ and of the unit vectors **i**, **j**, and **k**.

The cross product $\mathbf{A} \times \mathbf{B}$ is defined as

$$\mathbf{A} \times \mathbf{B} = \mathbf{n}AB \sin\theta \tag{1-16}$$

where A, B, and θ have the same meaning as above. The vector $\mathbf{n}$ is a unit vector perpendicular to *both* $\mathbf{A}$ and $\mathbf{B}$. If $\mathbf{A}$ and $\mathbf{B}$ are parallel, $\sin\theta = 0$ and $\mathbf{A} \times \mathbf{B} = 0$. Conversely, if $\mathbf{A} \times \mathbf{B} = 0$, the two vectors are parallel.

The "right-hand rule" must be used in evaluating the cross product. To use this rule, one places the bottom edge of the right palm along **A** and curls the fingers toward **B**. The thumb will then point in the direction of **n** as shown in Figure 1-6. As a result of this rule, it should be clear that

$$\mathbf{A} \times \mathbf{B} \neq \mathbf{B} \times \mathbf{A} \tag{1-17}$$

In fact,

$$\mathbf{A} \times \mathbf{B} = -(\mathbf{B} \times \mathbf{A})$$

When an equation such as Eq. 1-17 is true, it is said that the vectors **A** and **B** do not *commute*, or that they are not commutative. The multiplication of scalar quantities is always commutative; that is, the result does not depend

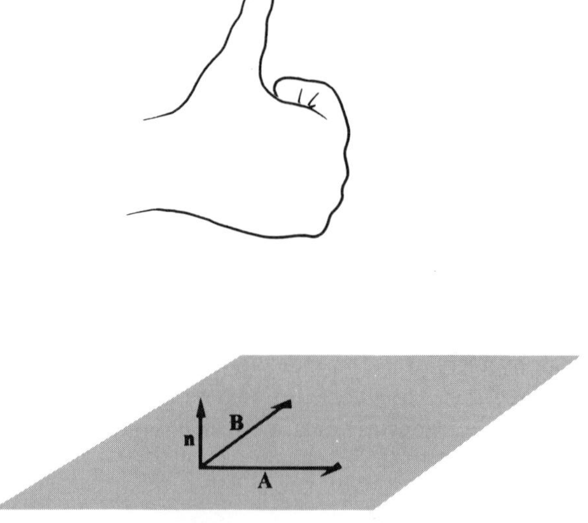

FIGURE 1-6 *The use of the right-hand rule to determine the direction of the vector* **A × B**. *In using the right-hand rule, the angle between* **A** *and* **B** *is always chosen to be less than* 180°.

on the order in which the multiplication is carried out. The commutative property does not hold, in general, for vector or matrix multiplication.

The geometrical interpretation of the cross product $\mathbf{A} \times \mathbf{B}$ is that of a vector perpendicular to both $\mathbf{A}$ and $\mathbf{B}$ whose length is equal to the area of the parallelogram defined by $\mathbf{A}$ and $\mathbf{B}$. A consideration of Figure 1-7 may help clarify this point. In terms of components, $\mathbf{A} \times \mathbf{B}$ is most conveniently written in the form of a determinant. Thus

$$\mathbf{A} \times \mathbf{B} = \begin{vmatrix} \mathbf{i} & \mathbf{j} & \mathbf{k} \\ A_x & A_y & A_z \\ B_x & B_y & B_z \end{vmatrix} \tag{1-18}$$

$$= \mathbf{i}(A_y B_z - A_z B_y) - \mathbf{j}(A_x B_z - B_x A_z) + \mathbf{k}(A_x B_y - B_x A_y)$$

For proof of Eq. 1-15, the reader is referred to [1–4] or to any standard work on vector analysis.

EXERCISE 1-5 Let $\mathbf{A} = 4\mathbf{i} + \mathbf{j} + 3\mathbf{k}$, $\mathbf{B} = \mathbf{i} - 3\mathbf{j} - \mathbf{k}$. Evaluate $\mathbf{A} + \mathbf{B}$, $\mathbf{A} - \mathbf{B}$, $\mathbf{A} \cdot \mathbf{B}$, and $\mathbf{A} \times \mathbf{B}$.

5. *Division* of vectors is not defined.
6. *Differentiation of vectors.* A vector is differentiated simply by differentiating its components. Thus

$$\mathbf{r} = x\mathbf{i} + y\mathbf{j} + z\mathbf{k}$$

$$\frac{d\mathbf{r}}{dt} = \frac{dx}{dt}\mathbf{i} + \frac{dy}{dt}\mathbf{j} + \frac{dz}{dt}\mathbf{k} \tag{1-19}$$

$$= v_x \mathbf{i} + v_y \mathbf{j} + v_z \mathbf{k}$$

$$= \mathbf{v}$$

where $\mathbf{v}$ is the velocity vector.

7. *Vector equations.* It should be noted that a vector equation is actually a summary of three scalar equations because, for two vectors to be equal, the

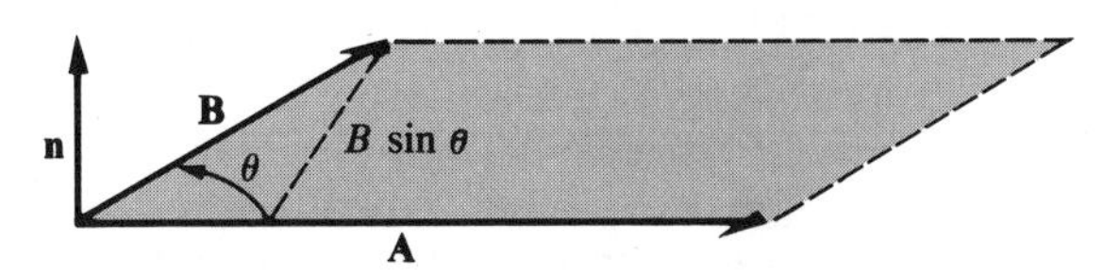

FIGURE 1-7 *The geometrical significance of* $\mathbf{A} \times \mathbf{B}$. *The quantity AB sin θ is the area of the parallelogram. The vector* **n** *is one unit in length.*

appropriate components on both sides of the equal sign must be equal. This point is illustrated in Exercise 1-6.

EXERCISE 1-6 In physics, the angular momentum $\mathbf{L}$ about a point is defined as

$$\mathbf{L} = \mathbf{r} \times \mathbf{p}$$

when $\mathbf{p}$ is the linear momentum at a point $\mathbf{r}$. The origin of the vector $\mathbf{r}$ is taken at the point about which the angular momentum is to be calculated. Write the equation for each component of angular momentum L_x, L_y, and L_z in terms of x, y, and z and the components of linear momentum p_x, p_y, and p_z.

EXERCISE 1-7 Referring to Figure 1-4 and using the additive property of vectors, derive an expression for r_A and r_B in terms of x, y, and z and R, the distance between the two foci of the ellipse.

EXERCISE 1-8 Using the results of Exercise 1-7, set μ equal to a constant in Eq. 1-3 and plot the intersection of the resulting surface with the XZ plane. This can be done by setting $y = 0$ and plotting z for different values of x. This procedure should result in an ellipse, as pointed out in Section 1-1. Repeat this procedure setting $\nu = \text{const.}$

1-5 COMPLEX NUMBERS

A complex number is one which contains $\sqrt{-1}$, or i as it is usually symbolized. Thus $A + iB$ is a complex number. We speak of the real (A) and the imaginary (B) part of a complex number. If

$$C = A + iB \tag{1-20}$$

then the complex conjugate of C, called C^*, is defined by replacing i wherever it appears by $-i$. Thus

$$C^* = A - iB \tag{1-21}$$

The magnitude or absolute value of a complex number is defined as

$$|C| \equiv (CC^*)^{\frac{1}{2}} = (A^2 + B^2)^{\frac{1}{2}} \tag{1-22}$$

Note that the magnitude of a complex number is always real. Two complex numbers are equal only if *both* their real and imaginary parts are equal. Addition and subtraction follow the same rules as for vectors. That is, the real and imaginary parts are added independently. Thus if $z_1 = x_1 + iy_1$ and $z_2 = x_2 + iy_2$, then $z_1 + z_2 = (x_1 + x_2) + i(y_1 + y_2)$.

An equation which will often be used in dealing with complex numbers is Euler's formula

$$e^{i\alpha} = \cos \alpha + i \sin \alpha \tag{1-23}$$

Euler's formula can be derived by expanding each of the quantities $e^{i\alpha}$, $\cos \alpha$, and $\sin \alpha$ in a Maclaurin series. By equating the power series expansions, it can easily be shown that Eq. 1-23 leads to an identity.

EXERCISE 1-9 Express the following quantities in the form $A + iB$:

(a) $(1 + i)^3$ (b) $(x + iy)(u + iv)$ (c) $e^{\pi i/2}$

What is the complex conjugate and magnitude of part (b)?

1-6 OPERATORS

In a study of quantum mechanics, it is necessary to utilize mathematical *operators*. An operator is nothing more than a symbol that tells one to do something to what follows the symbol. Thus in the expression $\sqrt{2}$, the $\sqrt{\ }$ is an operator telling one to take the square root of what follows, in this case 2. Likewise, in the expression

$$\frac{d}{dx}(x^2 + 5x + 1) \tag{1-24}$$

d/dx is an operator telling one to take the derivative with respect to x of what follows, that is, $x^2 + 5x + 1$. General operators will be indicated by a symbol with a caret over it, that is, $\hat{P}$ or $\hat{Q}$.

The algebra of operators follows definite mathematical procedures which the student is familiar with, although perhaps not consciously. Thus if

$$\hat{P} = \left(\frac{\partial}{\partial x}\right)_{yz} \qquad \hat{Q} = \left(\frac{\partial}{\partial y}\right)_{xz}$$

then (1-25)

$$\hat{P}\hat{Q} = \left|\frac{\partial}{\partial x}\left(\frac{\partial}{\partial y}\right)_{xz}\right|_{yz} = \frac{\partial^2}{\partial x\, \partial y}$$

When dealing with operators one must be careful of the order of operations because operations like vector multiplication are not necessarily commutative. By convention, one always begins with the operator on the right and works toward the left. For the particular operators given in Eq. 1-25, it turns out that $\hat{P}$ and $\hat{Q}$ do commute. That is,

$$\hat{P}\hat{Q} = \hat{Q}\hat{P} \quad \text{since} \quad \frac{\partial^2}{\partial x\, \partial y} = \frac{\partial^2}{\partial y\, \partial x} \tag{1-26}$$

but this, in general, will not be the case (see Exercises 1-10 and 1-11).

EXERCISE 1-10 Consider the function $f(x, y) = x^2 + y^2 + 2xy$. Let $\hat{P}$ and $\hat{Q}$ be the operators given in Eq. 1-25 and operate first on $f(x, y)$ with $\hat{P}\hat{Q}$, then with $\hat{Q}\hat{P}$. Note that the result is the same. What would be the result of operating on $f(x, y)$ with $\hat{P}\hat{Q} - \hat{Q}\hat{P}$?

EXERCISE 1-11 Let $\hat{P} = d/dx$, $\hat{O} = x$ (multiply by x), and $f(x) = x^2 + 2x + 1$. Show that $\hat{O}\hat{P}f(x) \neq \hat{P}\hat{O}f(x)$. For these two operators, can you derive a general expression for $\hat{P}\hat{O} - \hat{O}\hat{P}$?

The quantity $\hat{P}\hat{Q} - \hat{Q}\hat{P}$ is called the commutator of $\hat{P}$ and $\hat{Q}$ and is often symbolized $[\hat{P}, \hat{Q}]$. If $\hat{P}$ and $\hat{Q}$ commute, then the value of the commutator is zero. Conversely, if the value of the commutator is zero, the operators $\hat{P}$ and $\hat{Q}$ commute.

An operator can be a vector or a complex quantity. If an operator is a vector, one usually works with it in terms of its components. An example of a vector operator is "del."

$$\nabla = \frac{\partial}{\partial x}\mathbf{i} + \frac{\partial}{\partial y}\mathbf{j} + \frac{\partial}{\partial z}\mathbf{k} \tag{1-27}$$

The quantity ∇f, where f is some scalar function, is called the gradient of f. For example, suppose $f = x^2 + y^2 + z^2$; then the gradient of f is the vector

$$\nabla f = 2x\mathbf{i} + 2y\mathbf{j} + 2z\mathbf{k} \tag{1-28}$$

The gradient will be used in Chapter 2 in the discussion of classical mechanics. Since for a scalar function f, the quantities $\partial f/\partial x$, $\partial f/\partial y$, and $\partial f/\partial z$ are the rates of change of f with respect to distance in the x, y, and z directions, the gradient of f provides a means of evaluating the rate of change of f with distance in any direction.

If an operator $\hat{P}$ is complex, the complex conjugate $\hat{P}^*$ is formed by replacing i by $-i$ wherever it occurs. Thus, if $\hat{P} = i\, d/dx$, $\hat{P}^* = -i\, d/dx$.

In quantum mechanics, only linear operators are used. An operator is linear if it is true that

$$\hat{P}(f + g) = \hat{P}f + \hat{P}g$$

and

$$\hat{P}af = a\hat{P}f \tag{1-29}$$

where a is a constant. The student may easily verify that d/dx is a linear operator whereas $\sqrt{\ }$ is not.

EXERCISE 1-12 It is true in algebra that

$$(P + Q)(P - Q) = P^2 - Q^2$$

What is the value of $(P + Q)(P - Q)$ if P and Q are operators? Under what conditions will the first relation be true for operators?

An operator that will be used frequently in quantum mechanics is $\nabla^2 \equiv \nabla \cdot \nabla$. It will be shown in Chapter 3 that this operator is related to the kinetic energy. In Cartesian coordinates it can be seen using Eq. 1-15 and 1-27 that

$$\nabla^2 = \frac{\partial^2}{\partial x^2} + \frac{\partial^2}{\partial y^2} + \frac{\partial^2}{\partial z^2} \tag{1-30}$$

For many quantum chemical problems, spherical polar coordinates will be more appropriate than Cartesian coordinates and it will be necessary to express ∇^2 as a function of r, θ, and ϕ. This relation is given by

$$\nabla^2 = \frac{1}{r^2}\frac{\partial}{\partial r}\left(r^2\frac{\partial}{\partial r}\right) + \frac{1}{r^2 \sin\theta}\frac{\partial}{\partial\theta}\left(\sin\theta\frac{\partial}{\partial\theta}\right) + \frac{1}{r^2\sin^2\theta}\frac{\partial^2}{\partial\phi^2} \tag{1-31}$$

The derivation of the expression for ∇^2 in a general system of orthogonal coordinates as well as the derivation of Eq. 1-31 is given in [1].

1-7 EIGENVALUE EQUATIONS

An equation of the type

$$\hat{P}(q_i)G(q_i) = pG(q_i) \tag{1-32}$$

where $\hat{P}(q_i)$ is an operator, $G(q_i)$ is a function, both involving variable q_i, and p is a constant, is called an *eigenvalue equation*. When such an equation holds, $G(q_i)$ is called an eigenfunction of the operator $\hat{P}(q_i)$ and p is called the eigenvalue.

Eigenvalue equations play a major role in the mathematical formalism of quantum mechanics. In quantum mechanics, $\hat{P}$ is usually a differential operator and, therefore, the eigenvalue equation is a differential equation. The principal mathematical problem of quantum mechanics is to find the solution G and the eigenvalues p to these eigenvalue equations. The student must keep in mind that the mathematics of these equations were known long before quantum mechanics was developed. The mathematical properties of these equations should not be confused with the physical interpretation to be placed upon them in the discussion of quantum mechanics in Chapter 3.

To give an example of an eigenvalue equation, suppose that the operator of interest is $\hat{P} = d^2/dx^2$. We seek to find functions $G(x)$ such that when $\hat{P}$ operates on $G(x)$, $G(x)$ again results, multiplied by a constant. Looking at the specific form of the operator, it can be seen that what is required is a function which when differentiated twice results in the original function. There are a number of functions which have this property, for example, $\sin ax$, $\cos ax$, $e^{\pm ax}$. Let us choose $G(x) = \sin ax$ and evaluate

$$\hat{P}(x)G(x) = \frac{d^2}{dx} G(x) = \frac{d}{dx}\left(\frac{d}{dx} \sin ax\right) = \frac{d}{dx}(a \cos ax)$$

$$= -a^2 \sin ax \tag{1-33}$$

Thus if a is a constant, the function $G(x) = \sin ax$ is an eigenfunction of the operator d/dx^2 and the eigenvalue is $-a^2$.

EXERCISE 1-13 Show that the function Ae^{-ax} is an eigenfunction of the operator d^2/dx^2. What is the eigenvalue?

EXERCISE 1-14 Show that the function $\cos ax \cos by \cos cz$ is an eigenfunction of the operator ∇^2. What is the eigenvalue?

EXERCISE 1-15 Under what conditions is the function e^{-aq^2} an eigenfunction of the operator

$$\frac{d^2}{dq^2} - kq^2$$

where k is a constant. What is the eigenvalue under these conditions? *Hint*: the quantity a must be adjusted to eliminate terms involving q in the result.

1-8 *SOME COMMENTS ABOUT FUNCTIONS AND INTEGRALS*

The idea of a function is fundamental for algebra, although most of us would not be able to define a function precisely. (See [3] and [6] if you are interested in the definition.) The equations

$$y = x^3 - 2x + 5 \tag{1-34a}$$

$$z = 3 \sin x + 4i \cos x \tag{1-34b}$$

are both examples of functions; Eq. 1-34a is a real function; Eq. 1-34b is a complex function or, more precisely, a function of a complex variable.

The main property of functions that we wish to emphasize at this point has to do with the symmetry properties of certain simple functions. A function $f(x)$ is said to be an *even* function if

$$f(x) = f(-x) \tag{1-35}$$

A function is an *odd* function if

$$f(x) = -f(-x) \tag{1-36}$$

Thus, the function $y = x$ is an odd function because replacing x by $-x$ gives us a relation satisfying Eq. 1-36. The function $y = x^2$ is an even function because replacing x by $-x$ gives a result that satisfies Eq. 1-35.

EXERCISE 1-16 State whether each of the following functions of x are even or odd: x^3, x^4, $\sin x$, $\cos x$, $x \sin x$, $x \cos x$.

The last two parts of Exercise 1-16 are products of simple functions, and there are rules for determining the symmetry of a product function if one knows the symmetries of the simple functions. These rules are:

$$\begin{array}{ll} \text{even} \times \text{even} = \text{even} & \text{odd} \times \text{even} = \text{odd} \\ \text{even} \times \text{odd} = \text{odd} & \text{odd} \times \text{odd} = \text{even} \end{array}$$

One can see that these rules are identical to the prescriptions used to determine the sign of the answer when one multiplies $+1$ and -1. A $+1$ corresponds to an even function, and a -1 corresponds to an odd function. Similar rules hold for division.

EXERCISE 1-17 Give the symmetry of the following functions: $\tan x$, $\cos^2 x$, $\cos x \sin x$, $f(x) \sin x$ when $f(x)$ is given.

Some functions are unsymmetric and, therefore, a symmetry cannot be assigned to them. Sometimes they can be broken down into odd and even components, however. The function e^{iQ} has no symmetry, but since $e^{iQ} = \cos Q + i \sin Q$, one can see that the real part of e^{iQ} is even and the imaginary part is odd.

The principle reason for discussing symmetry properties of functions is that they often enable us to determine the value of an integral without specifically integrating the function. Consider the integral

$$y = \int_{-L}^{L} \sin x \, dx \tag{1-37}$$

We could integrate Eq. 1-37 to obtain

$$y = -\cos x \Big|_{-L}^{L} = -[\cos L - \cos(-L)] = 0$$

If we consider the geometric interpretation of an integral as an area under a curve, we can see by dividing Eq. 1-37 into two parts

$$y = \int_{-L}^{0} \sin x \, dx + \int_{0}^{L} \sin x \, dx \tag{1-38}$$

that since $\sin x$ is an *odd* function of x, the area indicated by the first integral in Eq. 1-38 will be negative and exactly compensated by the positive area indicated by the second integral. Thus, *all integrals between symmetric limits of an odd function must vanish by symmetry.*

EXERCISE 1-18 Plot the function $y = \sin x$ between $x = -180°$ ($-\pi$ radians) and $x = +180°$ ($+\pi$ radians). Label the areas corresponding to the two integrals in Eq. 1-38 when $L = 180°$. Note that if the areas were added, the area below the x axis (negative area) would exactly cancel the area above the x axis (positive area).

EXERCISE 1-19 Using the symmetry properties of functions and their effect on integrals, prove the following:

$$\int_{-\infty}^{\infty} f(t)e^{i\omega t} \, dt = 2\int_{0}^{\infty} f(t) \cos \omega t \, dt$$

if $f(t)$ is an even function of t. The integral on the left is called the Fourier transform of $f(t)$. The integral on the right is sometimes called the Fourier cosine transform.

1-9 SUMMARY

1. Some of the properties of coordinate systems, determinants, vectors, and complex numbers were reviewed.
2. Mathematical operators were introduced and some of their properties were discussed.
3. Several terms were introduced which will be important throughout the book. The student should be especially familiar with the meaning of the following terms: orthogonal, commute and commutator, complex conjugate, eigenvalue, eigenfunction, and even and odd functions.

REFERENCES

1. H. Margenau and G. M. Murphy, *The Mathematics of Physics and Chemistry* (D. Van Nostrand, Inc., Princeton, N.J., 1943).
2. J. M. Anderson, *Mathematics for Quantum Chemistry* (W. A. Benjamin, Inc., New York, 1966).
3. T. A. Bak and J. Lichtenberg, *Mathematics for Scientists* (W. A. Benjamin, Inc., New York, 1966).
4. C. A. Hollingsworth, *Vectors, Matrices and Group Theory for Scientists and Engineers* (McGraw-Hill Book Co., New York, 1967).
5. H. Eyring, J. Walter, and G. E. Kimball, *Quantum Chemistry* (J. Wiley & Sons, Inc., New York, 1949).

Chapter 2

CLASSICAL MECHANICS

BEFORE 1900, the science of mechanics, which dealt with the motions of bodies and with the forces that affected these motions, provided a powerful example of the ability of a mathematical, scientific theory to predict, correlate, and interpret observations on the nature of the physical world. For systems where the particles involved are considerably larger than atoms or molecules, this science still retains its power and usefulness. Mechanics, or classical mechanics as it is now called, was first based upon Newton's laws of motion. More general and powerful formulations were developed later by Lagrange and by Hamilton.

Since quantum mechanics developed out of classical mechanics, some of the ideas and nomenclature of the earlier science were retained. In fact one of the guideposts in the development of quantum mechanics was that in the limit of classical sized systems the quantum mechanical result had to go over into the classical mechanical one. This idea is referred to now as the "correspondence principle." It is for this reason that the student approaching quantum mechanics for the first time needs to have some knowledge about classical mechanics. This chapter provides a brief introduction to some of the necessary classical mechanical concepts. Students interested in a more complete study should see [1]. Good short introductions are provided in [2–4].

The fundamental problem of classical mechanics is to describe the motion of systems of particles under various kinds of forces and initial conditions. More practically, the problem is to solve the differential equations resulting from Newton's second law

$$\mathbf{F}_i = m\mathbf{a}_i \tag{2-1}$$

where $\mathbf{F}_i$ is the force acting on the ith particle in the system, and $\mathbf{a}_i$ is its acceleration.

2-1 CONSERVATIVE SYSTEMS

Before going into more detail concerning the solutions to Eq. 2-1, it is necessary to distinguish between two types of systems—conservative and nonconservative. There are a number of ways to define a conservative system. The word "conservative" is related to "conservation theorems." Conservation theorems state that, under certain conditions, a given quantity does not vary with time. A conservative system is any system of particles and forces whose total energy—the sum of its kinetic and potential energies—does not vary with time. Implicit in this definition is the idea that the forces must be such that the work done around a closed path must be zero. In mathematical language, it must be true that

$$\oint \mathbf{F}\cdot d\mathbf{s} = 0 \tag{2-2a}$$

When this relation holds, the forces are said to be conservative. Equation 2-2a means that there cannot be any friction or other dissipative forces in a conservative system.

A mathematical consequence of Eq. 2-2a allows us to write an equivalent definition of a conservative system. In this definition, both the internal and external forces can be represented as the negative gradient (see Section 1-4) of some potential function V. That is,

$$\mathbf{F}_i = -\nabla_i V \tag{2-2b}$$

To illustrate the equivalence of these two definitions, consider the case of a single particle constrained to move in one dimension, say, the x direction. Newton's second law for this case is

$$F_x = m\,\frac{d^2x}{dt^2} \tag{2-3}$$

and, if Eq. 2-2b holds,

$$F_x = -\frac{dV(x)}{dx} \tag{2-4}$$

Substituting Eq. 2-4 into 2-3 gives

$$-\frac{dV(x)}{dx} = m\ddot{x} = m\frac{d\dot{x}}{dt} \tag{2-5}$$

which, upon integrating over x, yields

$$-\int \frac{dV(x)}{dx}\,dx = -\int dV = m\int \frac{d\dot{x}}{dt}\,dx = m\int d\dot{x}\,\frac{dx}{dt} = m\int \dot{x}\,d\dot{x}$$

$$-V(x) + C = \tfrac{1}{2}m\dot{x}^2 \tag{2-6}$$

$$\tfrac{1}{2}m\dot{x}^2 + V(x) = C$$

where C is an arbitrary constant of integration. Thus if Eq. 2-2b is assumed, the sum of the potential and kinetic energies ($\frac{1}{2}m\dot{x}^2$) of the particle is independent of the time t. This is only an illustration of the equivalence of the two definitions of a conservative system. A general proof of the equivalence follows from Eq. 2-2a and Stokes's theorem. See [1] for details.

Any property of a mechanical system independent of time is called a constant of motion of the system. In this particular case, the constant of motion is the total energy of the particle, E. In what follows, the symbol T will be used for kinetic energy. Equation 2-6 then becomes $C = T + V$, and it becomes clear that the constant of integration in Eq. 2-6 is the total energy of the system.

2-2 *AN EXAMPLE OF NEWTONIAN MECHANICS*

The motion of a particle in which there is a restoring force proportional to the displacement of the particle from some point is called simple harmonic motion. This situation is applicable to the case of the stretching of an ideal spring (one which obeys Hooke's law) and, as will be seen in Chapter 5, is used as the lowest level of approximation in treating the vibration of diatomic molecules.

For motion in one dimension, say, the x direction, the force is

$$F_x = -kx \tag{2-7}$$

where k is usually referred to as the force constant and Newton's second law of motion becomes

$$-kx(t) = m\frac{d^2x(t)}{dt^2} \tag{2-8}$$

The classical mechanical problem is to find x as a function of t.

Rearranging Eq. 2-8 slightly, we obtain

$$\ddot{x} \equiv \frac{d^2x(t)}{dt^2} = -\frac{k}{m}x(t) \tag{2-9}$$

It is seen from Eq. 2-9 that what is required is a function $x(t)$ which when differentiated twice gives the same function back again multiplied by a constant. This is the same situation that was encountered in Section 1-7, and the solutions discussed there are also appropriate here. Remembering that general solutions to a second order differential equation have two undetermined constants, we try to find a solution to Eq. 2-9 of the form

$$x(t) = A \sin \alpha t \tag{2-10}$$

where A and α are the undetermined constants. Differentiating twice gives

$$\ddot{x}(t) = -\alpha^2 A \sin \alpha t \tag{2-11}$$

and we see that Eq. 2-10 is a solution to Eq. 2-8 if we make the identification

$$\alpha = \left(\frac{k}{m}\right)^{\frac{1}{2}} \tag{2-12}$$

The solution to the problem of one-dimensional simple harmonic motion then becomes

$$x(t) = A \sin \left(\frac{k}{m}\right)^{\frac{1}{2}} t \tag{2-13}$$

Since the sine function oscillates between -1 and $+1$, it can be seen that the constant A represents the maximum amplitude of displacement in the x direction.

This problem again illustrates the equivalence of the two definitions of a conservative system discussed in the previous section. Equation 2-2 holds, and therefore

$$-\frac{dV(x)}{dx} = -kx$$

$$\int dV(x) = \int kx\, dx \tag{2-14}$$

$$V(x) = \tfrac{1}{2}kx^2 + C$$

By choosing the initial conditions such that at $x = 0$, $V = 0$ we can set $C = 0$. Using Eq. 2-13 we see that the potential energy as a function of time is

$$V(t) = \frac{1}{2} kA^2 \sin^2\left(\frac{k}{m}\right)^{\frac{1}{2}} t \tag{2-15}$$

The kinetic energy of the particle is

$$\begin{aligned} T &= \frac{1}{2} mv^2 = \frac{1}{2} m\left(\frac{dx}{dt}\right)^2 = \frac{1}{2} m\left(\frac{k}{m}\right)A^2 \cos^2\left(\frac{k}{m}\right)^{\frac{1}{2}} t \\ &= \frac{1}{2} kA^2 \cos^2\left(\frac{k}{m}\right)^{\frac{1}{2}} t \end{aligned} \tag{2-16}$$

The total energy of the particle is then

$$\begin{aligned} E = T + V &= \frac{1}{2} kA^2\left[\sin^2\left(\frac{k}{m}\right)^{\frac{1}{2}} t + \cos^2\left(\frac{k}{m}\right)t\right] \\ &= \frac{1}{2} kA^2 \end{aligned} \tag{2-17}$$

a quantity independent of time.

EXERCISE 2-1 Consider the case of a particle of mass m moving in a gravitational field $V = mgz$. Write and solve the equations of motion starting with Newton's second law.

2-3 THE LAGRANGIAN AND HAMILTONIAN FORMS OF THE EQUATIONS OF MOTION

The equations of motion in Newtonian form are usually most convenient to solve if the physical problem is such as to make Cartesian coordinates appropriate. For problems in other coordinate systems it is often difficult even to write down the equations of motion. For this reason it would be more convenient if equations of motion could be derived whose form is independent of a coordinate system. Such general equations of motion were derived by two mathematicians, Joseph Lagrange and William Hamilton, and are called the Lagrangian and Hamiltonian forms of the equations of motion. The Hamiltonian form is especially important because it is directly used in the transformation from classical to quantum mechanics.

Before discussing these two forms of the equations of motion, it is necessary to introduce the idea of generalized coordinates, velocities, and momenta.

Consider a conservative system containing three particles. In order to specify completely the state of the system at a given time t, one would have to specify the positions and velocities of the three particles. To do this, one would have to specify nine coordinates, that is, $(x_1, y_1, z_1, \ldots, x_3, y_3, z_3)$ and nine velocities, that is, $(\dot{x}_1, \dot{y}_1, \dot{z}_1, \ldots, \dot{x}_3, \dot{y}_3, \dot{z}_3)$. In general, for a system containing N particles, one would have to specify $3N$ coordinates and $3N$ velocities. Such a system would have $6N$ degrees of freedom. This statement is true only if there are no constraints that make some of the $6N$ variables dependent inherent in the system. For instance, if the three particles were required to move on the spherical surface $x^2 + y^2 + z^2 = R^2$, then only 12 degrees of freedom (six coordinates and six velocities) rather than 18 would exist. We will always assume that the coordinates have been chosen so that the restraints have already been accounted for. To formulate classical mechanics in a general way, we introduce, for a system containing N particles, $3N$ generalized coordinates q_i and $3N$ generalized velocities $\dot{q}_i \equiv dq_i/dt$. We then derive the Lagrangian and Hamiltonian forms of the equations of motion in terms of these generalized coordinates and velocities. When working problems, we give these generalized quantities a specific form.

The Lagrangian function $L(\dot{q}, q)$ is defined as

$$L(\dot{q}, q, t) = T(\dot{q}, q) - V(q, t) \tag{2-18}$$

where T is the kinetic energy expressed as a function of the generalized velocities and coordinates, and V is the potential energy expressed as a function of the generalized coordinates and the time t. For conservative systems, the Lagrangian function L and the potential energy V will not depend explicitly on the time. For these systems, L is a function only of the $6N$ q_i and $\dot{q}_i$. Going through considerable algebra, one can show that the equations of motion in Lagrangian form are [2–4]

$$\frac{d}{dt}\left(\frac{\partial L}{\partial \dot{q}_i}\right)_{q_j, \dot{q}_{j \neq i}} = \left(\frac{\partial L}{\partial q_i}\right)_{q_{j \neq i}, \dot{q}_j} \tag{2-19}$$

These equations, since they are in generalized coordinates, hold for any system of coordinates. It should be pointed out that in the partial derivatives in Eq. 2-19, the other $6N - 1$ variables are held constant. From this point on, the constancy of these variables will be understood and the subscripts will not be used.

To illustrate the use of Eq. 2-18 and 2-19 let us return to the problem of simple harmonic motion discussed in the previous section. We first write down the appropriate quantities for this problem.

$$\begin{aligned} q_i &= x \\ \dot{q}_i &= \dot{x} \\ T(q_i, \dot{q}_i) &= \tfrac{1}{2}m\dot{x}^2 \\ V(q_i) &= \tfrac{1}{2}kx^2 \end{aligned} \tag{2-20}$$

The Lagrangian function $L(\dot{q}, q)$ is then

$$L(x, \dot{x}) = \tfrac{1}{2}m\dot{x}^2 - \tfrac{1}{2}kx^2 \tag{2-21}$$

To write the equations of motion we calculate

$$\left(\frac{\partial L}{\partial \dot{x}}\right) = m\dot{x} \qquad \left(\frac{\partial L}{\partial x}\right) = -kx$$

and substitute these quantities into Eq. 2-19 to obtain

$$\frac{d}{dt}(m\dot{x}) = m\ddot{x} = -kx \tag{2-22}$$

which is the same result as that obtained from Newton's second law.

The student can see that Lagrangian equations of motion are a set of $3N$ second order differential equations. To derive the equations of motion in Hamiltonian form, we transform these to a set of $6N$ first order equations. To do this, we first define the generalized momenta p_k as

$$p_k = \left(\frac{\partial L}{\partial \dot{q}_k}\right) \tag{2-23}$$

Applying Eq. 2-23 to the case of a free single particle whose motion is described by Cartesian coordinates gives the usual components of linear momentum $m\dot{x}$, $m\dot{y}$, and $m\dot{z}$. Note that with the definition 2-23 the equations of motion in Lagrangian form become simply

$$\dot{p}_i = \frac{\partial L}{\partial q_i} \tag{2-24}$$

EXERCISE 2-2 Write the Lagrangian function for a free (potential equal to a constant) particle moving in a three-dimensional space. Show that $p_x = m\dot{x}$, $p_y = m\dot{y}$, and $p_z = m\dot{z}$. Use Eq. 2-19 to write down the equations of motion. Solve them.

EXERCISE 2-3 Write the Lagrangian function for the particle of mass m moving in a gravitational field $V = mgz$. Derive the equations of motion from Eq. 2-19. Show that the equations obtained by this method are the same as those obtained from Newton's second law.

We next define a new function

$$\mathscr{H} = \sum_{i=1}^{3N} p_i \dot{q}_i - L \tag{2-25}$$

from which we can show that for a conservative system [2–4]

$$\begin{aligned} \frac{\partial \mathscr{H}}{\partial p_i} &= \dot{q}_i \\ \frac{\partial \mathscr{H}}{\partial q_i} &= -\frac{\partial L}{\partial q_i} = -\dot{p}_i \end{aligned} \tag{2-26}$$

EXERCISE 2-4 Show that Eq. 2-26 follows from Eq. 2-25. *Hint*: Write down the total differential for $\mathscr{H}$ using Eq. 2-25. Then compare it with the general expression for the total differential of $\mathscr{H}$ regarding $\mathscr{H}$ as a function of p_i and q_i. Use Eqs. 2-23 and 2-24 to prove the final result.

These are the equations of motion in Hamiltonian form. The quantity $\mathscr{H}$ is expressed as a function of the coordinates and the *momenta*, and is called the Hamiltonian function for the system. The Hamiltonian function for a conservative system has the property that it is equivalent to the total energy of the system. To show this, we substitute Eqs. 2-18 and 2-24 into Eq. 2-25 to obtain

$$\mathscr{H} = \sum_i \dot{q}_i \frac{\partial L}{\partial \dot{q}_i} - T + V \tag{2-27a}$$

$$= \sum_i \dot{q}_i \frac{\partial T}{\partial \dot{q}_i} - T + V \tag{2-27b}$$

Equation 2-27b follows from Eq. 2-27a because, for conservative systems, all of the dependence of L on $\dot{q}_i$ is in the T term. The first term in Eq. 2-27, however, is equal to $2T$. Rather than prove this general result, an example will be given. For more details, see [1–4]. Consider the example of a system in which one particle is constrained to move in one dimension. The kinetic energy for Cartesian coordinates is

$$T = \tfrac{1}{2} m \dot{q}_i^{\,2}$$

It is easily seen, then, that

$$\frac{\partial T}{\partial \dot{q}_i} = m\dot{q}_i$$

and

$$\dot{q}_i \frac{\partial T}{\partial \dot{q}_i} = m\dot{q}_i^2 = 2T$$

For the many-particle, many-dimensional system, using Cartesian coordinates $T = \frac{1}{2} \sum m_i \dot{q}_i$, and, following an argument similar to the above, we can show that

$$\sum_i \dot{q}_i \frac{\partial T}{\partial \dot{q}_i} = 2T \tag{2-28}$$

Using this result in Eq. 2-27b, it follows that

$$\mathscr{H} = 2T - T + V = T + V \tag{2-29}$$

Thus we see that Hamilton's function is identical with the total energy.

We now return to our one-dimensional harmonic oscillator problem to write it in the Hamiltonian formalism. The first step is to use L, Eq. 2-21, to define the appropriate momenta. Thus

$$p_x = \frac{\partial L}{\partial \dot{x}} = m\dot{x} \tag{2-30}$$

Then, in terms of *momenta*, T becomes

$$T = \frac{1}{2} m\dot{x}^2 = \frac{1}{2} m \left(\frac{p_x}{m}\right)^2 = \frac{1}{2m} p_x^2 \tag{2-31}$$

and the Hamiltonian is

$$\mathscr{H} = T + V = \frac{1}{2m} p_x^2 + \frac{1}{2} kx^2 \tag{2-32}$$

The student may easily show that applying Eqs. 2-26 through 2-32 gives the same equations of motion that were obtained previously.

EXERCISE 2-5 Write Hamilton's function for the case of a particle of mass m in a gravitational field. Write the equations of motion and show that they are identical with Newton's equation.

EXERCISE 2-6 A particle is constrained to move in the XY plane ($\theta = 90°$) under the potential $V = \frac{1}{2}k(x^2 + y^2)$.

1. Using Cartesian coordinates, write the equations of motion in Newtonian form.
2. Write the Lagrangian function L in Cartesian and polar coordinates.
3. Write the equations of motion in polar coordinates using the Lagrangian form.
4. Find the momenta p_r and p_ϕ.
5. Write the Hamiltonian function in both systems of coordinates.
6. What famous conservation law is obvious from the results of 3 above?

2-4 INTERNAL COORDINATES AND THE MOTION OF THE CENTER OF MASS

A specific problem that will be of great importance in quantum mechanics is that of two interacting particles of masses m_1 and m_2, where the potential is only a function of their distance apart, or, in other words, of their *relative* coordinates. The following treatment of this problem is essentially that of [3].

If the Cartesian coordinates of the two particles are x_1, y_1, z_1, and x_2, y_2, z_2, respectively, the square of their distance apart is

$$r_{12}^2 = (x_2 - x_1)^2 + (y_2 - y_1)^2 + (z_2 - z_1)^2 \tag{2-33}$$

The problem discussed above can be greatly simplified by transforming to new coordinates which involve the coordinates of the center of mass X, Y, Z and the "internal" or relative coordinates x, y, z. Thus we define

$$X = \frac{m_1 x_1 + m_2 x_2}{m_1 + m_2} \qquad Y = \frac{m_1 y_1 + m_2 y_2}{m_1 + m_2} \qquad Z = \frac{m_1 z_1 + m_2 z_2}{m_1 + m_2} \tag{2-34}$$

$$x = x_2 - x_1 \qquad y = y_2 - y_1 \qquad z = z_2 - z_1$$

EXERCISE 2-7 Write an expression for the kinetic energy of a system containing two particles of masses m_1 and m_2 moving in only two dimensions. Let the coordinates be x_1, y_1, x_2, y_2. Transform this expression to the new coordinate system X, Y, x, y making use of the relations 2-34. Show that the final expression is

$$T = \frac{1}{2}(m_1 + m_2)(\dot{X}^2 + \dot{Y}^2) + \frac{1}{2}\frac{m_1 m_2}{m_1 + m_2}(\dot{x}^2 + \dot{y}^2)$$

Write the corresponding equation for a system of two particles moving in three dimensions by analogy.

Making use of the results of Exercise 2-7, we can write

$$L = \frac{1}{2}(m_1 + m_2)(\dot{X}^2 + \dot{Y}^2 + \dot{Z}^2) + \frac{\mu}{2}(\dot{x}^2 + \dot{y}^2 + \dot{z}^2) - V(x, y, z) \tag{2-35}$$

where $\mu = m_1 m_2/(m_1 + m_2)$ and is called the reduced mass, and where V is a function only of x, y, z since, by hypothesis, the potential energy depends only on the internal or relative coordinates.

From the Lagrangian function in Eq. 2-35, the equations of motion for the six coordinates can be calculated using Eq. 2-19. These six equations of motion are

$$(m_1 + m_2)\ddot{X} = (m_1 + m_2)\ddot{Y} = (m_1 + m_2)\ddot{Z} = 0 \tag{2-36}$$

$$\mu\ddot{x} = -\frac{\partial V}{\partial x} \qquad \mu\ddot{y} = -\frac{\partial V}{\partial y} \qquad \mu\ddot{z} = -\frac{\partial V}{\partial z} \tag{2-37}$$

Equations 2-36 are identical to the equations of motion that are obtained if the problem of the motion of a free particle (see Exercise 2-2) of mass M is solved. Thus the motion of the center of mass of our two-particle system is the same as the motion of a free particle with mass equal to the total mass of the system. Equations 2-36 can be integrated to give

$$M\dot{q}_i = C \tag{2-38}$$

where $M = (m_1 + m_2)$, $\dot{q}_i$ can be $\dot{X}$, $\dot{Y}$, or $\dot{Z}$, and C is a constant. Equation 2-38 shows that the three components of the velocity of the center of mass are constants. The kinetic energy due to the motion of the center of mass, therefore, must also be a constant.

Equations 2-37 are identical to those that are obtained in solving the problem of the motion of a particle with mass μ subject to the potential function $V(x, y, z)$. The total energy of the system is the sum of the energies due to the motion of the center of mass and to the internal motion of the system. Since the translational energy of the center of mass adds only a constant to the total energy, it is usual to neglect its contribution and solve only the problem of the internal motion of the system. This example points up the power of expressing the laws of motion in generalized coordinates. In this case, we let $q_1 = X$, $q_2 = Y, \ldots, q_6 = z$, and so on, and we can immediately write down the equations of motion from the Lagrangian function.

The above example is an extremely important one that should be thoroughly understood. In general terms, its significance is that, as long as the potential energy depends only on the internal coordinates of the system, *the motion of the center of mass can always be separated from the internal motion of the system*, and the two problems can be solved independently.

2-5 *THE BASIC ASSUMPTIONS OF CLASSICAL MECHANICS*

At this point, it is good to think about the philosophical implications inherent in classical mechanics. First, it is implied that an experimentalist can precisely measure the positions and velocities of all of the particles in a system at some time t in order to describe the state of the system. Second, once this initial state is specified, the laws of mechanics and a knowledge of the forces acting on the system enable the system to be characterized at any later time. In principle, then, an experimentalist could measure the position, velocity, energy, momentum, and so on, of any particle at any time and compare it with the theoretical prediction. The following three statements summarize the assumptions inherent in this view.

1. There is no limit to the accuracy with which one or more of the dynamical variables of a classical system can be simultaneously measured *except* the limit imposed by the precision of the measuring instruments.
2. There is no restriction on the number of dynamical variables that can be accurately measured simultaneously.
3. Since the expressions for velocity are continuously varying functions of time, the velocity, and hence the kinetic energy, can vary continuously. That is, there are no restrictions on the values that a dynamical variable can have.

We shall see that when very small particles are involved, all three of these assumptions must be abandoned. For these systems, classical mechanics fails completely to describe their behavior. The new mechanics that was developed for these systems is called quantum mechanics.

2-6 *SUMMARY*

1. A conservative system was defined as a system in which the sum of the kinetic and potential energies remains constant with time, or one in which the forces are equal to the negative gradient of some potential function.
2. Newton's second law of motion was applied to the case of simple harmonic motion.
3. The Lagrangian and Hamiltonian forms of the equations of motion were introduced and shown to give the same results for a specific problem as Newton's second law. These equations are more general than Newton's because the forms of the equations are independent of the coordinate system.
4. For systems containing many particles, the motion of the center of mass was shown to be separable from the internal motion of the system as long as the potential energy depended only on the relative coordinates of the particles.
5. The basic assumptions of classical mechanics were discussed.
6. Terms which the student should understand are the following: conservative system, Lagrangian function, Hamiltonian function, internal coordinates, and reduced mass.

REFERENCES

1. H. J. Goldstein, *Classical Mechanics*, 2d ed. (Addison-Wesley, Reading, Mass., 1980).
2. J. M. Anderson, *Mathematics for Quantum Chemistry* (W. A. Benjamin, Inc., New York, 1962).
3. N. Davidson, *Statistical Mechanics* (McGraw-Hill Book Co., New York, 1922).
4. R. C. Tolman, *The Principles of Statistical Mechanics* (Oxford University Press, London, 1930).

Chapter 3

QUANTUM MECHANICS

BY THE end of the nineteenth century three types of observations made it apparent that classical mechanics could not give correct results when it was applied to molecular and atomic phenomena. These observations involved studies of atomic spectra, blackbody radiation, and the photoelectric effect. In the section that follows, each of these experiments is discussed, and it is shown how the basic assumptions of classical mechanics had to be abandoned. Following this, quantum mechanics is introduced by a series of postulates and these postulates are then applied to calculations on some simple systems.

3-1 ATOMIC SPECTRA, BLACKBODY RADIATION, AND THE PHOTOELECTRIC EFFECT

The discipline of spectroscopy began in the early part of the nineteenth century with the observation of the sunlight spectrum by Josef Fraunhofer. The study of the spectra of atoms was begun in 1861 by Kirchoff and by Bunsen, who extensively studied the spectrum of the alkali metals. In 1885, Balmer discovered the series of lines in the spectrum of atomic hydrogen that now bears his name, and found that he could write an empirical relationship that gave the positions of all the lines. This relationship was

$$\frac{1}{\lambda} = R\left(\frac{1}{2^2} - \frac{1}{n_2{}^2}\right) \qquad n_2 = 3, 4, 5 \cdots \tag{3-1}$$

where λ is the wavelength of the observed line and R is a constant called the Rydberg constant. This constant is one of the most accurately known physical constants and has the value 109,677.581 cm^{-1}.

EXERCISE 3-1 Equation 3-1 predicts the qualitative features of the spectra of other atoms besides hydrogen, especially the spectra of the alkali metals. In the latter case the constant R is replaced by other constants. Calculate the position of the lines in the hydrogen atom in units of R for $n_2 = 3, 4, 5, 10, 100, \infty$. Plot these results along a horizontal scale in $1/\lambda$. Note how the lines converge to a limiting value. This limiting value is called the series limit.

There were several striking features about atomic spectra. The first was the sharpness of the spectral lines. Apparently, energy was not emitted or absorbed by atoms in bands, or in a continuous fashion, but only at certain very precise frequencies. It was this sharpness of spectral lines that enabled the constant R to be determined with such accuracy. The second striking feature was the fact that the spectrum of each kind of atom was highly characteristic. In fact, the best proof for the presence of a particular element in a sample was the existence of its characteristic spectrum. It is clear that any theory of the structure of atoms would have to explain these two features of atomic spectra. In addition, for the hydrogen atom, the characteristic frequencies would have to be those predicted by Eq. 3-1. We see later how classical mechanics was at a loss to account for these facts if the Rutherford model of the atom were accepted.

Although the spectroscopic results had been accumulating for some time, the experiments that first led to questions about the validity of classical mechanics were those designed to measure the frequency (or wavelength, since the two are related by frequency $\times$ wavelength $=$ speed of light) dependence of blackbody radiation. A detailed discussion of blackbody radiation is beyond the scope of this text, but enough of the basic features can be outlined to acquaint the student with the experimental results and their significance.

Blackbody radiation is a familiar phenomenon even though the name may seem mysterious. When the heating element of an electric stove is turned on, it emits radiation. This radiation can be detected by placing one's hand at some distance above the heating element. If the stove is on low heat, the radiation can be detected by feeling only and not by sight. If the heat is turned up the stove element will begin to glow first red, then white, and, if the temperature could be raised high enough, even blue. This change in color is evidence that the frequency distribution of the radiation emitted by the hot body is changing with temperature.

In the laboratory, blackbody radiation is studied by constructing a cavity that is insulated so that the only energy that can be absorbed is energy added in the form of heat to raise the temperature of the cavity. The cavity is evacuated, and there is a small hole in one side of the apparatus through which radiation can pass. The approximation is made that the radiation coming from the hole in the cavity is a good sample of the radiation inside the cavity. The intensity of the radiation emitted from the hole is then studied as a function of wavelength at several different cavity temperatures. When this is done, data like those plotted in Figure 3-1 are obtained.

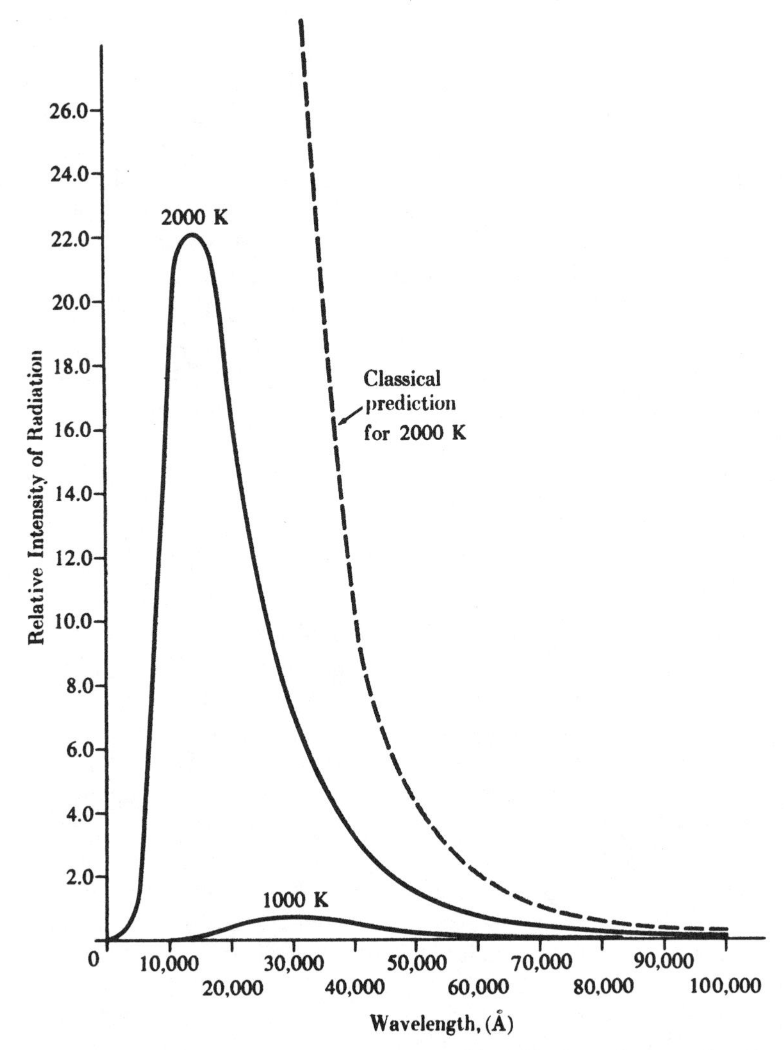

FIGURE 3-1 *Typical data for the intensity of blackbody radiation plotted as a function of wavelength. The student should note the falloff of intensity in the ultra-violet region of the spectrum as well as the "blue shift" of the maximum as the temperature increases. The dashed line is the prediction of the Rayleigh–Jeans formula at* 2000 K.

By applying classical statistical mechanics and the properties of wave motion, one can calculate the number of light waves between frequencies ν and $\nu + d\nu$ in a box of volume V [1,2]. This number is

$$g(\nu)\, d\nu = \frac{8\pi V \nu^2\, d\nu}{c^3} \tag{3-2}$$

To calculate the frequency distribution as a function of temperature, Lord Rayleigh and J. H. Jeans assumed that each electromagnetic wave had its classical value of the energy kT, where k is the Boltzmann constant. This result comes from the theorem of equipartition of energy in classical statistical mechanics. A loose statement of the theorem is that there is a contribution of $\frac{1}{2}kT$ to the energy of a system for each squared term that appears in the expression for the energy of the microscopic component making up the system. Since the energy of a classical light wave is proportional to $\mathscr{E}_0{}^2 + H_0{}^2$, where $\mathscr{E}_0$ and H_0 are the amplitudes of the oscillating electric and magnetic fields of the light wave respectively, there are two squared terms and the contribution to the system energy will then be kT. (See [1–3].)

For unit volume, this assumption, and Eq. 3-2, gives the Rayleigh–Jeans formula

$$\rho(\nu, T)\, d\nu = \frac{8\pi\nu^2 kT}{c^3}\, d\nu \tag{3-3}$$

where $\rho(\nu, T)\, d\nu$ is the energy density of radiation between frequencies ν and $\nu + d\nu$ at absolute temperature T and is proportional to the intensity of the light coming from the hole in the cavity at these frequencies. A plot of Eq. 3-3 for $T = 2000$ K is also shown in Figure 3-1. The student will notice that the fit of Eq. 3-3 is good at long wavelengths (low frequencies) but that the fit gets worse as the wavelength decreases and becomes qualitatively wrong at short wavelengths in that the Rayleigh–Jeans formula does not predict the maximum and sharp falloff. (Rayleigh and Jeans realized that this result was incorrect, but they did not know how to remedy the situation using the assumptions of classical mechanics.) The inability of classical mechanics to explain the falloff of radiation density in the ultraviolet region of the spectrum has been dubbed the "ultraviolet catastrophe."

Max Planck next attacked this problem. He used the fact, as did Rayleigh and Jeans, that the radiation on the inside of a heated body arises because of the emission from the vibrating constituents of the material. Rather than assume that these vibrations could emit energy of any value, he proposed that energy could only be emitted in discrete amounts, called quanta. Planck hypothesized that these fundamental quanta had an energy equivalent to $h\nu$, where ν was the fundamental frequency of the oscillator and where h is the now famous Planck's constant. At the time, this *ad hoc* assumption had no precedent and was justified only in that it gave the right answer. In this assumption, the idea of the existence of discrete energy states in matter was born.

Using Planck's hypothesis, the number of light waves between frequencies ν and $d\nu$ is not that given by Eq. 3-2 but is rather (see, for example, [3])

$$g(\nu)\,d\nu = \frac{8\pi\nu^2}{c^3}\,(e^{h\nu/kT} - 1)^{-1}\,d\nu \tag{3-4}$$

which, when combined with the energy of a light wave of this frequency ($h\nu$), gives the Planck radiation law

$$\rho'(\nu, T)\,d\nu = \frac{8\pi h\nu^3}{c^3}\,(e^{h\nu/kT} - 1)^{-1}\,d\nu \tag{3-5}$$

Equation 3-5 is in exact agreement with experiment (see Exercise 3-3).

EXERCISE 3-2 Show that in the limit of long wavelength (low frequency), Eq. 3-5 goes over into Eq. 3-3. *Hint*: Expand the exponential in a Maclaurin series and argue about what happens when $h\nu$ becomes less than kT.

EXERCISE 3-3 The function agreeing with experiment in Figure 3-1 is not the function $\rho'(\nu, T)d\nu$ but rather $\rho'(\lambda, T)d\lambda$. Using the fact that $\lambda\nu = c$, and therefore that $d\nu = -c/\lambda^2\,d\lambda$, show that

$$\rho(\lambda, T)\,d\lambda = \frac{8\pi hc}{\lambda^5}\,(e^{hc/\lambda kT} - 1)^{-1}\,d\lambda$$

This function agrees with the experimental results in Figure 3-1.

It was not long before Planck's quantity $h\nu$ had another application. In 1905, in order to explain the photoelectric effect, Albert Einstein postulated that light energy also had to be quantized. Planck originally thought that only the oscillators in a blackbody were quantized.

The photoelectric effect can best be illustrated by the description of an actual experiment. A schematic diagram of the apparatus is shown in Figure 3-2. The apparatus consists of a special cell C which contains a screen or grid and a receiving element constructed of a piece of metal which has been plated with a thin film of the metal to be studied. When the switch in Figure 3-2 is at position A, the screen is connected to the receiver element through a battery and a sensitive galvanometer G. The special cell is highly evacuated. Also included is a device R for changing the potential of the screen with respect to the receiving element. When light strikes the surface

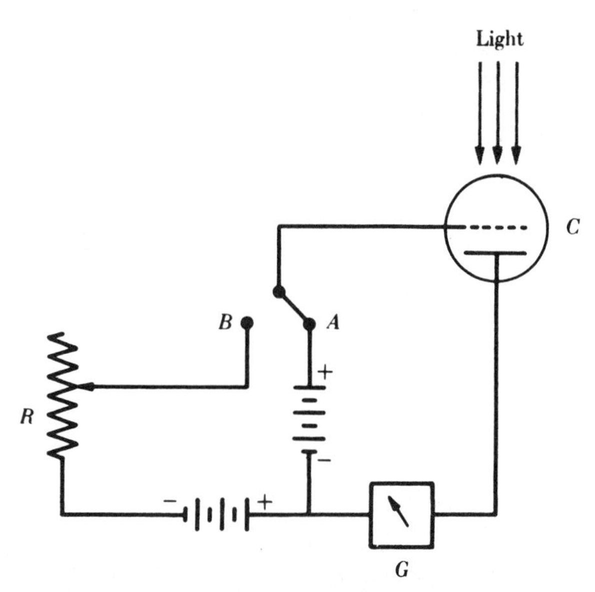

FIGURE 3-2 *Schematic drawing of an apparatus for studying the photoelectric effect. The apparatus consists of a special cell C, a galvanometer G, a variable resistor R, and a switch to connect the screen of the cell to either A or B. Note that when the switch is at B the screen is negative with respect to the plate.*

of the receiving element, electrons are ejected from the surface of the metal and are attracted to the positively charged screen. This causes current to flow in the galvonometer. The current registered by the galvanometer is proportional to the number of electrons striking the screen. When the switch is at *B*, a negative potential can be applied to the screen. When this negative potential energy just balances the kinetic energy of the emitted electrons, the current is reduced to zero. The voltage necessary to do this can then be recorded. Experimental observations in a typical experiment are the following.

1. No electrons are emitted until the frequency of the light becomes larger than a certain value. This minimum frequency is called the threshold frequency.

2. At light frequencies higher than the threshold frequency, the electrons are ejected with excess kinetic energy. This extra kinetic energy of the electrons is independent of the intensity of the incident light, but is directly proportional to its frequency.

3. When the frequency of the light is higher than the threshold frequency, the current flowing in the apparatus is dependent only on the intensity of the incident light. Below the threshold frequency no current flows.

These results can most easily be explained by the hypothesis that, when light is absorbed by a metal, it acts as though it were a stream of particles each of which has an energy proportional to its frequency. These light "particles," or quanta, are called photons, and the energy of a light quantum is $h\nu$. When a photon is absorbed by a metal, its total energy is given to one of the electrons in the conduction band of the metal. If the light quantum has enough energy, the electron can overcome the potential energy barrier at the surface of the metal and be attracted to the positively charged screen. The potential energy barrier of a metal is called its work function. The excess kinetic energy which an ejected electron has is then

$$\tfrac{1}{2}mv^2 = h\nu - h\nu_0 = h\nu - W \tag{3-6}$$

where ν_0 is the threshold frequency and W is the work function of the metal. When a potential is applied to the screen so as to stop the electron flow, the kinetic energy is just balanced by the potential energy of an electron in an electric field. Thus

$$\tfrac{1}{2}mv^2 = h\nu - h\nu_0 = -\mathscr{E}e \tag{3-7}$$

where e is the absolute value of the electronic charge and $\mathscr{E}$ is the potential difference between screen and receiver element. One can see that, by plotting $-\mathscr{E}$ versus ν one should get a straight line of slope h/e and intercept $h\nu_0/e$. Such a plot for lithium metal is shown in Figure 3-3 and the computation is illustrated by Exercise 3-4.

EXERCISE 3-4 An experiment was done on the emission of photoelectrons from a sodium surface by light of different wavelengths. The following values for the retarding potentials at which the photoelectric current was reduced to zero were obtained. Plot the voltage against frequency ($\nu = c/\lambda$)

λ, Å	$-\mathscr{E}$, volts
3125	2.128
3650	1.595
4047	1.215
4339	1.025
5461	0.467

and calculate (a) the threshold frequency and (b) Planck's constant. Note that 1 joule = 1 volt coulomb = 10^7 erg. Data from R. A. Millikan, *Phys. Rev.* **7**, 355 (1916).

We have seen how it was impossible to explain certain observations connected with the behavior of incandescent solids and of light striking a metal surface by a consistent application of classical mechanics. The final obstacle, which eventually led to the abandonment of the use of classical mechanics to describe microscopic phenomena, was its failure to accommodate the structure of atoms.

We have already pointed out that in atomic spectra the lines are unusually sharp and occur at definite frequencies characteristic of each atom. These results led to a peculiar paradox.

In 1911, Rutherford had enunciated the nuclear model of the atom. In

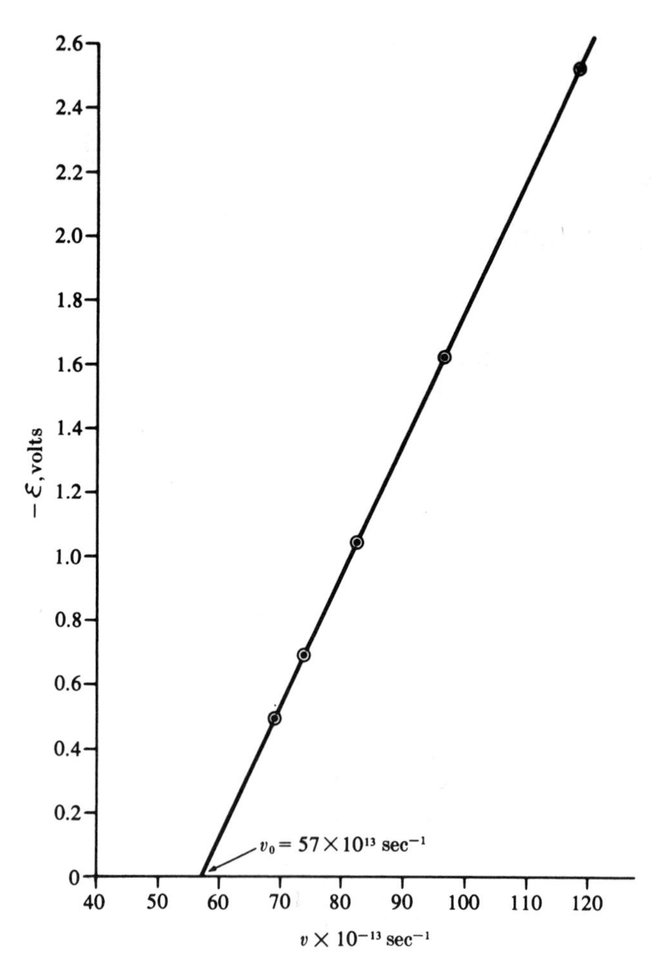

FIGURE 3-3 *Typical photoelectric effect data showing a plot of the retarding potential necessary to stop electron flow from lithium as a function of the frequency of the incident light. The slope of this line equals h/e. Plots such as the one shown here gave the first accurate determinations of Planck's constant. Data points are from R. A. Millikan, Phys. Rev.* 7, 355 (1916).

this model, the positive charge and most of the mass are concentrated in the center of the atom, called the nucleus. The electrons were postulated to revolve around it like planets around the sun. It is a simple consequence of Newton's laws of motion that, for an electron subject to the attractive force from the nucleus to move in a stable orbit, it must be accelerated. Further, according to classical electrodynamics, an accelerated charge must constantly lose energy by radiating. The paradox in this view of the structure of atoms is that an atom should not be stable. That is, as the electron radiates, it will lose energy and spiral inward, eventually colliding with the nucleus. Experimentally, it is known that atoms are stable indefinitely (unless they react to form molecules). Further, instead of continually radiating as the above picture predicts, atoms only radiate when excited by some means, and then only at a definite frequency. It is clear that the picture of the atom which arose from classical physics had some gross inadequacies.

In 1913, Niels Bohr proposed a hypothesis to try to explain these discrepancies between classical theory and experiment. Bohr's hypothesis was that the lines in an atomic spectrum come from a transition of an electron between two discrete states in an atom. The following discussion of Bohr's assumptions is an abbreviated form of that found in [4]. Bohr's assumptions were the following.

1. The Planck–Einstein relation, which relates an energy difference to the quantity $h\nu$, held for the emission and absorption of radiation from an atom. Thus, if the energies of the two discrete states of the electron in an atom are E_1 and E_2 respectively, the frequency of a spectral line due to a transition of an electron from state 1 to state 2 is

$$h\nu = E_2 - E_1 \tag{3-8}$$

2. In the discrete states, the magnitude of the angular momentum of the electron can only have the values

$$L\,(\text{angular momentum}) = n\left(\frac{h}{2\pi}\right) = n\hbar \tag{3-9}$$

where n is an integer and $\hbar$ is defined as $h/2\pi$.

3. The behavior of an electron during a transition cannot be visualized or explained classically.

To calculate the allowed orbits predicted by Bohr's theory for the hydrogen atom, we begin with Newton's second law

$$\mathbf{F} = m\mathbf{a} \tag{3-10}$$

The force is the Coulombic force between the positively charged nucleus and the electron;

$$\mathbf{F} = -\frac{(Ze)e}{r^2}\,\mathbf{u}_r \tag{3-11}$$

where Z is the nuclear charge and $\mathbf{u}_r$ is a unit vector along the direction from the nucleus to the electron. The acceleration in Eq. 3-10 is the centripetal acceleration and is given by

$$\mathbf{a} = -\frac{v^2}{r}\,\mathbf{u}_r \tag{3-12}$$

where v is the velocity of the electron and r is the radius of the orbit. This acceleration must also be directed inward or else the electron would not move in a stable orbit. Equation 3-10 then becomes

$$-\frac{(Ze)e}{r^2}\mathbf{u}_r = -\frac{mv^2}{r}\mathbf{u}_r \tag{3-13}$$

Equating the magnitudes leads to the scalar equation

$$\frac{Ze^2}{r^2} = \frac{mv^2}{r} \tag{3-14a}$$

$$mv^2 r = Ze^2 \tag{3-14b}$$

But, applying assumption 2 on page 41, we have

$$L \equiv mvr = \frac{nh}{2\pi} = n\hbar \tag{3-15}$$

Rearranging and combining Eq. 3-15 with Eq. 3-14b, one obtains

$$v^2 = \frac{n^2\hbar^2}{r^2 m^2} \tag{3-16a}$$

$$mr\frac{n^2\hbar^2}{m^2 r^2} = Ze^2 \tag{3-16b}$$

$$r = \frac{n^2\hbar^2}{me^2 Z} \tag{3-16c}$$

The student should notice that, because of assumption 2, the value of r is now restricted to certain orbits, that is, $\hbar^2/me^2Z$, $4\hbar^2/me^2Z$, ..., and so on. For the smallest allowed orbit in the hydrogen atom, $Z = 1$, $n = 1$, and

$$r_0{}^H \equiv a_0 = \frac{\hbar^2}{me^2} = 0.529 \text{ Å} \tag{3-17}$$

where a_0 is a symbol used for the radius of the first Bohr orbit.

The total energy of the atom is

$$E = T + V \tag{3-18}$$

where

$$T = \frac{1}{2} mv^2 = \frac{1}{2} \frac{Ze^2}{r} \tag{3-19}$$

and, since the system is conservative,

$$F_r = -\frac{dV}{dr}$$

and

$$V = -\int_\infty^r F_r \, dr = +\int_\infty^r \frac{Ze^2}{r^2} = -\frac{Ze^2}{r} \tag{3-20}$$

Therefore,

$$E = \frac{1}{2} \frac{Ze^2}{r} - \frac{Ze^2}{r} = -\frac{1}{2} \frac{Ze^2}{r} \tag{3-21}$$

Substituting Eq. 3-16c into Eq. 3-21, we obtain

$$E = -\frac{mZ^2e^4}{2n^2\hbar^2} \tag{3-22}$$

An important point in Eq. 3-22 is that only discrete energy levels appear because of the integer n^2 in the denominator.

It is of interest at this point to calculate the wavelength of the transitions in the hydrogen atom making use of assumption 1 above. For an absorption spectrum, we obtain

$$h\nu = E_2 - E_1 = \frac{me^4}{2\hbar^2} \left(\frac{1}{n_1{}^2} - \frac{1}{n_2{}^2} \right) \tag{3-23a}$$

or, since $\nu = c/\lambda$,

$$\frac{1}{\lambda} = \omega = \frac{2\pi^2 me^4}{h^3 c} \left(\frac{1}{n_1{}^2} - \frac{1}{n_2{}^2} \right) \tag{3-23b}$$

where ω is a commonly used spectroscopic unit called the wave number. Notice that ω has units of cm^{-1}, is proportional to the energy, and is the reciprocal of the wavelength of a spectral line. In this book ω always refers to an energy or an energy difference in wave numbers.

If one works out all of the constants in Eq. 3-14b, one obtains the value 109,737 cm^{-1}, which is in excellent agreement with the observed Rydberg constant. Setting $n_1 = 2$, the formula derived by Balmer for the lines in the Balmer series is obtained (Eq. 3-1). Other series of lines for atomic hydrogen were also found, and they were equally well correlated by Eq. 3-23b by setting $n_1 = 1$, 3, 4, and 5. These series are called the Lyman, Paschen, Brackett, and Pfund series, respectively, after their discoverers.

EXERCISE 3-5 Calculate the wavelength, frequency, and wave number for the first five lines in the Lyman series for atomic hydrogen. Sketch the spectrum on a piece of graph paper. What is the wave number of the series limit? What is the physical significance of the series limit?

It can be seen that the Bohr theory for the hydrogen atom worked very well indeed. Unfortunately, along with other difficulties, the theory *failed* to account for the spectrum of any atom having more than one electron. It was still necessary, therefore, to look for a more general form of mechanics for the treatment of atomic and molecular behavior.

The next step in the historical development of quantum mechanics was the suggestion in 1924 by Louis de Broglie [5] that if light, although usually regarded as a wave, sometimes could act like a particle, then electrons, although usually regarded as particles, sometimes could act like waves. De Broglie suggested that the bridge between the particle and wave descriptions of the electron was given by

$$\lambda = \frac{h}{p} = \frac{h}{mv} \tag{3-24}$$

where λ is the wavelength of the "electron wave" and m and v are its mass and velocity, respectively. De Broglie's suggestion received dramatic confirmation in the electron diffraction experiments of G. P. Thomson and of C. Davisson and L. H. Germer in 1927 [6]. Diffraction is a property that is only associated with wave motion, and the wavelength of the electrons involved was just that predicted by Eq. 3-24.

EXERCISE 3-6 Calculate (a) the wavelength of a beam of electrons accelerated by a voltage increment of 110 V and (b) the kinetic energy of an electron having a de Broglie wavelength of 1.5×10^{-8} cm.

Shortly after de Broglie's suggestion, quantum mechanics was founded practically simultaneously by Erwin Schrödinger and Werner Heisenberg.

3-2 THE FORMULATIONS OF QUANTUM MECHANICS

In its beginnings, quantum mechanics was approached in two completely different ways. Schrödinger, reasoning that electronic motions could be treated as waves, developed wave mechanics. In this treatment, he took over the great body of information from classical physics about wave motion and applied it to electronic and molecular motions. The stationary states that an electron or a molecule might have were analogous to standing waves (such as occur in a violin string) set up by applying appropriate boundary conditions. Later on, a mathematical formalism became associated with the Schrödinger method that related observable quantities to certain mathematical operations. Werner Heisenberg, independently and slightly earlier, had used the properties of matrices to get the same results as Schrödinger. This approach to quantum mechanics looked very different, but a little later M. Born and P. Jordan showed that they were equivalent. Later still, in the more general treatments of quantum mechanics by P. A. M. Dirac and by J. von Neumann, the Schrödinger and Heisenberg approaches were shown to be specific cases of a more general theory.

In chemistry texts, one finds many variations of these approaches. Usually, the time-independent Schrödinger equation (see below) is given in an *ad hoc* manner with the statement that the solutions to all atomic and molecular problems are the solutions to this second order differential equation. This approach leaves something to be desired, because the student usually wants to know *where* the Schrödinger equation came from and *why* its solutions give the answer to the allowed energy levels in atoms and molecules. A more rigorous mathematical approach is to press the analogy between the allowed states in atoms and molecules and standing and traveling waves. Difficulties arise in this approach because the average chemistry student has had little, if any, background in the physics and mathematics of wave motion. In the absence of a long introduction to wave mechanics, this approach becomes identical with the first in that the properties of wave motion must be introduced in an *ad hoc* way.

There is a formulation of quantum mechanics within the Schrödinger method that can be used in a consistent way in an introductory course. This approach treats quantum mechanics as a new subject with its own set of postulates analogous to the way in which Euclid's geometry is formulated. The basic ideas of quantum mechanics are introduced as postulates and these postulates are justified, in part, by the fact that the results that one gets by using them agree with the results of experiment. This is the same procedure that is followed in classical thermodynamics. The three laws of thermodynamics are postulates and they are justified by the fact that the conclusion that one draws from these laws are in accord with all experimental data presently at hand. For a more detailed discussion of this approach as well as discussion of the interrelation between the matrix and operator approach to quantum mechanics the student is referred to [7,8].

3-3 THE POSTULATES OF QUANTUM MECHANICS

The postulates of any theory are a set of fundamental statements that the student is asked to believe and draw conclusions from. These conclusions are then tested by experiment and if they are confirmed, the belief in the postulate is justified. The postulates can not be explained in terms of more fundamental concepts or else the more fundamental concepts would be used as postulates. For this reason, the student should not try to "understand" the postulates. A second point that should be kept in mind is that the ease with which a postulate may be made to appear reasonable depends on how readily it may be related to everyday experience. In quantum mechanics, the postulates are about atomic and molecular properties, and these are, in general, quite far from everyday experience. Consequently, the postulates may also, in this sense, be "difficult to understand." The main point to keep in mind is that *the postulates are justified only by their ability to predict and correlate experimental facts and by their general applicability.*

Before going to the postulates themselves, it is necessary to understand the meaning of the terms "dynamical variable" and "observable." Any property of a system of interest is called a dynamical variable. Thus the position $\mathbf{r}$, the energy E, the x component of linear momentum p_x, and so on, are all dynamical variables even though in a given system some of them may be constants. In general, any quantity of interest in classical mechanics is a dynamical variable. An example of dynamical variables that will be useful later is that of the three components of the momentum vector which a particle in a system has when it is at point P. An *observable* is any dynamical variable that can be measured. In classical mechanics all dynamical variables are observables, but, as we see later, there are certain fundamental restrictions placed upon simultaneously measurable quantities in quantum mechanics. In the case cited above, to measure the components of the momentum vector which the particle has at some point P, it is necessary to make a simultaneous measurement of the position and the momentum of the particle. As is seen later, there is an uncertainty relation that exists for such a simultaneous measurement on microscopic particles and the dynamical variable "the momentum at a point" is not an observable.

With this background in mind we introduce the basic postulates of quantum theory.

POSTULATE I

(a) *Any state of a dynamical system of N particles is described as fully as possible by a function* $\Psi(q_1, q_2, \ldots, q_{3N}, t)$ *such that*

(b) *the quantity* $\Psi^*\Psi\, d\tau$ *is proportional to the probability of finding* q_1 *between* q_1 *and* $q_1 + dq_1$, q_2 *between* q_2 *and* $q_2 + dq_2, \ldots, q_{3N}$ *between* q_{3N} *and* $q_{3N} + dq_{3N}$ *at a specific time t.*

What this postulate says (in less succinct form) is that all the information about the properties of a system is contained in a Ψ function (usually called a wave function) which is a function only of the coordinates of the N particles and the time t. If the wave function includes the time explicitly, it is called a time-dependent wave function. If the observable properties of a system do not change with time, the system is said to be in a *stationary* state. A Ψ function describing such a state is called a stationary state wave function, and the time dependence of such a wave function can be separated out. The second part of the postulate gives a physical interpretation of the Ψ function. This interpretation is easiest to visualize for a system containing a single particle constrained to move in one dimension. The quantity $\Psi^*\Psi\, dx$ is then the probability of finding the particle between x and $x + dx$ at a given time t. A Ψ function may be complex; hence the probability density $\Psi^*\Psi$ is a product of Ψ with its complex conjugate Ψ^*.

In order for these functions to be in accord with physical reality, they are subject to certain restrictions. These restrictions are the following.

1. The function should be continuous. This implies that its first and second derivatives will be continuous as well.
2. The function should be single valued.
3. The function should have an integrable square. For the cases encountered in this text, this requirement can be interpreted to mean that the function is everywhere finite. It also means that Ψ will go to zero at $\pm\infty$.

These restrictions all arise from the postulate that $\Psi^*\Psi\, d\tau$ represents a probability. The restriction of integrable squares is simply the requirement that the probability of finding the system in all space must be finite. A special case of this requirement is when the integral

$$\int_{\text{all space}} \Psi^*\Psi\, d\tau = 1 \tag{3-25}$$

When this is true, the function Ψ is said to be *normalized.* The physical meaning of this for a single-particle system is that the probability of finding the particle in some region in space must be 1. Normalized functions will always be used in this book.

EXERCISE 3-7 Which of the following functions meet the requirements for acceptable Ψ functions?

(a) $e^{\alpha x}$ $(\alpha > 0, 0 \leqq x \leqq \infty)$ (b) $e^{-\alpha x}$ $(\alpha > 0, 0 \leqq x \leqq \infty)$
(c) $e^{im\phi}$ $(0 \leqq \phi \leqq 2\pi$, m not an integer) (d) $e^{im\phi}$ $(0 \leqq \phi \leqq 2\pi$, m an integer).

EXERCISE 3-8 Given that $\Psi = Ae^{im\phi}$ for $0 \leqq \phi \leqq 2\pi$ is an acceptable function, find the value of the constant A to normalize Ψ.

POSTULATE II

For every observable property of a system, there exists a corresponding linear Hermitian operator, and the physical properties of the observable can be inferred from the mathematical properties of its associated operator.

The idea of a linear operator should be familiar to the student from Section 1-7. The only new property of operators in Postulate II is their Hermitian property. This property ensures that one always obtains real answers in the calculation of observables. A Hermitian operator is defined by the relation

$$\int_{\text{all space}} \Psi_i^* \hat{\alpha} \Psi_j \, d\tau = \int_{\text{all space}} \Psi_j \hat{\alpha}^* \Psi_i^* \, d\tau \tag{3-26}$$

where Ψ_i^* and Ψ_j are any two functions which satisfy the conditions for acceptability stated above, and where $\hat{\alpha}$ is the operator of interest.

EXERCISE 3-9 Making use of the expression for integrating by parts

$$\int u \, dv = uv - \int v \, du$$

show that the operator d/dx is not Hermitian, whereas the operator $i(d/dx)$ is Hermitian. *Hint:* you need to use the integrable squares condition to prove this.

At this point it is convenient to introduce a new notation for integrals of the type used in Eq. 3-26. This notation represents the integration over all space by parentheses or angular brackets. Thus

$$\int_{\text{all space}} \Psi_i^* \hat{\alpha} \Psi_j \, d\tau \equiv (\Psi_i | \hat{\alpha} | \Psi_j) \quad \text{or} \quad \langle \Psi_i | \hat{\alpha} | \Psi_j \rangle \tag{3-27}$$

and

$$\int \Psi^* \Psi \, d\tau \equiv (\Psi | \Psi) \quad \text{or} \quad \langle \Psi | \Psi \rangle$$

Equation 3-26, in this notation, becomes

$$(\Psi_i | \hat{\alpha} | \Psi_j) = (\Psi_j | \hat{\alpha} | \Psi_i)^*$$

Equation 3-26 is used to prove the theorem that *eigenvalues of a Hermitian operator must be real*, as they must be if they are to correspond to an observable.

Suppose we have a set of eigenfunctions Ψ_i of some Hermitian operator $\hat{\alpha}$. That is,

$$\hat{\alpha}\Psi_i = a_i \Psi_i \tag{3-28a}$$

The complex conjugate of this equation will also hold and is

$$\hat{\alpha}^*\Psi_i^* = a_i^*\Psi_i^* \tag{3-28b}$$

From the left, we now multiply Eq. 3-28a by Ψ_i^* and Eq. 3-28b by Ψ_i and integrate to obtain

$$(\Psi_i|\,\hat{\alpha}\,|\Psi_i) = (\Psi_i|\,a_i\,|\,\Psi_i) = a_i(\Psi_i\,|\,\Psi_i) \tag{3-29a}$$

and

$$(\Psi_i|\,\hat{\alpha}\,|\Psi_i)^* = (\Psi_i|\,a_i\,|\Psi_i)^* = a_i^*(\Psi_i\,|\,\Psi_i)^* \tag{3-29b}$$

Terms a_i and a_i^* can be brought outside the integral because they are constants. Since $\hat{\alpha}$ was postulated to be Hermitian, the left-hand sides of Eqs. 3-29a and 3-29b must be equal. Therefore,

$$a_i(\Psi_i\,|\,\Psi_i) = a_i^*(\Psi_i\,|\,\Psi_i)^*$$

Since Ψ_i^* and Ψ_i are functions (not operators), the order of multiplication is immaterial, and $(\Psi_i\,|\,\Psi_i) = (\Psi_i\,|\,\Psi_i)^*$. Therefore,

$$a_i = a_i^*$$

and the eigenvalue must be real because only real numbers equal their complex conjugates (Section 1-5). The Hermitian property of operators is also used in the proof of a number of important theorems about operators, eigenfunctions, and integrals. These theorems will be introduced at various points in the text, but as a sample of the kind of logic used, we will prove:

Theorem I. The product of two Hermitian operators is Hermitian only if the two operators commute.

Let $\hat{P}$ and $\hat{Q}$ be two Hermitian operators. These operators commute if $\hat{P}\hat{Q} - \hat{Q}\hat{P} = 0$. If Ψ_i and Ψ_j are any two acceptable functions, we wish to investigate the conditions for the relation

$$(\Psi_j\,|\,\hat{P}\hat{Q}\,|\,\Psi_i) = (\Psi_i\,|\,\hat{P}\hat{Q}\,|\,\Psi_j)^* \tag{3-30}$$

to hold. Since the following argument involves keeping careful track of the quantities that are starred, we will write out the integrals in two notations.

We begin with the integral on the left-hand side of Eq. 3-30 and make use of the facts that $\hat{P}$ is Hermitian and that $\hat{Q}\Psi_i$ is simply another function to write

$$(\Psi_j|\hat{P}\hat{Q}|\Psi_i) \equiv \int \Psi_j^*\hat{P}(\hat{Q}\Psi_i)\,d\tau = \int (\hat{Q}\Psi_i)\hat{P}^*\Psi_j^*\,d\tau \equiv (\hat{Q}\Psi_i|\hat{P}|\Psi_j)^*$$

But $\hat{P}^*\Psi_j^*$ is also a function, not an operator. Since the order of multiplication of functions is not important, we can write

$$(\hat{Q}\Psi_i|\hat{P}|\Psi_j)^* \equiv (\hat{Q}\Psi_i|\hat{P}\Psi_j)^* \equiv \int (\hat{Q}\Psi_i)(\hat{P}\Psi_j)^*\,d\tau$$

$$= \int (\hat{P}\Psi_j)^*(\hat{Q}\Psi_i)\,d\tau \equiv (\hat{P}\Psi_j|\hat{Q}\Psi_i)$$

Now we can make use of the fact that $\hat{Q}$ is Hermitian (the student should keep track of the stars this time) to write

$$(\hat{P}\Psi_j|\hat{Q}\Psi_i) = (\hat{P}\Psi_j|\hat{Q}|\Psi_i) = (\Psi_i|\hat{Q}|\hat{P}\Psi_j)^* = (\Psi_i|\hat{Q}\hat{P}|\Psi_j)^*$$

Summarizing the entire argument, we have shown that

$$(\Psi_j|\hat{P}\hat{Q}|\Psi_i) = (\Psi_i|\hat{Q}\hat{P}|\Psi_j)^* \tag{3-31}$$

Equation 3-31 will be true for any two Hermitian operators, but Eq. 3-30 will hold only if $\hat{Q}\hat{P}$ can be replaced by $\hat{P}\hat{Q}$, and this replacement is possible only if $\hat{P}\hat{Q} - \hat{Q}\hat{P} = 0$, that is, if $\hat{P}$ and $\hat{Q}$ commute.

The question naturally arises as to how one gets the operators for a given observable. A rigorous way to do this is to relate the commutator of two operators to a classical quantity called a Poisson bracket [7], but in this book we employ the following prescription. First, the classical expression for the observable of interest is written down in terms of coordinates, momenta, and the time. Next, the following replacements are made.

1. The time and all coordinates are left just as they are.
2. For Cartesian coordinates the momenta p_q are replaced by the differential operator $-i\hbar(\partial/\partial q)$.

As an example of this prescription, let us construct the quantum mechanical operator for the kinetic energy T. The classical expression for the kinetic energy of a particle in Cartesian coordinates is

$$T = \frac{1}{2m}(p_x{}^2 + p_y{}^2 + p_z{}^2) \tag{3-32}$$

Using step 2 in the above prescription, and being careful of the order of operations, this becomes

$$\hat{T} = \frac{1}{2m}\left[\left(-i\hbar\frac{\partial}{\partial x}\right)\left(-i\hbar\frac{\partial}{\partial x}\right) + \left(-i\hbar\frac{\partial}{\partial y}\right)\left(-i\hbar\frac{\partial}{\partial y}\right) + \left(-i\hbar\frac{\partial}{\partial z}\right)\left(-i\hbar\frac{\partial}{\partial z}\right)\right] \tag{3-33}$$

or

$$\hat{T} = -\frac{\hbar^2}{2m}\left(\frac{\partial^2}{\partial x^2} + \frac{\partial^2}{\partial y^2} + \frac{\partial^2}{\partial z^2}\right) = -\frac{\hbar^2}{2m}\nabla^2 \tag{3-34}$$

Even though the last form of Eq. 3-34, $\hat{T} = -(\hbar^2/2m)\nabla^2$, was derived using Cartesian coordinates, it is completely general and holds for any coordinate system.

Perhaps the most important operator that will concern us is the operator connected with the total energy E of a system. The classical expression for the total energy is Hamilton's function, and the corresponding operator is called the Hamiltonian. The expression for the Hamiltonian for a single-particle system is

$$\hat{\mathscr{H}} = \hat{T} + \hat{V}$$

But $\hat{T}$ is given by Eq. 3-34, and $\hat{V}$ is only a function of the coordinates q that, according to our prescription, remain the same. Therefore,

$$\hat{\mathscr{H}} = -\frac{\hbar^2}{2m}\nabla^2 + V(q) \tag{3-35}$$

EXERCISE 3-10 Derive the quantum mechanical operators for the three components of angular momentum. For the classical expression, see Exercise 1-6.

POSTULATE III

Suppose that $\hat{\alpha}$ is an operator corresponding to an observable and that there is a set of identical systems in state Ψ_s. Suppose further that Ψ_s is an eigenfunction of $\hat{\alpha}$. That is, $\hat{\alpha}\Psi_s = a_s\Psi_s$, where a_s is a number. Then, if an experimentalist makes a series of measurements of the quantity corresponding to $\hat{\alpha}$ on different members of the set, the result will always be a_s. It is only when Ψ_s and $\hat{\alpha}$ satisfy this condition that an experiment will give the same result on each measurement.

This is one of the postulates that bridges the gap between the mathematical formalism of quantum mechanics and experimental measurements in the laboratory. For example, suppose one is interested in calculating the allowed energies in a molecular or atomic system and in comparing them with the result of experiment. Postulate III states that, for an energy measurement on a series of identical systems to be exactly reproducible (that is, precise), the state of the system must be described by a function Ψ which is an eigenfunction of the operator corresponding to the total energy, the Hamiltonian. The problem of computing the allowed energies then is reduced to finding the Ψ_n and E_n that satisfy the eigenvalue equation

$$\mathscr{H}\Psi_n = E_n \Psi_n \tag{3-36a}$$

Substituting Eq. 3-35 into Eq. 3-36a, one obtains for a single-particle system

$$-\frac{\hbar^2}{2m}\nabla^2\Psi + V\Psi = E\Psi \tag{3-36b}$$

or

$$+\frac{\hbar^2}{2m}\nabla^2\Psi + (E - V)\Psi = 0 \tag{3-36c}$$

Equation 3-36c is Schrödinger's wave equation for a single particle in a stationary state.

If one is interested in calculating other properties of the system, such as the value of the angular momentum about the z axis, the procedure is the same, but the appropriate operator must be used in deriving the eigenvalue equation.

Many times we will wish to know the behavior of a property for a system not characterized by an eigenfunction appropriate to the operator corresponding to that property. This is the purpose of Postulate IV.

POSTULATE IV

Given an operator $\hat{\alpha}$ and a set of identical systems characterized by a function Ψ_s that is not an eigenfunction of $\hat{\alpha}$, a series of measurements of the property corresponding to $\hat{\alpha}$ on different members of the set will not give the same result. Rather, a distribution of results will be obtained, the average of which will be

$$\langle\hat{\alpha}\rangle = \frac{(\Psi_s|\hat{\alpha}|\Psi_s)}{(\Psi_s|\Psi_s)} \tag{3-37}$$

This is the so-called "mean-value" theorem that tells what the experimental result will be when a system is not described by an eigenfunction of the operator involved. The symbol $\langle\hat{\alpha}\rangle$ is called the *average* or *expectation* value of the quantity associated with $\hat{\alpha}$. The "average value" in quantum mechanics should not be confused with a time average in classical mechanics. Rather, it is the number average of a large number of measurements of the property corresponding to $\hat{\alpha}$. Obviously, if Ψ_s is an eigenfunction of $\hat{\alpha}$, the average value will be the same as the eigenvalue.

Much modern research in quantum chemistry and spectroscopy is concerned with time-dependent phenomena. In this case the problem is to know how the state function $\Psi(q, t)$ develops in time. We therefore introduce Postulate V.

POSTULATE V
The evolution of a state vector $\Psi(q, t)$ in time is given by the relation

$$i\hbar \frac{\partial \Psi}{\partial t} = \mathscr{H}\Psi \tag{3-38}$$

where $\mathscr{H}$ is the Hamiltonian operator for the system.

Equation 3-38 is called the *time-dependent* Schrödinger equation.

If the Hamiltonian operator $\mathscr{H}$ does not depend explicitly on the time, then it is always possible to find a formal solution to Eq. 3-38 of the form

$$\Psi(q, t) = \Psi_0(q)e^{-(i/\hbar)At} \tag{3-39}$$

To show that this is the case, one substitutes Eq. 3-39 into Eq. 3-38 to obtain

$$i\hbar \frac{\partial \Psi(q, t)}{\partial t} = i\hbar\Psi_0(q)\left(-\frac{i}{\hbar}\right)Ae^{-(i/\hbar)At} = \mathscr{H}[\Psi_0(q)\bar{e}^{(i/\hbar)At}] \tag{3-40}$$

If $\mathscr{H}$ does not depend on the time, then the exponential time dependence can be brought through the operator to give

$$i\hbar\left(-\frac{i}{\hbar}\right)A\Psi_0(q)e^{-(i/\hbar)At} = e^{-(i/\hbar)At}\mathscr{H}\Psi_0(q)$$

$$A\Psi_0(q) = \mathscr{H}\Psi_0(q) \tag{3-41}$$

which is the stationary state Schrödinger equation given in Eq. 3-36. Note, on comparing Eqs. 3-41 and 3-36, that the constant A must be the stationary state energy E. It will be necessary to bring in some features of time dependence when transition probabilities are discussed in Chapter 4.

3-4 APPLICATIONS OF THE POSTULATES TO SIMPLE SYSTEMS

As an application of the above postulates, we now discuss a simple problem—that of a particle constrained to move in a one-dimensional box. This problem is an excellent one because it illustrates a number of quantum mechanical principles, and at the same time shows how discrete energy levels inevitably arise whenever a small particle is confined to a region in space. Consider the situation shown in Figure 3-4. The particle is constrained to move in a one-dimensional box of length a. This constraint is brought about by making the potential energy inside the box zero, and everywhere outside the box infinity. The observable that we are interested in is the energy of a particle; therefore, the quantum mechanical operator that will be appropriate to the problem is the Hamiltonian $\hat{\mathscr{H}}$. According to Postulate III, if the same result is to be obtained when a series of measurements of the energy on a set of identical systems is made, the state of each particle must be described by an eigenfunction of $\hat{\mathscr{H}}$. Therefore, to find the allowed energies and wave functions for this particle, the eigenvalue equation $\hat{\mathscr{H}}\Psi_n = E_n \Psi_n$ must be solved. The solution is most conveniently divided into two parts corresponding to the regions outside and inside the box. Outside the box, T is the first term in Eq. 3-35, and $V = \infty$. Equation 3-36a then becomes

$$\frac{\hbar^2}{2m}\frac{d^2\Psi}{dx^2} + (E - \infty)\Psi = 0$$

or

$$\frac{d^2\Psi}{dx^2} = \infty\Psi \tag{3-42}$$

since E is small compared with infinity, and $(2m/\hbar^2)\infty$ is still infinity. A little thought should convince the student that there is no function subject to

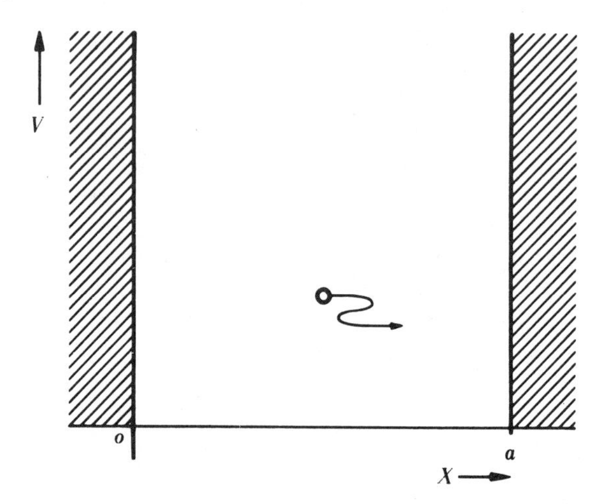

FIGURE 3-4 *Quantities characterizing the problem of a particle moving in a one-dimensional box. The potential energy is zero between $x = 0$ and $x = a$, and is infinity everywhere else. This is not the only choice of coordinates for this problem. Solutions with the origin at the middle of the box are discussed in Exercises 3-17 and 3-18.*

the restrictions discussed under Postulate I that, when differentiated twice, will give infinity times itself. Any function satisfying Eq. 3-42 would of necessity have a nonintegrable square. Therefore, the only solution outside the box is $\Psi = 0$. This means, according to Postulate I(b), that the probability of finding the particle somewhere outside the box is zero, which is the result we would expect. Inside the box, the eigenvalue equation is (since $V = 0$)

$$\frac{\hbar^2}{2m}\frac{d^2\Psi}{dx^2} + E\Psi = 0 \tag{3-43a}$$

or, by rearranging,

$$\frac{d^2\Psi}{dx^2} = -\frac{2mE}{\hbar^2}\Psi \tag{3-43b}$$

This is a second order differential equation whose solutions are functions that, when differentiated twice, will give the same function back multiplied by a constant. Recalling past discussions of this type of function in Chapters 1 and 2, we first try a function of the type

$$\Psi = A \sin \alpha x \tag{3-44}$$

(A second order differential equation will, in general, contain two arbitrary constants. These constants can be determined in the integrated form from the appropriate boundary conditions and the normalization requirement. Equation 3-44 only has one arbitrary constant A, and is not the most general solution. For the most general solution see Exercise 3-17.)

Differentiating Eq. 3-44 twice, we obtain

$$\frac{d\Psi}{dx} = \alpha A \cos \alpha x$$

$$\frac{d^2\Psi}{dx^2} = -\alpha^2 A \sin \alpha x = -\alpha^2\Psi \tag{3-45}$$

which is identical to Eq. 3-43b if we identify the constant α^2 with $2mE/\hbar^2$. Thus Eq. 3-44 is a solution to Eq. 3-43. So far in the solution, however, there is nothing which restricts the values which E can have.

We now apply the boundary conditions. The requirements that Ψ must be single valued means that it must become zero at the edges of the box. That is,

$$\Psi(0) = \Psi(a) = 0 \tag{3-46}$$

The quantity $\Psi(0)$ is clearly equal to zero, but $\Psi(a) = A \sin \alpha a$. Equation 3-46 requires that

$$\Psi(a) = A \sin \alpha a = 0$$

and this is true only if $\alpha a = n\pi$, where n is an integer. We thus have the requirement that

$$\alpha = \frac{n\pi}{a} \qquad (n = 1, 2, 3, \cdots)$$

$$\alpha^2 = \frac{2mE}{\hbar^2} = \frac{n^2\pi^2}{a^2} \tag{3-47}$$

Thus the energies which the particle can have become

$$E_n = \frac{\hbar^2 n^2 \pi^2}{2ma^2} = n^2\left(\frac{h^2}{8ma^2}\right) \qquad (n = 1, 2, 3, \cdots) \tag{3-48}$$

The important point to notice is that the imposition of the boundary conditions has restricted the energy to discrete values.

To complete the calculation of the wave functions, they should be normalized. This requires that

$$\int_0^a \left(A \sin \frac{n\pi x}{a}\right)^2 dx = 1 \tag{3-49}$$

Equation 3-49, when the integral is evaluated, requires that A must have the value

$$A = \left(\frac{2}{a}\right)^{\frac{1}{2}}$$

To summarize, the allowed wave functions and energies for the particle in a box are

$$\Psi_n = \left(\frac{2}{a}\right)^{\frac{1}{2}} \sin \frac{n\pi x}{a} \qquad E_n = \frac{n^2h^2}{8ma^2} \tag{3-50}$$

These results are represented diagrammatically in Figure 3-5.

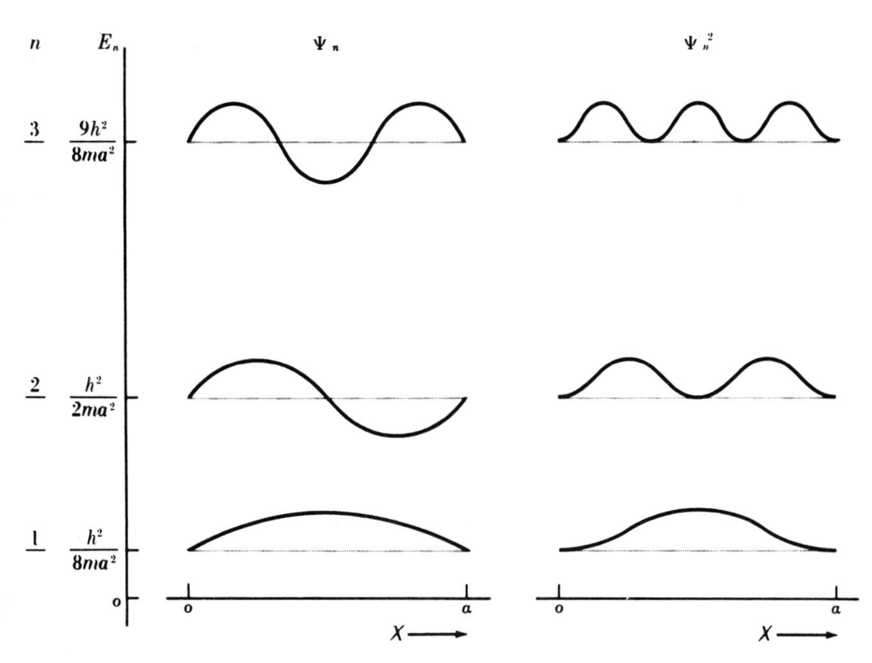

FIGURE 3-5 *Schematic drawing of E_n, Ψ_n, and Ψ_n^2 for the case of a particle moving in a one-dimensional box. The plot of Ψ_n as a function of x is a light line; the darker line is a plot of Ψ_n^2. Note that Ψ_n changes sign at each node while Ψ_n^2 always remains positive. Also note that the energy spacing between levels diverges as n increases.*

This example has several instructive features. For the same value of the quantum number n, the energy is inversely proportional to the mass of the particle and to the square of the length of the box. Thus, as the particle becomes heavier and the box larger, the energy levels become more closely spaced. It is only when the quantity ma^2 is of the same order as h^2 that quantized energy levels become important in experimental measurements. When dealing with dimensions of 1 g and 1 cm, the energy levels become so closely spaced that they seem to us to be continuous. The quantum mechanical formula, therefore, gives the classical result for systems with dimensions such that $ma^2 \gg h^2$. This is an illustration of the "correspondence principle" that states that the quantum mechanical result must become identical with the classical one in the limit where the quantum numbers describing the system become very large. These statements are illustrated by Exercises 3-11 and 3-12.

EXERCISE 3-11 Calculate the energy in cm^{-1} of the first two energy levels of a particle in a box, and the energy difference $\Delta E_{2-1} = E_2 - E_1$ for (a) an electron in a box 2 Å in length and (b) a ball bearing of mass 1 g in a box 10 cm long.

EXERCISE 3-12 Calculate the value of n necessary to give an energy equal to kT at room temperature for a 1-g ball bearing in a 10-cm long box. k is Boltzmann's constant and T is the absolute temperature.

The second feature of the solutions to the particle-in-a-box problem that should be pointed out is the relationship between the energy of a state and the number of nodes in the wave function. A node is a point where the wave function becomes zero. Neglecting the nodes at the end of the box, in the state $n = 2$ there is one node, in $n = 3$, two nodes, and in state n, $n - 1$ nodes. It is a general property of wave functions that the greater the number of nodes in a wave function, the higher the energy of the corresponding state. This is shown in Figure 3-5, and is reasonable when considered along with the de Broglie relationship, Eq. 3-24. The greater the number of nodes in the length of the box, the shorter the wavelength must be. According to Eq. 3-24, if the wavelength becomes shorter, the momentum, and hence the kinetic energy of the particle, must become greater.

A third important feature of the solutions to the particle-in-a-box problem is illustrated by the integral

$$(\Psi_1 | \Psi_2) = \frac{2}{a} \int_0^a \sin \frac{\pi x}{a} \sin \frac{2\pi x}{a} \, dx \tag{3-51a}$$

To evaluate integrals such as Eq. 3-51a, we make the substitutions

$$y = \frac{\pi x}{a} \qquad dy = \frac{\pi}{a} dx$$

and the integral then becomes

$$(\Psi_1 | \Psi_2) = \frac{2}{a} \left(\frac{a}{\pi}\right)^2 \int_0^\pi \sin y \sin 2y \, dy \tag{3-51b}$$

If we then use the trigonometric substitution

$$\sin 2y = 2 \sin y \cos y$$

followed by

$$u = \sin y \qquad du = \cos y \, dy$$

our integral becomes

$$(\Psi_1 | \Psi_2) = \frac{2a}{\pi^2} \cdot 2 \int_0^\pi \sin^2 y \cos y \, dy = \frac{4a}{\pi^2} \int_0^0 u^2 \, du = 0$$

Whenever an integral of the type $(\Psi_i|\Psi_j)$ equals zero, the two orbitals Ψ_i and Ψ_j are said to be *orthogonal*. Thus, our evaluation of $(\Psi_1|\Psi_2)$ in Eq. 3-51a shows us that Ψ_1 and Ψ_2 are orthogonal. In fact, we can show that all integrals of the type $(\Psi_i|\Psi_j)$ where $i \neq j$ equal zero for the particle in a box by proving the following general theorem about eigenfunctions of a Hermitian operator.

Theorem II. Eigenfunctions belonging to different eigenvalues of a Hermitian operator are orthogonal.

Let $\hat{F}$ be a Hermitian operator with eigenfunctions Ψ_i and Ψ_j such that

$$\hat{F}\Psi_i = f_i\Psi_i \qquad \hat{F}\Psi_j = f_j\Psi_j \tag{3-52a and b}$$

and where $f_i \neq f_j$. We next carry out the following steps:

1. Multiply Eq. 3-52a from the left by Ψ_j^* and integrate.
2. Take the complex conjugate of Eq. 3-52b, multiply from the left by Ψ_i, and integrate. This gives

$$(\Psi_j|\hat{F}|\Psi_i) = f_i(\Psi_j|\Psi_i) \tag{3-53a}$$

$$(\Psi_i|\hat{F}|\Psi_j)^* = f_j^*(\Psi_i|\Psi_j)^* = f_j(\Psi_i|\Psi_j)^* = f_j(\Psi_j|\Psi_i) \tag{3-53b}$$

In the next to last step in Eq. 3-53b we have used the fact that eigenfunctions of a Hermitian operator must be real. Subtracting Eq. 3-53b from 3-53a, we obtain

$$(\Psi_j|\hat{F}|\Psi_i) - (\Psi_i|\hat{F}|\Psi_j)^* = (f_i - f_j)(\Psi_j|\Psi_i) \tag{3-54}$$

Since $\hat{F}$ is Hermitian, the left-hand side of Eq. 3-54 is zero. The quantity $(f_i - f_j)$ cannot be zero and, therefore, $(\Psi_j|\Psi_i) = 0$ and the two functions Ψ_j and Ψ_i are orthogonal.

In the particle-in-a-box problem, the Hamiltonian is a Hermitian operator. All of the eigenvalues E_i are different and, thus, all of the eigenfunctions are orthogonal. There is a neat mathematical way of summarizing the values of integrals over any pair of functions for the particle in a one-dimensional box. We write

$$(\Psi_i|\Psi_j) = \delta_{ij}, \qquad \delta_{ij} = 1 \quad \text{for } i = j \tag{3-55}$$

$$\delta_{ij} = 0 \quad \text{for } i \neq j$$

where δ_{ij} is called the *Kronecker delta*. When Eq. 3-55 is true for a set of quantum mechanical functions the set is said to be *orthonormal.*

Orthogonality of functions plays a very important role in the mathematical formalism and manipulations of quantum mechanics and many qualitative conceptual arguments as well.

We next inquire about some other properties of the particle in a box. Suppose we are interested in measuring the component of momentum in the x direction for a set of identical systems in which the particle is known to be in the lowest energy state. The appropriate operator to use in the calculation of the expected result is $-i\hbar(d/dx)$, and we obtain

$$\hat{p}_x \Psi_1 = -i\hbar \frac{d}{dx}\left(A \sin \frac{\pi x}{a}\right) = -i\hbar A \frac{\pi}{a} \cos \frac{\pi x}{a} \tag{3-56a}$$

It is clear that Ψ_1 is not an eigenfunction of $\hat{p}_x$; therefore, according to Postulate IV, a series of measurements of $\hat{p}_x$ will not yield the same result. We must use the average-value theorem to calculate the expectation value of $\hat{p}_x$. This gives

$$\langle \hat{p}_x \rangle_1 = \int_0^a \Psi_1 \hat{p}_x \Psi_1 \, dx \Big/ \int_0^a \Psi_1{}^2 \, dx = \left[\frac{2}{a} \int_0^a \sin \frac{\pi x}{a} \left(-i\hbar \frac{\pi}{a}\right) \cos \frac{\pi x}{a} \, dx\right] \Big/ 1$$
$$= 0 \tag{3-56b}$$

Accordingly, the average of a large number of measurements of $\hat{p}_x$ on the set of identical systems is zero.

Suppose one now considers the square of the momentum in the x direction. The appropriate operator is $-\hbar^2(d^2/dx^2)$ and, applying this operator to Ψ_1, we obtain

$$-\hbar^2 \frac{d^2}{dx^2} A \sin \frac{\pi x}{a} = +\hbar^2 \frac{\pi^2}{a^2} A \sin \frac{\pi x}{a}$$

Ψ_1 is thus an eigenfunction of $p_x{}^2$, and a series of measurements of $p_x{}^2$ on a set of identical systems will always give the same result, namely, the eigenvalue

$$(p_x{}^2)_1 = \hbar^2 \frac{\pi^2}{a^2} = 2mE_1$$

Taking the square root, we obtain

$$(p_x)_1 = \pm(2mE_1)^{\frac{1}{2}} \tag{3-57}$$

The results calculated in Eqs. 3-56b and 3-57 present an interesting dilemma. The results of Eq. 3-56b indicate that the average value of $\langle p_x \rangle_1$ is zero. The results of Eq. 3-57 indicate that $(p_x)_1$ must be either $\pm(2mE_1)^{\frac{1}{2}}$. The apparent contradiction is resolved by considering the meaning of Postulates III and IV. Since a measurement of p_x^2 always gives the result $2mE_1$, the momentum p_x must always be either plus or minus $(2mE_1)^{\frac{1}{2}}$. A single measurement of p_x will give one of these values. What the mean-value postulate states is that, if one makes a large number of measurements of p_x, we end up with $(p_x)_1 = -(2mE_1)^{\frac{1}{2}}$ as often as $+(2mE_1)^{\frac{1}{2}}$, and the average value of p_x will be zero. The important point is that we never know in advance whether an experimental result will give plus or minus $(2mE_1)^{\frac{1}{2}}$. It can therefore be said that an uncertainty exists in our knowledge of the momentum, and the magnitude of this uncertainty is equal to $2(2mE_1)^{\frac{1}{2}}$.

In a similar manner, we can argue that if we know that the particle is in state Ψ_n, the only thing that can be said about the position of the particle is that it is somewhere in the box. That is, our uncertainty in the x coordinate of the particle is the length of the box, a. It is of interest to calculate the product of our uncertainties in the position and the momentum of a particle in a box. This is

$$\begin{aligned} \Delta x \, \Delta p_x &\geqq a \times 2(2mE_n)^{\frac{1}{2}} \geqq 2a \times \frac{n\pi\hbar}{a} \\ &\geqq nh \end{aligned} \tag{3-58a}$$

This will have its smallest value when $n = 1$, and thus we obtain the result that

$$\Delta x \, \Delta p_x \approx h \tag{3-58b}$$

This is a form of the *Heisenberg uncertainty principle*, which states that the simultaneous measurement of both the position and momentum of a particle cannot be made to an accuracy greater than Planck's constant h. Planck's constant is a very small number, however, and it is clear why the uncertainty principle is of no consequence in measurements on systems of large dimensions and/or containing particles with large masses.

EXERCISE 3-13 Calculate the uncertainty in momentum and velocity of an electron in a 1-Å box, a hydrogen atom in a 10-Å box, and a 1-g ball bearing in a 10-cm box.

The student may wonder why, given that the particle is described by state function Ψ_1, it is not possible to locate the particle more precisely than "somewhere in the box." To answer this question, suppose that we want to narrow the limits of uncertainty on the position of the particle, say to between $\frac{1}{4}a$ and $\frac{3}{4}a$. This means that the probability of finding the particle somewhere in this region must be *one* and outside of this region *zero*. But that is simply the problem of a particle in a smaller box (length $= \frac{1}{2}a$). As a consequence of the smaller box E_1 has been raised by a factor of 4 and our uncertainty in the momentum has been raised by 2, exactly compensating for the increased accuracy with which the particle can be located.

It should be emphasized that Eq. 3-58b is much more general than our derivation from the properties of a particle in a box would indicate. A completely general derivation gives the result

$$\Delta x \, \Delta p \geq \tfrac{1}{2}\hbar \tag{3-59}$$

where Δx and Δp are defined as root mean square derivations from the average value;

$$\Delta x \equiv (\langle x^2 \rangle - \langle x \rangle^2)^{\frac{1}{2}} \tag{3-60a}$$

$$\Delta p \equiv (\langle p^2 \rangle - \langle p \rangle^2)^{\frac{1}{2}} \tag{3-60b}$$

Many physicists believe that Eq. 3-59 is a fundamental property of nature. Professor Dirac, who developed one of the general statements of quantum mechanics, has a contrary opinion [9]. It should be emphasized, however, that the uncertainty relations are more than statements that presently available measuring instruments are inadequate.

Commuting operators are important in quantum theory because there is a link between the uncertainty principle and the commutation properties of operators. Before proceeding with this idea, we need to prove another theorem.

Theorem III. If two operators $\hat{F}$ and $\hat{G}$ commute, then there exists a set of functions that are simultaneous eigenfunctions of both operators.

Suppose we have a set of eigenfunctions of the operator $\hat{F}$. That is,

$$\hat{F}\Psi_i = f_i \Psi_i$$

We have the further postulate that $\hat{F}\hat{G} - \hat{G}\hat{F} = 0$. We will restrict ourselves to the case that none of the f_i are equal. The theorem can also be proved without this restriction. Operating on the above equation from the left by $\hat{G}$ and taking advantage of the commutative property of $\hat{F}$ and $\hat{G}$, we write

$$\hat{G}\hat{F}\Psi_i = \hat{F}(\hat{G}\Psi_i) = f_i(\hat{G}\Psi_i)$$

From the last equality, it is seen that the function $(\hat{G}\Psi_i)$ must also be an eigenfunction of $\hat{F}$ with eigenvalue f_i. The only way that this result can be consistent with our assumption that Ψ_i was the only function with eigenvalue

f_i is if the function $\hat{G}\Psi_i$ is a simple multiple of Ψ_i. Any other functional form of $\hat{G}\Psi_i$ would be a new function. Thus,

$$\hat{G}\Psi_i = g_i \Psi_i$$

must hold, and we have the result that Ψ_i is also an eigenfunction of the operator $\hat{G}$ with eigenvalue g_i. If there is more than one function Ψ_i with eigenvalue f_i, it is always possible to construct a linear combination of these functions which is an eigenfunction of the operator $\hat{G}$. This will not be proved, however.

This theorem has important consequences when it is considered along with Postulate III. If a set of functions that are simultaneous eigenfunctions of two operators exists, then it is possible to measure simultaneously and precisely the values of the observables corresponding to these operators. Theorem III tells us that such a condition exists if the two operators commute. Thus, there will be no uncertainty relationship between observables corresponding to commuting operators. The more general derivation of the uncertainty principle just mentioned relates the product of the uncertainties of two quantities to the commutator of the two corresponding quantum mechanical operators. Since the operators $\hat{P}_x$ and $\hat{x}$ do not commute (see Exercise 1-11), it will be impossible to measure both quantities simultaneously and precisely.

Theorem IV. Given a pair of commuting Hermitian operators, $\hat{F}$ and $\hat{G}$, and a set of functions such that

$$\hat{F}\Psi_i = f_i \Psi_i$$

then all integrals of the type $(\Psi_i|\hat{G}|\Psi_j) = 0$ unless $f_i = f_j$.

To prove this theorem, we consider the integral

$$(\Psi_i|\hat{F}\hat{G} - \hat{G}\hat{F}|\Psi_j)$$

Because $\hat{F}$ and $\hat{G}$ commute, this integral must vanish. Expanding we obtain

$$(\Psi_i|\hat{F}\hat{G}|\Psi_j) - (\Psi_i|\hat{G}\hat{F}|\Psi_j) = (\Psi_i|\hat{F}\hat{G}|\Psi_j) - f_j(\Psi_i|\hat{G}|\Psi_j) = 0$$

But because $\hat{F}$ and $\hat{G}$ are Hermitian and eigenvalues of a Hermitian operator must be real, we have

$$\begin{aligned}(\Psi_i|\hat{F}\hat{G}|\Psi_j) &= (\Psi_j|\hat{G}\hat{F}|\Psi_i)^* = f_i^*(\Psi_j|\hat{G}|\Psi_i)^* \\ &= f_i(\Psi_j|\hat{G}|\Psi_i)^* = f_i(\Psi_i|\hat{G}|\Psi_j)\end{aligned}$$

Substituting this equation into the one above gives

$$0 = f_i(\Psi_i|\hat{G}|\Psi_j) - f_j(\Psi_i|\hat{G}|\Psi_j) = (\Psi_i|\hat{G}|\Psi_j)(f_i - f_j)$$

If $f_i \neq f_j$, the second term cannot be zero and therefore,

$$(\Psi_i | \hat{G} | \Psi_j) = 0$$

Theorem IV is extremely valuable in quantum mechanical problems because it allows one to set many integrals equal to zero without evaluating them. Much use of this theorem will be made when symmetry is discussed in Chapter 8.

EXERCISE 3-14 Calculate the expectation value of the x position of a particle known to be in the state $n = 2$ of a one-dimensional box. What is the probability of finding the particle in a small unit of length dx at this position? Can you rationalize these results?

EXERCISE 3-15 Given a particle in the state $n = 1$ of a one-dimensional box, what is the probability of finding it somewhere in the region between 0 and $\frac{1}{4}a$?

EXERCISE 3-16 Evaluate for a particle in a one-dimensional box the integral $(\Psi_1 | x - \frac{1}{2}a | \Psi_2)$. Use the orthogonality condition to simplify if possible.

There are, of course, other solutions to Eq. 3-43. The most general are

$$\Psi = A \sin \alpha x + B \cos \alpha x \tag{3-61a}$$

and

$$= Ae^{-i\alpha x} + Be^{+i\alpha x} \tag{3-61b}$$

The use and interpretation of these solutions are illustrated in Exercises 3-17 and 3-18.

EXERCISE 3-17 Show that Eq. 3-61a is a solution to Eq. 3-43, with the same value for α. Use the appropriate boundary conditions to evaluate the constants A and B. Suppose, instead of the coordinates of Figure 3-4, one were to let the box go from $x = -\frac{1}{2}a$ to $x = +\frac{1}{2}a$. That is, put the origin in the middle of the box. What values will the constants A and B in Eq. 3-61a have then?

EXERCISE 3-18 Show that Eq. 3-61b is a solution of Eq. 3-43. Evaluate the constants A and B (see Eq. 1-23). Reevaluate A and B, placing the origin at the middle of the box. What happens when p_x operates on either half of Eq. 3-61b? What does a solution in the form of Eq. 3-61b imply about measurements of p_x?

It is instructive also to consider the problem of a particle in a three-dimensional box because it illustrates a particular technique which will be useful in solving other quantum mechanical problems. The eigenvalue equation for inside the box becomes

$$-\frac{\hbar^2}{2m}\nabla^2\Psi = E\Psi$$

or

$$\frac{\partial^2\Psi}{\partial x^2} + \frac{\partial^2\Psi}{\partial y^2} + \frac{\partial^2\Psi}{\partial z^2} = -\frac{2mE}{\hbar^2}\Psi \tag{3-62}$$

To solve Eq. 3-62, the technique of separation of variables is used. To separate the variables in a differential equation, we try to find a solution of the form

$$\Psi = X(x)Y(y)Z(z) \tag{3-63}$$

where $X(x)$, $Y(y)$, and $Z(z)$ are functions of only x, y, or z, respectively. Substituting Eq. 3-63 into Eq. 3-62, and performing the indicated partial differentiations, we obtain

$$YZ\frac{\partial^2 X}{\partial x^2} + XZ\frac{\partial^2 Y}{\partial y^2} + XY\frac{\partial^2 Z}{\partial z^2} = -\frac{2mE}{\hbar^2}XYZ \tag{3-64}$$

Dividing both sides of Eq. 3-64 by XYZ and rearranging, we obtain

$$\frac{1}{X}\frac{\partial^2 X}{\partial x^2} + \frac{1}{Y}\frac{\partial^2 Y}{\partial y^2} + \frac{2mE}{\hbar^2} = -\frac{1}{Z}\frac{\partial^2 Z}{\partial z^2} \tag{3-65}$$

Equation 3-65 must hold for all values of x, y, and z. The only way that this can be true is if both sides of the equation are equal to a constant. We arbitrarily (at this point) call this constant $2mE_z/\hbar^2$. Equation 3-65 then becomes two equations,

$$-\frac{1}{Z}\frac{\partial^2 Z}{\partial z^2} = \frac{2mE_z}{\hbar^2} \tag{3-66}$$

and

$$\frac{1}{X}\frac{\partial^2 X}{\partial x^2} + \frac{1}{Y}\frac{\partial^2 Y}{\partial y^2} + \frac{2mE}{\hbar^2} = \frac{2mE_z}{\hbar^2} \tag{3-67}$$

But Eq. 3-67 can be rearranged to give

$$\frac{1}{X}\frac{\partial^2 X}{\partial x^2} + \frac{2m}{\hbar^2}(E - E_z) = -\frac{1}{Y}\frac{\partial^2 Y}{\partial y^2} \tag{3-68}$$

and, applying the same argument as above, both sides of this equation must equal a constant, which we call $2mE_y/\hbar^2$. We are then left with

$$-\frac{1}{Y}\frac{\partial^2 Y}{\partial y^2} = \frac{2mE_y}{\hbar^2} \tag{3-69}$$

and

$$\frac{1}{X}\frac{\partial^2 X}{\partial x^2} = -\frac{2m}{\hbar^2}(E - E_z - E_y) = -\frac{2m}{\hbar^2}E_x \tag{3-70}$$

Equations 3-66, 3-69, and 3-70 are exactly the same as the equation for a particle in a one-dimensional box except that Ψ has been replaced by X, Y, and Z, and E has been replaced by E_x, E_y, and E_z. The one-dimensional problem has already been solved, though, and we can write down the answer immediately. If a, b, and c are the lengths of the box in the x, y, and z directions, respectively, then

$$\begin{aligned} X &= \left(\frac{2}{a}\right)^{\frac{1}{2}} \sin\frac{n_x \pi x}{a} \qquad & E_x &= \frac{n_x^2 h^2}{8ma^2} \\ Y &= \left(\frac{2}{b}\right)^{\frac{1}{2}} \sin\frac{n_y \pi y}{b} \qquad & E_y &= \frac{n_y^2 h^2}{8mb^2} \\ Z &= \left(\frac{2}{c}\right)^{\frac{1}{2}} \sin\frac{n_z \pi z}{c} \qquad & E_z &= \frac{n_z^2 h^2}{8mc^2} \end{aligned} \tag{3-71}$$

and

$$\Psi = XYZ = \left(\frac{8}{abc}\right)^{\frac{1}{2}} \sin\frac{n_x \pi x}{a} \sin\frac{n_y \pi y}{b} \sin\frac{n_z \pi z}{c} \tag{3-72}$$

$$\begin{aligned} E &= E_x + E_y + E_z \\ &= \frac{h^2}{8m}\left(\frac{n_x^2}{a^2} + \frac{n_y^2}{b^2} + \frac{n_z^2}{c^2}\right) \end{aligned} \tag{3-73}$$

One further point of interest arises in a three-dimensional box if all three sides have equal lengths. Then $a = b = c$ and Eq. 3-73 becomes

$$E = \frac{h^2}{8ma^2}(n_x^{\,2} + n_y^{\,2} + n_z^{\,2}) \tag{3-74}$$

Suppose we consider the state with next to lowest energy. This state arises when one of the quantum numbers is 2 and the other two are 1 and $E = \frac{3}{4}(h^2/ma^2)$. There are three different combinations of quantum numbers which will give this energy, however. If the values of three quantum numbers n_x, n_y, and n_z are listed in parentheses after the energy, we can express this as

$$E(2, 1, 1) = E(1, 2, 1) = E(1, 1, 2) = \frac{3}{4}\frac{h^2}{ma^2} \tag{3-75}$$

When more than one state has the same energy, the states are said to be *degenerate*. The number of states with the same energy is the degree of degeneracy. Thus the state with next to lowest energy for a particle in a cubical box is *threefold degenerate*.

The problem of a particle in a box has been used in this chapter to illustrate some of the fundamental ideas of quantum mechanics. This problem has practical applications as well, however. A one-dimensional box model has been quite successful in correlating the wavelengths of maximum absorption of certain long conjugated dye molecules [10]. A two-dimensional box model gives quite good electron density distribution maps in aromatic systems [11]. The solutions of the three-dimensional box problem are used in the derivation of the expression for the translational partition function in statistical mechanics [1–3]. The first two of these applications are illustrated in Exercises 3-19 through 3-23.

EXERCISE 3-19 Consider the case of hexatriene,

$CH_2{=}CH{-}CH{=}CH{-}CH{=}CH_2$

and assume that the π electrons (see Section 7-8 if the idea of π electrons is not familiar) are free to move along the length of the molecule. Approximate the energy levels of this system by using a one-dimensional box model whose length is the length of the molecule plus one C—C single bond length. Use 1.54 Å as a C—C and 1.35 Å as a C═C length. There are six π electrons and only two of them may be placed in each energy level. Using these six electrons to fill the three lowest energy levels calculate the following quantities.

a. The energy of the highest filled level.

b. The energy of the lowest unfilled level.

c. The difference in energy between the highest energy filled and the lowest energy unfilled levels. This energy difference should be approximately equal to the energy of the longest wavelength absorption band in the ultraviolet visible spectrum of this dye. Compare your calculated energy with the experimental wavelength of maximum absorption $\lambda_{max} = 268$ mμ (millimicrons, 1 $m\mu = 10^{-7}$ cm).

d. Derive an expression for λ_{max} for the general polyene H—(CH═CH)$_k$—H. Leave the length of the chain as an undetermined parameter in the expression for λ. See [10] for help.

EXERCISE 3-20 Assume that there are two particles in each of the two lowest energy levels of a one-dimensional box of length a. Compute the total particle density at the points $x = \frac{1}{8}a, \frac{1}{4}a, \frac{3}{8}a$, and $\frac{1}{2}a$. Since all functions are symmetric about $x = \frac{1}{2}a$, the particle densities at $\frac{5}{8}a, \frac{3}{4}a, \frac{7}{8}a$ will also be known. Plot the particle density as a function of x.

EXERCISE 3-21 Calculate the allowed energies and wave functions for a particle constrained to move on a rectangular surface of side lengths a and b.

EXERCISE 3-22 Benzene has its carbon atoms arranged in a regular hexagon with side length 1.39 Å. It has six π electrons. Adapt the solutions from Exercise 3-21 to the situation of a square and assume that there are six particles filling the three lowest energy levels. Take a section at $y = \frac{1}{2}a$ and plot the particle density as a function of x (See Exercise 3-20). A section at $x = \frac{1}{2}a$ will be identical by symmetry. Next take a section along $x = y$ and plot the particle density. Draw a contour map on a piece of graph paper showing particle density in the xy plane. Note that the electron density has a maximum in a circle. Using 4×1.39 Å as the length of a side, this model provides a good representation of the π electron distribution of benzene. This works for naphthalene and higher aromatics as well. See [11].

EXERCISE 3-23 Calculate the allowed energies and wave functions for a particle constrained to move on a circle of radius R. What causes the energy levels to be quantized in this case? What is the lowest value that the quantum number can have? Why is this lowest value different for a particle on a circle and a particle in a box?

The separation of variables in the three-dimensional box problem is an example of a procedure that is frequently used in quantum chemistry. There is a general rule about when such a separation can be made. When the Hamiltonian operator $\mathscr{H}$ can be written as a sum of terms, each of which is a function of only one variable, then it is always possible to find a solution to Eq. 3-36 of the form in which Ψ is a product of single coordinate functions. In more mathematical terms, when

$$\mathscr{H} = \sum_i \hat{h}_i \tag{3-76}$$

where each $\hat{h}_i$ can be expressed as a function of a single coordinate and derivatives with respect to that coordinate, then Ψ can always be written in the form

$$\Psi = \prod_i \phi_i(q_i) \tag{3-77}$$

where $\phi_i(q_i)$ depends only on the single coordinate q_i. It is true, furthermore, that the total energy E can be expressed as a sum of single orbital energies ε_i, where each orbital energy arises from the motion with respect to one coordinate q_i. Thus

$$E = \sum_i \varepsilon_i \tag{3-78}$$

where the index i goes over all of the coordinates of all of the particles.

The separation of variables is used widely in "many-electron" problems in quantum mechanics. In these problems, the assumption is often made that the total Hamiltonian $\mathscr{H}$ can be written as a sum of single particle or "one-electron" operators, that is, operators that depend only on the coordinates of a single electron or particle. If this is true, then the total wave function of the system can be written as a product of single-electron functions, and the energy is the sum of the single-electron energies. This assumption about many-electron systems is called the independent-particle model. It is usually a poor assumption because all of the interparticle interaction terms must be left out of the Hamiltonian. This point is discussed at some length in a later chapter. Since the procedure of the separation of variables will be used many times in the material that follows, it is important that the student have a good understanding of this procedure.

A comment should be made on some of the nomenclature introduced in the above discussion of separation of variables. In that discussion and in the material that follows, a single-particle wave function and energy are designated by lower case Greek letters, usually ψ, ϕ, χ for wave functions and ε for energy. A many-particle wave function will be designated by a capital Greek letter and a many-particle energy by a capital E. A small $\hat{h}$ will designate a single-particle *approximate* Hamiltonian. The capital script $\mathscr{H}$ will designate either an exact or a many-particle Hamiltonian.

3-5 PERTURBATION THEORY

In practice it is possible to find exact solutions to the eigenvalue equation (Eq. 3-36) for only a very few systems. For other problems it is necessary to try and find approximate solutions. Two main approximation methods are used for this purpose: the linear variation method to be discussed in Chapter 7 and perturbation theory to be discussed in this section. The following material is somewhat more difficult than the preceding, and the remainder of the book does not require a mastery of this section.

Perturbation theory is useful when the problem that one is trying to solve is almost like a problem that can be solved exactly. In mathematical terms, this means that the solutions of some zero-order problem

$$\mathscr{H}^0\Psi_m{}^0 = E_m{}^0\Psi_m{}^0 \tag{3-79}$$

are known and that we wish to solve a new problem

$$\mathscr{H}\Psi_m = E_m\Psi_m \tag{3-80}$$

where $\mathscr{H}$ is only slightly different from $\mathscr{H}^0$ and consequently where E_m and Ψ_m will not differ too much from $E_m{}^0$ and $\Psi_m{}^0$, respectively. To do this we write the Hamiltonian as

$$\mathscr{H} = \mathscr{H}^0 + \lambda\mathscr{H}' \tag{3-81a}$$

where the term $\lambda\mathscr{H}'$ represents a small perturbation on $\mathscr{H}^0$. The quantity λ is an arbitrary multiplier which is introduced for reasons that will soon become apparent. It is clear, however, that what is being sought are solutions of Eq. 3-80 such that

$$E_m{}^0 = \lim_{\lambda\to 0} E_m \tag{3-81b}$$

$$\Psi_m^0 = \lim_{\lambda\to 0} \Psi_m \tag{3-81c}$$

The mathematical manipulations to be used in the following, while not difficult, may be unfamiliar to many students. Since it is the expression for the perturbation corrections to the energy and wave functions that will be useful in some later parts of the book, the student may want to skip directly to these results, Eqs. 3-92 and 3-93. Students who want to know where results come from should keep reading.

It is assumed that both Ψ_m and E_m can be expanded in a power series in λ. Thus

$$\Psi_m = \Psi_m{}^0 + \lambda\Psi_m^{(1)} + \lambda^2\Psi_m^{(2)} + \cdots$$

or

$$\Psi_m = \sum_{n=0}^{\infty} \lambda^n\Psi_m^{(n)} \tag{3-82a}$$

and

$$E_m = E_m{}^0 + \lambda E_m^{(1)} + \lambda^2 E_m^{(2)} + \cdots$$

or

$$E_m = \sum_{n=0}^{\infty} \lambda^n E_m^{(n)} \tag{3-82b}$$

Equations 3-82 satisfy the conditions of Eqs. 3-81. The problem is to evaluate the so-called first, second, and so on, order corrections to the energy E_m and wave function Ψ_m. These are the coefficients of the corresponding powers of λ in Eqs. 3-82. Thus $E_m^{(1)}$ is the first order correction to the zero order energy $E_m{}^0$.

To evaluate these coefficients, we substitute Eqs. 3-82 and 3-81 into Eq. 3-80 to obtain

$$(\mathscr{H}^0 + \lambda\mathscr{H}')(\Psi_m{}^0 + \lambda\Psi_m^{(1)} + \lambda^2\Psi_m^{(2)} + \cdots) = (E_m{}^0 + \lambda E_m^{(1)} + \cdots)(\Psi_m{}^0 + \lambda\Psi_m^{(1)} + \lambda^2 \cdots) \quad (3\text{-}83)$$

Collecting terms according to powers of λ gives

$$\mathscr{H}^0 \sum_{n=0}^{\infty} \lambda^n \Psi_m^{(n)} + \mathscr{H}' \sum_{n=0} \lambda^{n+1}\Psi_m^{(n)} = E_m{}^0\Psi_m{}^0 + \lambda(E_m^{(1)}\Psi_m{}^0 + E_m{}^0\Psi_m^{(1)}) + \lambda^2(\cdots) + \cdots \quad (3\text{-}84)$$

If it can be assumed that Ψ_m and E_m are continuous functions of λ, then it follows that for Eq. 3-84 to be true for all values of λ, the coefficients of each power of λ on the left- and right-hand sides of the equation must be equal. Using this fact, we obtain the results

$$\mathscr{H}^0\Psi_m{}^0 = E_m{}^0\Psi_m{}^0 \quad (3\text{-}85a)$$

$$\mathscr{H}^0\Psi_m^{(1)} + \mathscr{H}'\Psi_m{}^0 = E_m^{(1)}\Psi_m{}^0 + E_m{}^0\Psi_m^{(1)} \quad (3\text{-}85b)$$

and so on.

EXERCISE 3-24 Write down the next equation in the series using Eqs. 3-83 and 3-84.

Equations 3-85 are the zero and first order perturbation equations. Since it is our postulate that the zero order solutions are available, they can be used to find the first order correction to the energy $E_m^{(1)}$ and to the wave function $\Psi_m^{(1)}$.

Rewriting Eq. 3-85b, we obtain

$$(\mathscr{H}^0 - E_m{}^0)\Psi_m^{(1)} + \mathscr{H}'\Psi_m{}^0 = E_m^{(1)}\Psi_m{}^0 \quad (3\text{-}86)$$

Multiplying Eq. 3-86 from the left by Ψ_m^{0*} and integrating gives

$$(\Psi_m{}^0 | \mathscr{H}^0 - E_m{}^0 | \Psi_m^{(1)}) + (\Psi_m{}^0 | \mathscr{H}' | \Psi_m{}^0) = E_m^{(1)} \quad (3\text{-}87)$$

The first term in Eq. 3-87 is zero because of the Hermitian property of $\mathscr{H}^0$, and therefore the first order correction to the energy is

$$E_m^{(1)} = (\Psi_m{}^0 | \mathscr{H}' | \Psi_m{}^0) \equiv H'_{mm} \tag{3-88}$$

EXERCISE 3-25 Use the fact that $\mathscr{H}^0$ is Hermitian to show that

$$(\Psi_m{}^0 | \mathscr{H}^0 - E_m{}^0 | \Psi_m^{(1)}) = 0$$

To obtain $\Psi_m^{(1)}$ we make use of the fact, to be introduced here without proof, that the function $\Psi_m^{(1)}$ can be written as a linear combination of the zero order functions $\Psi_i{}^0$. Thus

$$\Psi_m^{(1)} = \sum_i a_{im} \Psi_i{}^0 \tag{3-89}$$

where the a_{im} are coefficients that need to be determined in order to specify $\Psi_m^{(1)}$. Substituting Eq. 3-89 into Eq. 3-85b, multiplying from the left by Ψ_k^{0*}, and integrating, we obtain

$$(\Psi_k{}^0 | \mathscr{H}^0 - E_m{}^0 | \sum a_{im} \Psi_i{}^0) + (\Psi_k{}^0 | \mathscr{H}' | \Psi_m{}^0) = E_m^{(1)}(\Psi_k{}^0 | \Psi_m{}^0) \tag{3-90}$$

For $k = m$, one obtains Eq. 3-88. For all other k, the right-hand side of Eq. 3-90 vanishes since we can take the $\Psi_k{}^0$ to be an orthonormal set. The only nonvanishing part of the first term on the left-hand side of Eq. 3-90 is the $a_{km} \Psi_k{}^0$ term, and we are therefore left with

$$(\Psi_k{}^0 | a_{km}(E_k{}^0 - E_m{}^0) | \Psi_k{}^0) = -(\Psi_k{}^0 | \mathscr{H}' | \Psi_m{}^0) \tag{3-91a}$$

or

$$a_{km} = -\frac{(\Psi_k{}^0 | \mathscr{H}' | \Psi_m{}^0)}{E_k{}^0 - E_m{}^0} \tag{3-91b}$$

$$= \frac{H'_{km}}{E_m{}^0 - E_k{}^0} \qquad k \neq m \tag{3-91c}$$

Equation 3-91c determines all of the a_{km} except for a_{mm}. Applying the normalization condition and only keeping terms to first order in λ it can be shown that $a_{mm} = 0$. The final results for E_m and Ψ_m correct to first order are

$$E_m = E_m{}^0 + \lambda H'_{mm} \tag{3-92a}$$

$$\Psi_m = \Psi_m{}^0 + \lambda \sum_k{}' \frac{H'_{km}}{E_m{}^0 - E_k{}^0} \Psi_k{}^0 \tag{3-92b}$$

where the prime on the sum sign in Eq. 3-92b means that the term $k = m$ is omitted. When $\mathscr{H}'$ itself is small it is usual to set $\lambda = 1$ in Eqs. 3-92. $\mathscr{H}'$ is considered small if the integrals H'_{km} are considerably less than the energy differences $E_m{}^0 - E_k{}^0$. This means that the mixing coefficients a_{km} must be on the order of 0.1 or less.

The second order correction to the energy can be obtained by substituting Eqs. 3-92 into the second order perturbation equation worked out in Exercise 3-24. Only the result will be given here.

$$E_m^{(2)} = \sum_k{}' \frac{H'_{mk} H'_{km}}{E_m{}^0 - E_k{}^0} \tag{3-93}$$

The second order correction to the wave function and higher order corrections to both E_m and Ψ_m are seldom used. Students may wish to test their understanding of the material in this section by deriving Eq. 3-93.

The derivation in this section only holds for cases where there are no degenerate levels. It is clear that if states k and m have the same energy, then the denominators in Eqs. 3-91, 3-92b, and 3-93 become zero and the expansion blows up. Cases where there is degeneracy can be handled, and the interested student is referred to some of the standard quantum chemistry texts for more details. There is an especially good discussion of perturbation theory in [7].

As an example of the application of perturbation theory, consider the problem of a particle in a box with a sloping bottom shown in Figure 3-6. The zero order problem is the particle in a one-dimensional box whose solutions are given by Eqs. 3-50. The perturbation Hamiltonian $\mathscr{H}'$ is

$$\mathscr{H}' = \frac{x}{a} V_1 \tag{3-94}$$

where V_1 is the height of the potential at $x = a$. The first order correction to the energy of the nth level is from Eq. 3-88.

$$\begin{aligned} E_n^{(1)} &= \left(\Psi_n \middle| x \frac{V_1}{a} \middle| \Psi_n\right) \\ &= \left(\frac{V_1}{a}\right) \int_0^a \left(\frac{2}{a}\right)^{\frac{1}{2}} \left(\sin \frac{n\pi x}{a}\right)(x) \left(\frac{2}{a}\right)^{\frac{1}{2}} \left(\sin \frac{n\pi x}{a}\right) dx \qquad (3\text{-}95) \\ &= \frac{2V_1}{a^2} \int_0^a x \sin^2 \frac{n\pi x}{a}\, dx \end{aligned}$$

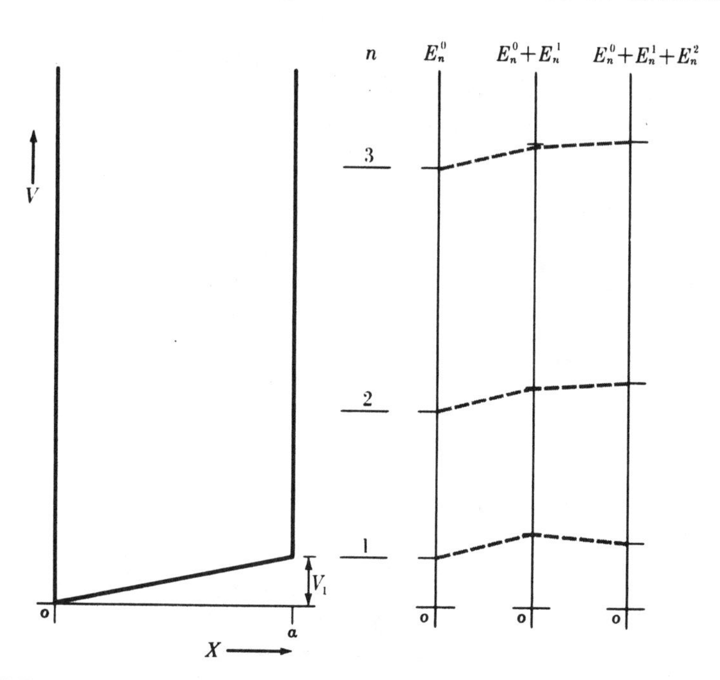

FIGURE 3-6 *Quantities characterizing the problem of a particle moving in a one-dimensional box with a linear potential gradient added as a perturbation. The first order energy correction calculated from perturbation theory raises the energy of all levels by an equal amount. The second order correction changes the energy of each level by a different amount, but the differences are too small to show in the figure.*

Evaluation of the integral in Eq. 3-95 (see Exercise 3-26) gives

$$E_n^{(1)} = \tfrac{1}{2}V_1 \tag{3-96}$$

It is seen that to first order the perturbation raises the energies of all of the levels by a constant amount $\frac{1}{2}V_1$. This is shown in Figure 3-6.

EXERCISE 3-26 Evaluate the integral in Eq. 3-95 and show that the result given in Eq. 3-96 is correct. *Hint*: First make the substitution $y = n\pi x/a$. The resulting integral is

$$\int y \sin^2 y \, dy = \frac{y^2}{4} - \frac{y \sin 2y}{4} - \frac{\cos 2y}{8}$$

The first order corrections to the wave function are given by Eq. 3-91c, and for this problem they are

$$a_{km} = \left(\frac{2}{a^2} V_1 \int_0^a x \sin \frac{k\pi x}{a} \sin \frac{m\pi x}{a} \, dx\right) \bigg/ (E_m{}^0 - E_k{}^0) \tag{3-97}$$

The denominator in Eq. 3-97 is equal to $(h^2/8ma^2)(m^2 - k^2) = (m^2 - k^2)E_1$; the numerator is a general form of the integral evaluated in Exercise 3-16. To evaluate this integral we use the trigonometric relation

$$\sin\theta \sin\phi = \tfrac{1}{2}[\cos(\theta - \phi) - \cos(\theta + \phi)] \tag{3-98}$$

which, when substituted into the expression for H'_{km}, gives

$$\begin{aligned}
H'_{km} &= \frac{1}{\pi^2} V_1 \int_0^{\pi} y[\cos(k-m)y - \cos(k+m)y]\, dy \\
&= \frac{1}{\pi^2} V_1 \left(\frac{1}{(k-m)^2} \int_0^{(k-m)\pi} z \cos z dz - \frac{1}{(k+m)^2} \int_0^{(k+m)\pi} y \cos y\, dy \right) \\
&= \frac{1}{\pi^2} V_1 \left(-\frac{2}{(k-m)^2} + \frac{2}{(k+m)^2} \right) \qquad (k-m,\ k+m \text{ odd}) \\
&= \frac{2}{\pi^2} V_1 \left(\frac{1}{(k+m)^2} - \frac{1}{(k-m)^2} \right) \qquad (3\text{-}99)
\end{aligned}$$

$$H'_{km} = 0 \qquad (k-m,\ k+m \text{ even}) \tag{3-100}$$

EXERCISE 3-27 Work out the details of getting to Eq. 3-99.

This result shows that not all of the zero order states are mixed into any given state. In fact, if the state of interest (m) is characterized by an even quantum number, then only states with odd quantum numbers will be mixed in and vice versa. Furthermore, for any k and m the $(k - m)$ term will make a larger contribution to Eq. 3-99 than the $(k + m)$ term, and the former will make its biggest contribution when $k - m = \pm 1$. Thus adjacent energy states in the zero order problem are most strongly mixed.

To get an idea of the magnitudes of the quantities involved, some mixing coefficients will be evaulated. Let state m be the lowest energy state Ψ_1. The index k then goes all over the other states $(2, 3, 4, 5, \cdots)$. Using Eqs. 3-99 and 3-100, one obtains

$$H'_{21} = \frac{2}{\pi^2} V_1 \left(\frac{1}{3^2} - \frac{1}{1^2} \right) = -\frac{16}{9} \frac{V_1}{\pi^2} = -0.18 V_1 \tag{3-101}$$

$$H'_{31} = 0$$

$$H'_{41} = \frac{2}{\pi^2} V_1 \left(\frac{1}{5^2} - \frac{1}{3^2} \right) = -\frac{32}{225} \frac{V_1}{\pi^2} = -0.0144 V_1 \tag{3-102}$$

$$H'_{51} = 0$$

and so on. Since H'_{41} is less than a tenth as large as H'_{21}, the perturbation sum is converging very rapidly and it is not necessary to include higher terms.

We next need to examine V_1 to see how large it can be for perturbation theory to be valid. We see that the integral $H'_{21} = -0.18V_1$ while the denominator in Eq. 3-97 for this case would be

$$E_1{}^0 - E_2{}^0 = (1^2 - 2^2)E_1 = -3E_1 \tag{3-103}$$

Thus V_1 can be as large as E_1 and still have the largest mixing coefficient only about 0.06.

We can now write the mixing coefficients in terms of the ratio V_1/E_1. They are

$$a_{12} = \frac{0.18}{3}\left(\frac{V_1}{E_1}\right) = 0.06\left(\frac{V_1}{E_1}\right) \tag{3-104}$$

$$a_{13} = a_{15} = \cdots = 0$$

$$a_{14} = \frac{0.0144}{15}\left(\frac{V_1}{E_1}\right) = 0.00096\left(\frac{V_1}{E_1}\right) \tag{3-105}$$

The perturbed wave function for Ψ_1 is, to a high degree of approximation,

$$\Psi_1 = \Psi_1{}^0 + 0.06\left(\frac{V_1}{E_1}\right)\Psi_2{}^0 + 0.00096\left(\frac{V_1}{E_1}\right)\Psi_4{}^0 \tag{3-106}$$

EXERCISE 3-28 Calculate the perturbation corrections to $\Psi_2{}^0$.

If the functions $\Psi_k{}^0$ and $\Psi_m{}^0$ are real, the integrals H'_{km} and H'_{mk} are equal and the expression for the second order correction to the energy is

$$E_m^{(2)} = \sum_k{}' \frac{H_{mk}^2}{E_m{}^0 - E_k{}^0} \tag{3-107}$$

Using the values of the matrix elements determined above we have for the second order correction to E_1 (keeping only the first two terms in the sum)

$$E_1^{(2)} = \frac{(-0.180V_1)^2}{-3E_1} + \frac{(-0.0144V_1)^2}{-15E_1}$$

$$= -0.0109\frac{V_1{}^2}{E_1} - 0.0000139\frac{V_1{}^2}{E_1}$$

$$\approx -0.0109\frac{V_1{}^2}{E_1} \tag{3-108}$$

The energy E_1 is thus correct to second order.

$$E_1 = E_1^{\,0} + 0.500V_1 - 0.0109\,\frac{V_1^{\,2}}{E_1} \qquad \text{(3-109)}$$

It should be pointed out that the first order correction to E is independent of m but the second order correction depends on m. Thus any property of interest that depends on the spacing between levels will not be affected by the first order correction but will be affected by the second order correction.

EXERCISE 3-29 Calculate the perturbation corrections to E_2.

EXERCISE 3-30 (Suitable for a take-home exam.) Use perturbation theory to calculate the energy levels of a particle in a box with a cosine function bottom. Let the box extend from $0 \geqq x \geqq a$ and let the perturbing potential be $\mathscr{H}' = V_0[1 + \cos(2\pi mx/a)]$. This potential oscillates between $2V_0$ and 0. The number of oscillations is determined by m. Discuss restrictions on the possible values of $k + 1$ and $k - 1$ in the integrals H'_{kl}.

3-6 SUMMARY

1. The historical development of quantum mechanics was traced, and a discussion of three significant experiments in this development was given. These three experiments involved the study of blackbody radiation, the photoelectric effect, and atomic spectra.
2. Bohr's theory of the energy levels of the hydrogen atom was presented.
3. Quantum mechanics was introduced with five postulates that relate a mathematical formalism used in the theory to experimental measurements performed in the laboratory.
4. The problem of a particle in a one-dimensional box was discussed. It was shown that quantized energy levels arise when the boundary conditions are applied to the solutions of the appropriate eigenvalue equation. The various postulates were illustrated with this simple problem.
5. The Heisenberg uncertainty principle was introduced using the solutions of a particle in a box as an example.
6. The method of the separation of variables for solving differential equations was illustrated for the problem of the energy levels in a three-dimensional box.
7. Perturbation theory was introduced and applied to the problem of a one-dimensional potential well with a sloping bottom.
8. Many new terms were introduced. The student should study the meanings of the following terms: Hermitian operator, normalized function, orthogonal function, stationary state, degenerate orbitals, Kronecker delta, correspondence principle, and first and second order perturbation corrections.

REFERENCES

1. N. Davidson, *Statistical Mechanics* (McGraw-Hill Book Co., New York, 1962), Chap. 12.
2. H. Eyring, D. Henderson, B. J. Stoner, and E. M. Eyring, *Statistical Mechanics and Dynamics* (J. Wiley & Sons, Inc., New York, 1964), pp. 185ff.
3. R. E. Dickerson, *Molecular Thermodynamics* (W. A. Benjamin, Inc., New York, 1969).
4. W. Moore, *Physical Chemistry* (Prentice-Hall Inc., Englewood Cliffs, N. J., 1962) 3rd ed., p. 473.
5. L. de Broglie, *Phil. Mag.* **47**, 446 (1924).
6. C. Davisson and L. H. Germer, *Nature* **119**, 558 (1927).
7. J. M. Anderson, *Introduction to Quantum Chemistry* (W. A. Benjamin, Inc., New York, 1969), Chap. 1.
8. A. Messiah, *Quantum Mechanics* (J. Wiley & Sons, Inc., New York, 1961), Vol. I, Chaps. 5 and 7 .
9. P. A. M. Dirac, *Sci. Am.* **208**, 45 (1963).
10. H. Kuhn, *Helv. Chim. Acta* **31**, 1441 (1948); **34**, 1308 (1951); *J. Chem. Phys.* **29**, 958 (1958).
11. J. R. Platt, *J. Chem. Phys.* **22**, 1448 (1954).

Chapter 4

SPECTROSCOPY AND SPECTROSCOPIC MEASUREMENTS

IN THE last chapter, it was shown that placing boundary conditions on the motion of a particle requires that the energy levels for the particle be quantized. This result is a completely general one even though the last chapter dealt only with a particle in an infinite, square potential well. The task of the quantum theorist is to calculate the energies and state functions associated with the quantized levels. Where possible, these calculations are carried out with no approximations on the actual physical system. This type of calculation is called an "exact" one. Exact calculations are only possible for a few simple systems, however, and much quantum theory is devoted to approximate calculations. These approximate calculations can involve either approximations in the mathematical apparatus used to treat the actual system or approximations to the physical system that produce a mathematically treatable problem, or both. In all cases, a key question is: How can quantum theorists check the validity of their computations?

The principal source of *experimental information* about the allowed energy states in atoms and molecules is *spectroscopy*. This fact has led to a symbiotic relationship between the quantum theorist and the spectroscopist. The quantum theorist uses spectroscopic results to test the validity of his or her calculations, and the spectroscopist uses quantum theory to interpret experimental spectra and plan new experiments.

There are many different kinds of spectroscopy, but all have in common the fact that electromagnetic radiation is allowed to interact with a sample of matter. This radiation can be absorbed, emitted, or scattered by the sample. Most students will have already encountered infrared, ultraviolet and visible, and nuclear magnetic resonance spectra in their organic courses. In these cases, electromagnetic radiation is absorbed and one can correlate *frequencies* and *intensities* of the absorbed radiation with structural features of the molecules in the sample of matter. Biochemists and some organic students may also have dealt with *fluorescence* or *phosphorescence*. In this case, radiation is emitted, and again the frequency and intensity of the emitted radiation are correlated with structural features of the sample. These correlations are possible because of the fundamental link between the energy of absorbed or emitted radiation and the allowed energy states of the molecules in the sample. This link is the Planck relation introduced in the last chapter,

$$h\nu = E_j = E_i = \Delta E \tag{4-1}$$

In Eq. 4-1, ν is the frequency of the electromagnetic radiation impinging on the sample, and E_j and E_i are allowed energy states in the molecules or atoms making up the sample. The spectroscopist measures ν and can thus calculate an experimental value of $E_j - E_i$. The quantum theorist calculates E_j and E_i and compares the calculated values with the experimental ones.

The intensities of radiation absorbed, emitted, or scattered by a sample of matter can also be measured, and these quantities are related to the state functions Ψ_j and Ψ_i associated with the energy levels E_j and E_i. Thus, intensities present another opportunity for comparison between theory and experiment. In the discussion that follows, Section 4-1 summarizes the different units used to specify the energies of photons or of energy states in matter. Section 4-2 discusses the relation between measured intensities and state functions. Section 4-3 briefly surveys the various types of spectroscopy. The student interested in more experimental details is referred to [1].

4-1 UNITS

The energy of the electromagnetic radiation absorbed by an atomic or molecular system is always equal to an energy *difference* between two allowed states of the system. Any quantity related to this energy difference can be used as a unit. Units used in spectroscopy can be divided into two classes depending on whether they are directly or inversely proportional to the energy.

The International Committee on Weights and Measures has recommended the adoption of a standard set of units called SI units. In several cases, these units are different from those commonly employed by theoretical chemists, and students will, therefore, need to become familiar with both sets of units in order to compare the older literature with that which is still to come.

The quantity of most concern to the theoretical chemist is *energy*. The SI unit of energy is the joule. The cgs unit, used extensively in theoretical calculations, is the erg. The two units are related by the conversion factor 1 J = 10^7 erg. Conversion is accomplished simply by changing the exponent. A more difficult problem has to do with the *calorie*. Many molecular energies have been expressed in units of kilocalories per mole, but in the SI system, the use of the calorie is to be progressively discouraged. The calorie is now defined by the relation 1 cal = 4.184 J. In this text, we will try to eliminate the use of calories and kilocalories, but we will retain the use of cgs units.

Theoretical chemists will continue to use other units for energy, such as the electron volt (eV) and wave number (cm^{-1}). In addition, the Planck relation, $h\nu = \Delta E$, relates the frequency of electromagnetic radiation to the energy and, therefore, the frequency units, hertz (Hz or s^{-1}) and megahertz (MHz), can be used as energy units.

The Planck relation also allows the energy of absorbed radiation to be expressed in wavelength units. Since the frequency of electromagnetic radiation is related to the wavelength by the relation $\nu\lambda = c$, where c is the speed of light, the wavelength of absorbed radiation is *inversely* proportional to the energy of the transition. The older common wavelength units are angstroms (Å), millimicrons (mμ), and microns (μ), which are 10^{-8}, 10^{-7}, and 10^{-4} cm, respectively. Here again, using SI units requires some changes. The SI unit of length is the *meter*, m, and wavelength should be expressed in *nanometers*, nm, or micrometers, μm. Note that nanometers and micrometers are equivalent to the older millimicron and micron, respectively. Use of angstroms is supposed to be progressively discouraged, but because it is such a natural unit of atomic and molecular dimensions, it will be retained in this text.

The reciprocal of the wavelength $1/\lambda$ is directly proportional to the energy and is the wave number unit referred to above. It has dimensions of cm^{-1}. The student should verify that an expression for the energy in ergs can be converted to an expression in wave numbers by dividing the energy in ergs by the quantity hc.

In the material that follows the symbol ω will always designate an energy in wave number units. The symbol ε will usually designate an energy in ergs although it may occasionally be used for an energy in electron volts.

Some conversion factors between these units are given in Table 4-1 (see also the table at the back of the book).

EXERCISE 4-1 A blue dye has an absorption band in the visible region of the spectrum at 700 nm. Calculate the energy of the transition in Hz, cm^{-1}, eV, kJ $mole^{-1}$, and erg $molecule^{-1}$.

TABLE 4-1 CONVERSION FACTORS FOR USE IN CHANGING ENERGY UNITS[a]

To convert from energy in	*To an energy in*				
	erg molecule^{-1}	eV	cm^{-1}	kJ mole^{-1}	MHz
	multiply by				
erg molecule^{-1}	–	6.242×10^{11}	5.035×10^{15}	6.023×10^{13}	1.509×10^{20}
eV	1.602×10^{-12}	–	8.067×10^{3}	9.649×10^{1}	2.418×10^{8}
cm^{-1}	1.986×10^{-16}	1.240×10^{-4}	–	1.196×10^{-2}	2.998×10^{4}
kJ mole^{-1}	1.660×10^{-14}	1.036×10^{-2}	8.359×10^{1}	–	2.506×10^{6}
MHz	6.626×10^{-21}	4.136×10^{-9}	3.336×10^{-5}	3.990×10^{-7}	–

[a] To obtain the wavelength equivalent, in cm, of an energy difference, express the energy in wave numbers and take the reciprocal.

4-2 *SOME FACTS ABOUT ABSORPTION STRENGTHS*

The absorption strength of a spectral band or line can be described in a number of ways. Experimentally, absorption strengths are measured using the Lambert–Beer law. To derive the Lambert–Beer law, consider the situation shown in Figure 4-1. (This derivation is essentially that of [2]. For a more rigorous treatment see [3].) In this figure, I_1 is the intensity of an incident beam of monochromatic radiation, I_0 is the intensity of the beam after it has passed through one cell wall, I is the intensity after it has passed through the absorbing medium, and I_2 is the intensity after the beam has traversed the last cell wall. The transmittance T is defined as

$$T = \frac{I_2}{I_1} \tag{4-2a}$$

and is the quantity that is usually measured in spectrophotometers. The quantity of interest to the physical chemist is the internal transmittance of the system

$$T_i = \frac{I}{I_0} \tag{4-2b}$$

Usually T and T_i are not very different because cells are constructed of materials that do not absorb or scatter much light. Even this difference can be eliminated by using matched cells, one containing the sample of interest and one containing a suitable reference material (usually air or solvent). If T is set at 100% for the reference cell, then a measurement of T of the sample gives T_i.

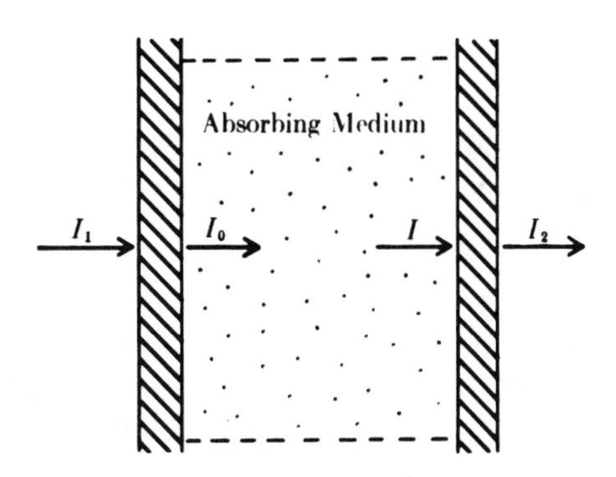

FIGURE 4-1 *Intensities of a light beam at different points in an absorbing system. [Reprinted by permission from J. Waser, "Quantitative Chemistry" (W. A. Benjamin, Inc., New York,* 1962).]

The quantity I/I_0 is related to the length of the absorbing medium (Bouguer–Lambert law) and to the concentration of absorbing medium (Beer's law). The Bouguer–Lambert law states that, for a thin layer of absorbing material, the decrease in intensity of a monochromatic light beam in an absorbing material is directly proportional to the incident intensity and the length of the layer. In differential form, this law becomes

$$-dI = kI\,dx \tag{4-3}$$

where k is called the absorption coefficient. Separating variables and integrating, we obtain

$$-\ln I \Big|_{I(x=0)}^{I(x)} = kx \Big|_{x=0}^{x}$$

$$-\ln \frac{I}{I_0} = kx \tag{4-4}$$

since, by definition, $I = I_0$ at $x = 0$. Since it is easier to use base 10 than natural logarithms, Equation 4-4 is usually written

$$-\log \frac{I}{I_0} = \left(\frac{k}{2{\cdot}303}\right)x = ax \tag{4-5}$$

where a is called the absorbancy index of the medium. Experimentally, the sample is usually contained in a cell of fixed length l, and Eq. 4-5 becomes

$$-\log \frac{I}{I_0} = al$$

or

$$I = I_0 10^{-al} \tag{4-6}$$

In 1852, Beer reasoned that, for a solution in which the solvent was transparent ($a = 0$), the exponent al could be written as

$$al = C\varepsilon l \tag{4-7}$$

where C is the concentration of solute in moles per liter and ε is a property of the absorbing molecules called the molar absorptivity, molar absorption coefficient, or molar extinction coefficient. Combining Eqs. 4-7 and 4-5, we obtain the Lambert–Beer law

$$-\log \frac{I}{I_0} = C\varepsilon l \tag{4-8}$$

Sometimes spectrophotometers are calibrated in units of log I_0/I. This quantity is called the absorbancy or optical density of the medium.

When more than one absorbing species is present, the molar extinction coefficient is given by the sum

$$C\varepsilon = \sum_j C_j \varepsilon_j \tag{4-9}$$

where $C = \sum C_j$ is the total concentration of all absorbing species, and C_j and ε_j are the concentration and molar extinction coefficients of the substance j, respectively. This additivity law is of great importance in that it permits spectroscopic analysis of solutions containing several components.

EXERCISE 4-2 Benzene and toluene both absorb light at 255 and 269 nm. For benzene, the molar extinction coefficient at these two wavelengths is 234 and 12.5 l mole^{-1} cm^{-1}; for toluene, the molar absorbancy indices are 210 and 267 l mole^{-1} cm^{-1}. When a spectrum of the mixture of benzene and toluene was run, the optical density at 255 nm was 0.800 and the optical density at 269 nm was 0.267. Calculate the concentrations of benzene and toluene in the mixture. Assume a 1.0 cm cell.

The molar extinction coefficient in Eq. 4-8 is a property characteristic of the absorbing species, and it is a measure of that species' ability to absorb light of a given frequency. It should be related to the theoretical absorption strength of a spectral line, and this strength should, at least in principle, be calculable by the methods of quantum mechanics if the wave functions characterizing the two energy states involved in a transition are known. Some of the factors involved in a calculation of absorption strengths will now be considered.

All spectroscopic measurements depend on the presence of some type of interaction between the particle, atom, or molecule being studied and an electromagnetic wave. The absorption strength of a spectral line will depend on two factors.

1. The magnitude of the interaction with the light wave.
2. The difference in population between the initial and final states of the transition.

An electromagnetic wave can be regarded as oscillating electric and magnetic field vectors propagating through space, as in Figure 4-2. For simplicity, a light wave that is plane polarized is shown. For an unpolarized wave, the electric field (or magnetic field) vector is not confined to a single plane. Before a particle or molecule can absorb energy from such a wave, there must be some means by which it can interact with either the electric or magnetic field. In most types of spectroscopy, the interaction of interest is with the electric field of the electromagnetic wave. For the various kinds of magnetic

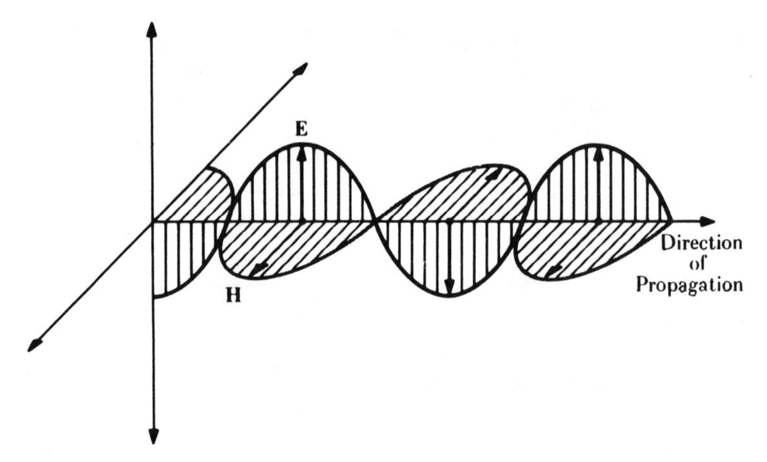

FIGURE 4-2 *Oscillating electric* (**E**) *and magnetic* (**H**) *field vectors in a plane polarized electromagnetic wave. The interaction of the oscillating electric field with a changing dipole moment of an atom or molecule causes a transition between allowed energy states.*

resonance spectroscopy (see below and Chapter 9) the system being studied interacts with the magnetic field.

A rigorous mathematical treatment of the interaction of a light wave with an atom or molecule is beyond the scope of this book, and therefore in what follows only a few of the basic ideas will be given. Suppose we have a system that at time $t = 0$ is known to be in the state Ψ_n. It is also assumed that the stationary state eigenvalue problem $\Psi_m = E_m \Psi_m$ has been solved. Suppose that the system is now subjected to a light wave whose electric field component can be represented by the equation

$$\mathscr{E} = \mathscr{E}_0 \cos 2\pi \nu t \tag{4-10}$$

where ν is the frequency of the light wave. For purposes of discussion Eq. 4-10 will be broken down into two parts—the electric field part $\mathscr{E}_0$ and the time-dependent part, $\cos 2\pi\nu t$.

An atom or a molecule consists of an assembly of charged particles and at any moment in time has an instantaneous dipole moment

$$\boldsymbol{\mu} = \sum_i q_i \mathbf{r}_i \tag{4-11}$$

where q_i is the charge on particle i and $\mathbf{r}_i$ is the vector position of particle i with respect to some origin. For neutral species $\boldsymbol{\mu}$ does not depend on origin and the following discussion is restricted to this case. The energy of interaction between an electric field and a dipole moment is known from classical physics to be

$$E = -\boldsymbol{\mu} \cdot \mathscr{E} = -\sum_i q_i \mathbf{r}_i \cdot \mathscr{E} \tag{4-12}$$

Thus the presence of an electromagnetic wave adds a term to the Hamiltonian operator for our problem that will depend on the instantaneous dipole moments in the system.

To take the time dependence into account, time-dependent perturbation theory is used. This is mathematically more complicated than stationary state perturbation theory discussed in Section 3-5, but the basic ideas are the same. Beginning with our system in state $\Psi_n{}^0$ and subject to the perturbing Hamiltonian $\mathscr{H}' = -\boldsymbol{\mu} \cdot \mathscr{E}_0 \cos 2\pi\nu t$ we wish to calculate the probability that at some later time t the state will be in state $\Psi_k{}^0$. This probability is equal to $a^2{}_k$, where a_k is the mixing coefficient of $\Psi_k{}^0$ in the perturbation expansion for the time-dependent wave function. Since $\mathscr{H}$ contains the dipole moment of the system, $\boldsymbol{\mu}$, the integrals H'_{nk} that appear in the expression for a_k will be of the form $(\Psi_k{}^0 \,|\boldsymbol{\mu}|\, \Psi_n{}^0)$ (see Eq. 3-91b). This particular integral is called the transition moment and is given the symbol $\mathbf{R}^{kn}$. Also, the time-dependent perturbation calculation shows that a_k will be small unless the υ in Eq. 4-10 is equal to the energy difference between the two states $\Psi_k{}^0$ and $\Psi_n{}^0$. Let this frequency be designated

$$\nu_{kn} \equiv \frac{E_k{}^0 - E_n{}^0}{h} \tag{4-13}$$

Since the probability that the system will make a transition from state n to state k depends on $a_k{}^2$, it will be proportional to $|\mathbf{R}^{kn}|^2$ as well as to the density of radiation at ν_{kn}. The quantum mechanical theory was worked out by Einstein, who showed that

$$P_{kn} = B_{kn}\, \rho(\nu_{kn}) \tag{4-14}$$

where

$$B_{kn} = \frac{8\pi^3}{3h^2} |\mathbf{R}^{kn}|^2 \tag{4-15}$$

is called the Einstein coefficient of induced absorption. In Eq. 4-14, P_{kn} is the probability that in one second of exposure of a system to electromagnetic radiation of frequency ν_{kn}, the system will absorb a photon of energy $h\nu_{kn}$ and make a transition from state n to state k. The quantity $\rho(\nu_{kn})$ is the density of radiation at frequency ν_{kn}. For a more complete discussion of this subject, the student is referred to [4].

Following a treatment due to Strickler and Berg [5], we wish now to relate the probability of absorption in unit time, P_{kn}, to the observed molar extinction coefficient. The Lambert–Beer law can be written in the differential form

$$-dI = 2.303\varepsilon(\nu)IC\,dx \tag{4-16}$$

The radiation density $\rho(\nu)$ is related to the radiation intensity I by the relationship

$$I = \frac{c}{\eta}\,\rho(\nu) \tag{4-17}$$

where c is the speed of light and η is the refractive index of the medium. Equation 4-17 is reasonable when it is recalled that the intensity of light is defined as the energy flowing through a 1 cm^2 cross-sectional area in 1 sec. The energy density ρ is the energy contained in 1 cm^3 per unit frequency range and c/η is the speed of light in the absorbing medium. Since the intensity of the incident light varies with both the frequency of the light and with the location x in the cell, the energy density must be a function of both ν and x, so we shall write it as $\rho(\nu, x)$. Making use of Eq. 4-17, the differential form of the Lambert–Beer Law may be written

$$-d\rho(\nu, x) = 2.303\varepsilon(\nu)\rho(\nu, x)C\,dx \tag{4-18}$$

where $-d\rho(\nu, x)$ is the decrease in energy density when the light moves a distance dx through the medium. If N' is the number of absorbing molecules per cubic centimeter, then the number of molecules in an element of volume 1 cm^2 in cross section and dx in width is $N'dx$, so that

$$C\,dx = 1000N'\,\frac{dx}{N_0} \tag{4-19}$$

where N_0 is Avogadro's number. In addition, if P_{kn} is the transition probability for a single molecule in 1 sec, then $P_{kn}N'\,dx$ is the number of molecules excited in the layer in 1 sec, with an energy absorption of $h\nu_{kn}$ for each molecule. Therefore, the loss of intensity in the light beam in passing through the layer is

$$-dI = P_{kn}N'\,dx\,h\nu_{kn} \tag{4-20}$$

The corresponding decrease in radiation density is given by

$$-\frac{c}{\eta}\,d\rho = P_{kn} N'\,dx h\nu_{kn} \tag{4-21}$$

which, on making use of Eqs. 4-18 and 4-19, gives

$$P_{kn} = \frac{2303c\varepsilon(\nu_{kn})}{\eta N_0 h\nu_{kn}}\,\rho(\nu_{kn}, x) \tag{4-22}$$

Electronic transitions in molecules do not, however, occur at a single frequency, but over a range of frequencies, so that Eq. 4-22 must be integrated over this range of frequencies to get the total probability of absorption from state $\Psi_n{}^0$ to state $\Psi_k{}^0$. In doing this the energy density $\rho(\nu)$ at a single frequency must be replaced by the energy density $\rho'(\nu)\,d\nu$ in the frequency range between ν and $\nu + d\nu$. We then obtain for the total probability that a molecule will be excited when exposed to radiation covering the entire absorption band

$$P'_{kn} = \int_{\text{band}} P_{kn}\,d\nu = \frac{2303c}{h\eta N_0}\int_{\text{band}} \varepsilon(\nu)\rho'(\nu)\,d\ln\nu \tag{4-23a}$$

If the radiation density $\rho'(\nu)$ is constant throughout the band (this is usually a safe assumption), and if we write this constant as ρ, then Eq. 4-23 can be written as

$$P'_{kn} = \left(\frac{2303c}{h\eta N_0}\int_{\text{band}} \varepsilon(\nu)\,d\ln\nu\right)\rho \tag{4-23b}$$

If Eqs. 4-23b and 4-14 are compared, we see that

$$B_{kn} = \frac{2303c}{h\eta N_0}\int_{\text{band}} \varepsilon(\nu)\,d\ln\nu \tag{4-24}$$

This equation is the desired link between the theoretically computed absorption intensities (using Eq. 4-15) and experimental absorption measurements. For narrow absorption bands, it is good approximation to take the center frequency of the band out of the integral in Eq. 4-24, giving

$$B_{kn} = \frac{2303c}{h\nu_{kn}\eta N_0}\int_{\text{band}} \varepsilon(\nu)\,d\nu \tag{4-25}$$

Equation 4-25 is the formula used in most elementary texts and involves the integrated absorption intensity $\int \varepsilon(\nu)\,d\nu$.

The second factor involved in absorption strengths arises from the difference in population of states Ψ_k^0 and Ψ_n^0. Obviously, before molecules can be excited from state Ψ_n^0, this state must be populated. In addition, a treatment similar to that discussed above shows that the probability of induced *emission* (that is, a transition from state Ψ_k^0 to state Ψ_n^0) is equal to the probability of induced absorption. The net absorption of energy from the sample will depend, therefore, on the difference in population of the two states involved in the transition. If the populations of the two states are equal, no absorption will be observed. Experimental conditions that equalize the populations in two states involved in a transition occur quite often in magnetic resonance spectroscopy, and this gives rise to the phenomenon of *saturation*. Under certain conditions, it is possible to make the population of an excited state temporarily larger than that of the ground state. If light of the appropriate energy then strikes the crystal, the number of ions that can emit radiation exceeds the number that can absorb, and an emission is obtained which is more intense than the exciting light. Under these conditions, the sample acts as an optical amplifier or *laser*.

To calculate the population differences between various states of a system, use is made of the Boltzmann distribution law [6]. This law has the form

$$N_j = Nq^{-1}g_j e^{-\varepsilon_j/kT} \tag{4-26}$$

where

N is the total number of systems
N_j is the number of systems in state j
q is the partition function of the system, a constant for our purposes
g_j is the degeneracy of state j
ε_j is the energy of state j in joules
k is Boltzmann's constant $= 1.3804 \times 10^{-23}$ J deg^{-1} (molecule)$^{-1}$
T is the absolute temperature.

If the lowest state of the system is arbitrarily assigned a population of 1.00, Eq. 4-26 can be used to calculate the relative populations of the higher states. Such a calculation is illustrated in Exercise 4-3.

EXERCISE 4-3 For a set of identical systems, each containing a helium atom in a one-dimensional box of length 0.5 nm, calculate the relative populations of the states $n = 2$ and $n = 1$ at 300 K.

EXERCISE 4-4 Calculate kT at 300 K (approximately room temperature) in J molecule^{-1}, cm^{-1}, eV, and kJ mole^{-1}.

4-3 SURVEY OF TYPES OF SPECTROSCOPY

In this section, the various types of spectroscopy will be listed in the order of increasing size of the quantum of radiation used. A brief description will be given of each type. Reference should be made to Figure 4-3 for information about various units and positions in the electromagnetic spectrum.

1. *Nuclear magnetic and nuclear quadrupole resonance spectroscopy.* This type of spectroscopy utilizes a portion of the radiofrequency region of the electromagnetic spectrum, usually 5–100 MHz. Nuclear magnetic resonance spectroscopy detects transitions between nuclear spin states in an applied magnetic field and is discussed in more detail in Chapter 9. Quadrupole resonance spectroscopy detects the splitting in the nuclear spin levels arising from the interaction of an unsymmetrical charge distribution in certain nuclei with an electric field gradient. For more details, the student should consult [6] after studying Chapter 9. The quantum size in this region of the spectrum is very small. A frequency of 100 MHz corresponds to an energy of only 0.0033 cm^{-1} and a wavelength of 300 cm or 3 m. This energy is much less than kT at room temperature (see Exercise 4-4) and all of the nuclear spin states are almost equally populated. Detection of resonance signals is quite difficult and requires fairly large and/or highly concentrated samples.

2. *Electron spin resonance spectroscopy.* If a sample containing unpaired electrons—for example, organic free radicals or certain transition metal ions—is placed in an applied magnetic field, transitions between different *electron spin* states can be induced by radiation in the microwave range. Microwaves are characterized by the fact that they are generated by klystrons and magnetons instead of *LC* circuits. Usually, they are conducted by waveguides (hollow pieces of metal tubing with either a circular or rectangular cross section) instead of wires. Experiments are usually performed at 9,500 MHz (*X* band) or 12,000 MHz (*K* band) but, in general, any frequency in the range of 2,000 MHz to 36,000 MHz can be used. A frequency of 10,000 MHz corresponds to a wavelength of 3 cm and an energy of 0.3 cm^{-1}, still considerably less than kT at room temperature. The different electron spin states are still almost equally populated, but there is a larger excess population in the lowest electron spin state than there is in the corresponding nuclear spin state. Consequently, saturation effects, when the populations of the lowest and excited states become equal, are not as pronounced as with nuclear magnetic resonance. Electron spin resonance spectroscopy is discussed in Chapter 9.

3. *Pure rotational spectroscopy.* In this type of spectroscopy, transitions are observed between different rotational states of a molecule. Most of these transitions also occur in the microwave region except those for light molecules (HCl, HF, and so on) which take place in the far infrared. The remarks made about quantum size and populations under electron spin resonance

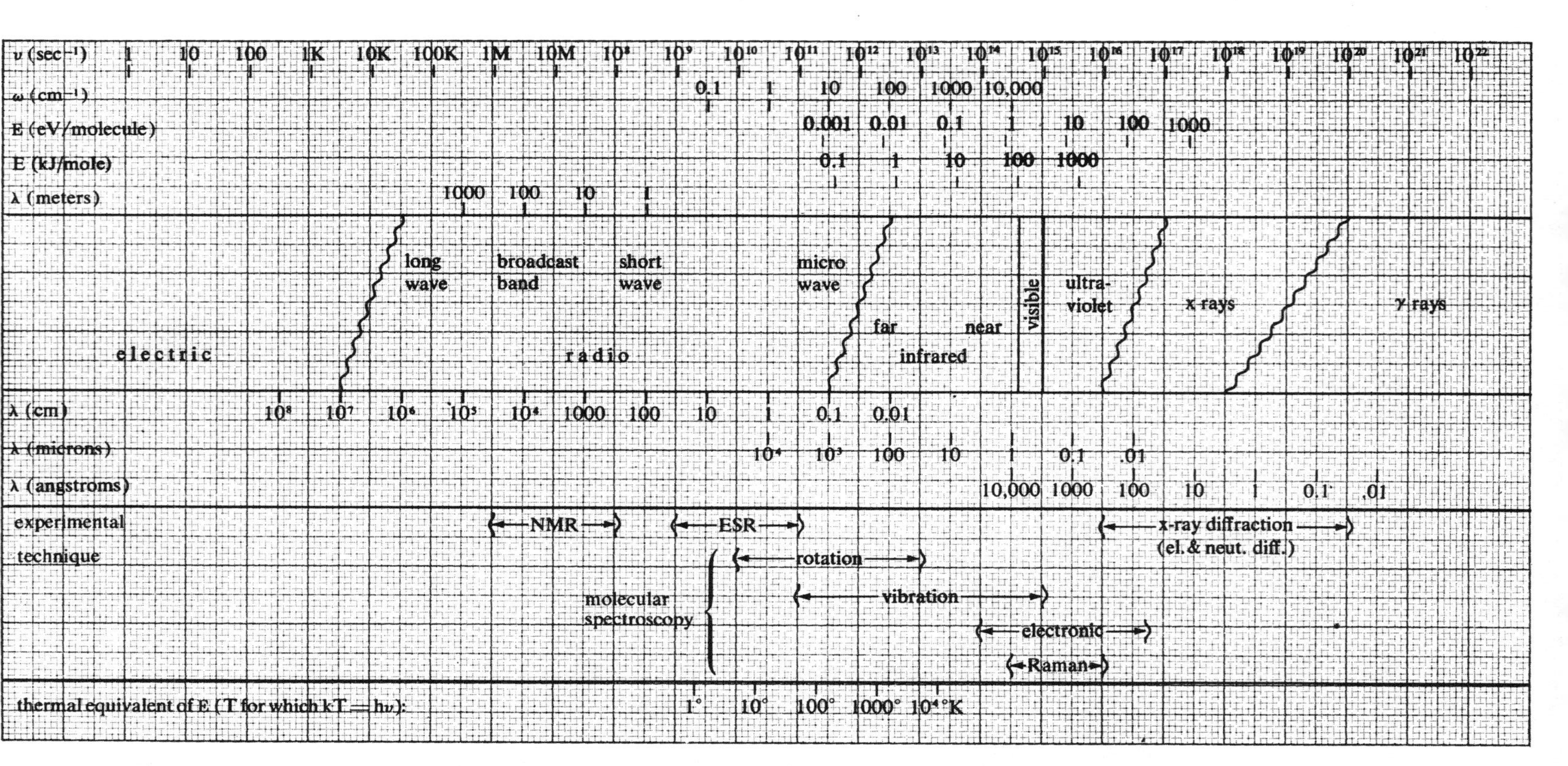

FIGURE 4-3 *Schematic diagram of the electromagnetic spectrum showing the relationships between frequency, wavelength and various other energy units. The various kinds of spectroscopy are placed in the appropriate regions and important characteristic temperatures* ($= h\nu/k$) *are also given.*

spectroscopy also apply here. Pure rotational spectroscopy is discussed in Chapter 5.

4. *Vibrational (infrared) and vibration-rotation spectroscopy.* Transitions between vibrational states in molecules absorb energy in the infrared region of the electromagnetic spectrum. Conventional infrared spectrometers usually scan the range 200–4,000 cm^{-1}. These energies correspond to a wavelength range of 50–2.5 μm. Radiation in this region is produced by a hot glowing wire and is an example of a practical use of blackbody radiation. The optics of infrared spectrometers are constructed from large single crystals of sodium chloride, lithium chloride, or potassium bromide. The quantum size in this region, except at the low energy end of the range, is greater than kT at room temperature and, as a result of this, only the lowest vibrational state has appreciable population.

Overtones (see Section 5-6 for a definition of "overtone") of high energy vibrational transitions are found in the near infrared region which extends from 2.5 μm to 800 nm (8,000 Å). Some low energy electronic transitions are also found in the high energy part of this region. Vibration spectroscopy is also discussed in Chapter 5.

5. *Raman spectroscopy.* Raman spectroscopy measures vibrational transition energies, but does so by observing the frequency of scattered light rather than that of absorbed light. An intense beam of monochromatic light, usually in the visible region of the spectrum, is allowed to strike a sample, and the intensity of scattered light is observed at right angles to the incident beam. Most of the scattered light will have the same frequency as the incident beam, but a small amount of light will have frequencies different from that of the incident beam. The energy differences between these weak lines and the main line correspond to vibrational and/or rotational transitions in the system being studied.

6. *Electronic spectroscopy.* In electronic spectroscopy, transitions between allowed electronic states of atoms and molecules are observed. It is electronic transitions that give rise to the atomic line spectra discussed in Chapter 3. These transitions occur over a wide range of energies encompassing the visible, ultraviolet, and vacuum ultraviolet regions. The extensions of each of these regions as well as the energy ranges of the corresponding quanta are given in Figure 4-3.

7. *Ultraviolet and x-ray photoelectron spectroscopy.* If a light quantum possesses enough energy, it can cause an electron to be completely ejected from an atom or a molecule. This process is called ionization. To the extent that the energy of the photon is larger than the ionization energy, the ejected electron will possess excess kinetic energy. In photoelectron spectroscopy, one allows a beam of photons of known energy to interact with a sample and measures the excess kinetic energy that the ejected electrons have. By subtracting this excess kinetic energy from the photon energy, one can calculate

the binding energy of the electron. Ultraviolet photons are used to study valence electron states of molecules, and x-ray photons are used to study core electrons. Electronic energies are then correlated with structural features of the molecule or compared with theoretical calculations. Recent articles that are good starting points for reading in this field are [7, 8].

8. *Gamma ray or Mössbauer spectroscopy.* This type of spectroscopy measures the transition energy between allowed states of certain nuclei. Chemical applications of Mössbauer spectroscopy have mainly used compounds containing ^{57}Fe or ^{119}Sn. Gamma-ray emissions from these isotopes are characterized by being extremely narrow. This narrowness can be used to construct an extremely sensitive time measuring device, and such a device has recently been used to test the general theory of relativity. Energies in the gamma-ray regions are extremely large, ranging from 8×10^3 to 1×10^5 eV. For more details see [9].

4-4 SUMMARY

1. Any unit related to an energy difference can be used as a spectroscopic unit. Some common units used in spectroscopic measurements are cm^{-1}, eV, kJ $mole^{-1}$, nm, Å, and MHz.
2. The Bouguer–Lambert–Beer law was derived. This law relates the absorbance of the sample to its length and to the concentration and the molar absorbancy index of the absorbing species.
3. Factors affecting the intensity of a transition were discussed. Intensities are determined by the magnitude of the interaction between the sample being studied and the light wave and by the relative populations of the states involved in the transitions.
4. The Boltzmann distribution law was introduced to make it possible to calculate relative populations.
5. Various types of spectroscopy were surveyed briefly.
6. The student should be familiar with the following terms: molar absorbancy index. transition moment, populations of a state, integrated absorption intensity, dipole moment, and Einstein's coefficient of induced absorption.

REFERENCES

1. *Chemical Applications of Spectroscopy*, W. West, Ed., Vol. 9 of *Techniques of Organic Chemistry*, A. Weisburger, Ed. (Wiley-Interscience, Inc., New York, 1963).
2. J. Waser, *Quantitative Chemistry* (W. A. Benjamin, Inc., New York, 1962), pp. 155ff.
3. D. F. Swinehart, J. *Chem. Ed.* **39**, 333 (1962).

4. G. Barrow, *An Introduction to Molecular Spectroscopy* (McGraw-Hill Book Co., New York, 1962), Chap. 4.
5. S. J. Strickler and R. A. Berg, *J. Chem. Phys.* **37**, 814 (1962).
6. W. Kauzmann, *Quantum Chemistry* (Academic Press, Inc., New York, 1957), pp. 484ff.
7. H. Bock and P. D. Mollère, *J. Chem. Ed.* **51**, 506 (1974).
8. F. O. Ellison and M. G. White, *J. Chem. Ed.* **53**, 430 (1976).
9. G. K. Wertheim, *Science* **144**, 253 (1964).

Chapter 5

ROTATION AND VIBRATION SPECTROSCOPY

STUDENTS WILL have already encountered data on the structure of molecules, such as bond lengths, bond angles, and dipole moments, in their freshman and sophomore chemistry courses. Some may have encountered *force constants*: parameters related to the difficulty of causing bonded atoms to vibrate. Almost all students will have used the infrared spectrum of a molecule to identify it or, at least, to tell whether certain functional groups are present. This chapter shows how studies of the rotational and vibrational motions of molecules are related to these kinds of molecular structure parameters.

Bond lengths and bond angles in molecules can be determined by x-ray diffraction techniques as long as single crystals of the material can be obtained. Many molecules cannot be conveniently studied as solids because of their low freezing points, however, and other techniques must be used to determine structural features in these cases. Rotational spectroscopy is a technique that can be used to study molecules in the gas phase. The allowed rotational energy states of a molecule depend on a parameter called the *moment of inertia*, which is related to the bond distances and bond angles within a molecule. The problem for the quantum chemist is to determine how the allowed energy levels for the rotation of a molecule depend on these structural features. Once this has been accomplished, structural data can be determined from experimentally measured transition energies. Calculations related to the intensities of these measurements show that these intensities are determined, in part, by the dipole moment of a molecule. Thus, one can measure dipole moments from rotational spectra studies. Differences between rotational energy levels lie in the microwave region of the spectrum, so this form of spectroscopy is often called microwave spectroscopy.

In addition to the rotation of a whole molecule, individual atoms in a molecule undergo vibratory motion. For a diatomic molecule, only one type of vibration, corresponding to the stretching and compression of the bond, is possible. Quantized energy levels are associated with this vibratory motion, and transitions between these levels are studied by infrared and Raman spectroscopy. The quantum theorist relates the observed frequencies to force constants characteristic of the bonds and the masses of the atoms. The force constants can in turn be related theoretically to the electronic state functions for the molecule, again demonstrating the symbiotic relationship between experimental spectroscopy and quantum theory.

The introduction to Chapter 4 states that approximations must often be made to treat a complex physical system theoretically. In the discussion that follows, we will use an approximation that chooses a simple physical model that can be treated mathematically to approximate the rotational and vibrational motion of a diatomic molecule. Then corrections will be added to the results using perturbation theory to make the model more appropriate to an actual molecule. The simple model that we will begin with assumes first of all that the rotation and vibrations can be treated independently. The rotation of a diatomic molecule is then approximated by assuming that the atoms are two mass points held together by a rigid rod. This is called the rigid rotor approximation. The vibration of diatomic molecules is approximated by assuming that the atoms are two mass points held together by a Hooke's law spring. This is called the harmonic oscillator approximation.

The following sections are devoted to the quantum mechanical solutions for these two approximate models, and the energy levels will be corrected to take into account the fact that real molecules do not behave quite like rigid rotors and harmonic oscillators. Finally, we will give a brief discussion of the vibrational motion of polyatomic molecules.

5-1 THE RIGID ROTOR APPROXIMATION

We have shown in Chapter 2 that, in problems where the potential energy is only a function of the internal coordinates, the motion of the center of mass can be separated from the internal motion of the molecule. In the rigid rotor approximation, we assume that a diatomic molecule can be regarded as a dumbbell with atoms of masses m_A and m_B at the ends held together by a massless bar of length r, the bond length. Since there is no potential energy in rotational motion (in the absence of electric and magnetic fields), the separation of the motion of the center of mass can be made, and the internal motion of the molecule can be treated as a separate problem.

The problem of rotational motion is most conveniently solved using spherical polar coordinates. The appropriate quantities are shown in Figure 5-1, where the origin has been placed at the center of mass of the system. From the definition of the center of mass, we can write

$$m_B r_B = m_A r_A \qquad r_B = \left(\frac{m_A}{m_B}\right) r_A \tag{5-1}$$

Using the fact that the bond length $r = r_A + r_B$, the student can show that

$$r_A = \frac{m_B}{m_A + m_B} r \tag{5-2}$$

$$r_B = \frac{m_A}{m_A + m_B} r$$

The moment of inertia I about an axis is defined as

$$I = \sum_i m_i r_i^2 \tag{5-3}$$

where m_i is the mass of the ith particle and r_i is the distance of the ith particle from the axis. For a diatomic rotor, the moment of inertia about an axis

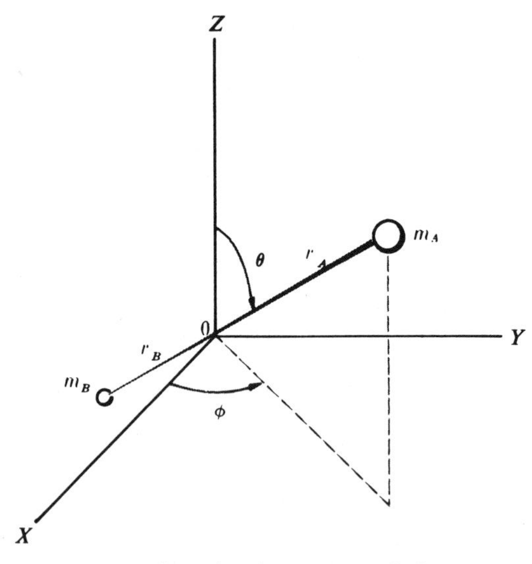

FIGURE 5-1 *Quantities used in the description of the rotational motion of a diatomic molecule. The origin is taken at the center of mass.*

through the center of mass and perpendicular to the molecular axis is

$$I = m_A r_A^{\,2} + m_B r_B^{\,2} \tag{5-4}$$

After a little algebra, the student can show that Eq. 5-4 becomes

$$I = \left(\frac{m_A m_B}{m_A + m_B}\right) r^2 = \mu r^2 \tag{5-5}$$

where μ is the reduced mass of the system.

The general method for solving quantum mechanical problems is now followed. This includes:

1. Writing down the classical Hamiltonian.
2. Transforming to the corresponding quantum mechanical operator.
3. Finding the solutions to the eigenvalue equation $\hat{\mathscr{H}}\Psi_i = E_i \Psi_i$. From the discussion of the separation of the center of mass in Chapter 2 we can write the classical Hamiltonian for the internal motion

$$\mathscr{H} = \frac{1}{2\mu}(p_x^{\,2} + p_y^{\,2} + p_z^{\,2}) \tag{5-6}$$

where x, y, and z are the internal coordinates $x_B - x_A$, $y_B - y_A$, $z_B - z_A$, respectively. Using Eq. 3-34 to transform to the corresponding quantum mechanical operator gives

$$\hat{\mathscr{H}} = -\frac{\hbar^2}{2\mu}\nabla^2$$

which in spherical polar coordinates is (Eq. 1-31)

$$\hat{\mathscr{H}} = -\frac{\hbar^2}{2\mu}\left[\frac{1}{r^2}\frac{\partial}{\partial r}\left(r^2\frac{\partial}{\partial r}\right) + \frac{1}{r^2 \sin\theta}\frac{\partial}{\partial\theta}\left(\sin\theta\frac{\partial}{\partial\theta}\right) + \frac{1}{r^2\sin^2\theta}\frac{\partial^2}{\partial\phi^2}\right] \tag{5-7}$$

For the rigid rotor r is a constant and Eq. 5-7 then becomes

$$\hat{\mathscr{H}} = -\frac{\hbar^2}{2\mu r^2}\left[\frac{1}{\sin\theta}\frac{\partial}{\partial\theta}\left(\sin\theta\frac{\partial}{\partial\theta}\right) + \frac{1}{\sin^2\theta}\frac{\partial^2}{\partial\phi^2}\right] \tag{5-8}$$

(The student should note that this problem in spherical coordinates is only a two-dimensional problem. This is obscured in Cartesian coordinates, where three variables are apparently involved.) The student should note that μr^2 in Eq. 5-8 is the moment of inertia I according to Eq. 5-5.

The final step necessary to find the allowed rotational energy levels is to find the functions and energies that make the equation

$$\mathscr{H}\psi_i = E_i\psi_i \tag{5-9}$$

true for the $\mathscr{H}$ given in Eq. 5-8.

This approach is similar to that of the problem of the particle in a three-dimensional box. That is, we try to reduce the two-dimensional problem to two one-dimensional problems. To do this, we attempt to separate the variables θ and ϕ by trying solutions of the form

$$\Psi = \Theta(\theta)\Phi(\phi) \tag{5-10}$$

where Θ is a function of θ only and Φ is a function of ϕ only. Substituting Eqs. 5-10 and 5-8 into Eq. 5-9, and performing some straightforward algebra, one obtains

$$\frac{\sin^2\theta}{\Theta}\left[\frac{1}{\sin\theta}\frac{\partial}{\partial\theta}\left(\sin\theta\frac{\partial\Theta}{\partial\theta}\right) + \frac{2IE\Theta}{\hbar^2}\right] + \frac{1}{\Phi}\frac{\partial^2\Phi}{\partial\phi^2} = 0$$

Note that μr^2 has been replaced by I. Since Θ and Φ are functions of θ and ϕ only, the partial derivatives can be replaced by total derivatives. Rearranging, one then obtains

$$\frac{\sin^2\theta}{\Theta}\left[\frac{1}{\sin\theta}\frac{d}{d\theta}\left(\sin\theta\frac{d\Theta}{d\theta}\right) + \frac{2IE\Theta}{\hbar^2}\right] = -\frac{1}{\Phi}\frac{d^2\Phi}{d\phi^2} \tag{5-11}$$

Once again we apply the argument of Chapter 3. If Eq. 5-11 is to hold for all values of θ and ϕ, it must be true that both sides are equal to a constant. Calling this constant β, Eq. 5-11 becomes two equations

$$-\frac{1}{\Phi}\frac{d^2\Phi}{d\phi^2} = \beta \tag{5-12a}$$

$$\frac{\sin^2\theta}{\Theta}\left[\frac{1}{\sin\theta}\frac{d}{d\theta}\left(\sin\theta\frac{d\Theta}{d\theta}\right) + \frac{2IE\Theta}{\hbar^2}\right] = \beta \tag{5-12b}$$

Equation 5-12a is by now a familiar one, and the solutions can be immediately written down. For reasons illustrated in Exercises 5-1 and 5-2, we choose the solutions in the form

$$\Phi = Ae^{im\phi} \tag{5-13}$$

By differentiating twice, the constant β is shown to be equal to m^2, and by applying the "single valuedness" condition, m is restricted to the values $0, \pm 1, \pm 2, \ldots$.

EXERCISE 5-1 Find the allowed energies and wave functions for a particle constrained to move on the surface of a right circular cylinder of radius R and height b. (The particle is not allowed to move on the bottom and top surfaces of the cylinder.) This problem involves setting up the Hamiltonian in cylindrical coordinates, separating variables, and showing that the eigenvalue equation reduces to two problems, one solved in Chapter 3 and one discussed above.

EXERCISE 5-2 Show that in spherical coordinates the operator for the z component of angular momentum becomes

$$\hat{L}_z = -i\hbar \frac{\partial}{\partial \phi}$$

Show that the functions 5-13 are eigenfunctions of $\hat{L}_z$ while the functions $\Phi = A \sin m\phi$, or $\Phi = A \cos m\phi$ are not. Previous information on angular momentum is given in Exercises 1-6 and 3-10.

EXERCISE 5-3 Evaluate the normalization constant A in Eq. 5-13.

If Eq. 5-12b is expanded, and if the substitution $x = \cos \theta$ is made, one obtains

$$(1 - x^2)\frac{d^2\Theta}{dx^2} - 2x\frac{d\Theta}{dx} + \left(\frac{2IE}{\hbar^2} - \frac{\beta}{1 - x^2}\right)\Theta = 0 \tag{5-14}$$

EXERCISE 5-4 Starting with Eq. 5-12b, derive Eq. 5-14. *Hint*: You will need to use the fact that $d/d\theta = (dx/d\theta)\, d/dx$ and $d^2/d\theta^2 = [d/d\theta\,(dx/d\theta)]\, d/dx + (dx/d\theta)\, d/d\theta\, d/dx$. Using these relations you should get $d/d\theta = -\sin\theta\, d/dx$ and $d^2/d\theta^2 = \sin^2\theta\, d^2/dx^2 - \cos\theta\, d/dx$. From here derivation of Eq. 5-14 is straightforward.

Equation 5-14 is an equation of the form

$$(1 - x^2)z'' - 2xz' + \left(l(l + 1) - \frac{m^2}{1 - x^2}\right)z = 0 \tag{5-15}$$

This equation is a well-known equation of physics called the associated Legendre equation. Functions z that are finite, have integrable squares, and are single valued, exist only for the conditions that l is a positive integer or zero, and that $|m| \leq l$. The solutions to Eq. 5-14 are called the associated Legendre polynomials. The solutions for $l = 0$ to $l = 2$ are given in Table 5-1. Students interested in more detailed information about the solution to Eq. 5-14 are referred to [1, 2].

TABLE 5-1 SOME NORMALIZED ASSOCIATED LEGENDRE POLYNOMIALS Θ(1, m)

l	m	$\Theta(l, m)$
0	0	$\frac{1}{2}\sqrt{2}$
1	0	$(3/2)^{\frac{1}{2}} \cos \theta$
1	± 1	$(3/4)^{\frac{1}{2}} \sin \theta$
2	0	$(5/8)^{\frac{1}{2}} (3 \cos^2 \theta - 1)$
2	± 1	$(15/4)^{\frac{1}{2}} \sin \theta \cos \theta$
2	± 2	$(15/16)^{\frac{1}{2}} \sin^2 \theta$

EXERCISE 5-5 By direct substitution, show that the functions $\Theta(1, 0)$, $\Theta(1, 1)$, and $\Theta(2, 0)$ are solutions of Eq. 5-12b.

The product functions $\theta(l, m)\phi(m)$ are called spherical harmonics, and in many books they are given the symbol $Y_{l,m}$. In addition to being the solution to the rigid rotor problem, spherical harmonics appear in the quantum mechanical treatment of atoms where they are associated with the angular momentum of the electron. The $Y_{l,m}$ form a complete orthonormal set of functions and, because of this, they are used in the evaluation of integrals that appear in calculations involving several particles. To form the $Y_{l,m}$ from the $\theta(l, m)$ functions in Table 5-1, multiply the $\theta(l, m)$ by $(2\pi)^{-\frac{1}{2}}e^{im\phi}$.

Comparing Eq. 5-14 with Eq. 5-15, one obtains the result that quantum mechanically acceptable solutions to Eq. 5-14 exist only if

$$\frac{2IE}{\hbar^2} = l(l + 1) \tag{5-16}$$

Thus the allowed energies for the rigid rotor are

$$E = J(J + 1)\frac{\hbar^2}{2I} \qquad (J = 0, 1, 2, 3, \ldots) \tag{5-17}$$

The symbol J has been substituted for l in Eq. 5-16 because l is reserved for the quantum number of electronic angular momentum (see Chapter 6).

The allowed wave functions are those given by Eq. 5-10 and depend on two quantum numbers, J and m. Thus

$$\Psi(J, m) = \Theta(J, m)\Phi(m)$$

It should be noted that, for every value of J, there will be $2J + 1$ values of m. Thus if $J = 2$, m can have the five values $\pm 2, \pm 1, 0$. Because the energy depends only on J, each energy level will be $(2J + 1)$fold degenerate. In the presence of an electric or magnetic field, this degeneracy is removed if the molecule contains an electric or magnetic dipole moment, and the energy of the state will depend on m also.

5-2 *PURE ROTATIONAL SPECTROSCOPY*

In pure rotational spectroscopy, allowed transitions between rotational states are studied. As was pointed out in Chapter 4, these transitions occur in the far infrared or microwave region of the electromagnetic spectrum. By convention, spectroscopists designate quantities referring to the upper and lower states in a transition by single and double primes, respectively. The energy of the transition $\Psi(J'', m'') \rightarrow \Psi(J', m')$ is then

$$E' - E'' = \frac{\hbar^2}{2I}[J'(J' + 1) - J''(J'' + 1)] \tag{5-18}$$

It is shown below that only transitions for which $\Delta J = \pm 1$ are allowed. If this selection rule is used in Eq. 5-18, the expression for the energy of a rotational transition becomes

$$E' - E'' = \frac{\hbar^2}{2I} 2(J'' + 1) \tag{5-19}$$

Spectroscopists usually use wave number units. In these units, the energies of the allowed rotational transitions are

$$\omega_R(\text{cm}^{-1}) = \frac{E' - E''}{hc} = \frac{h}{8\pi^2 cI} 2(J'' + 1) = 2B(J'' + 1) \tag{5-20}$$

where $B = h/8\pi^2 cI$ is in wave number units and is called the rotational constant of the molecule. At this level of approximation, it is predicted that the pure rotational spectrum of a diatomic molecule should consist of a series of equally spaced lines with spacing $2B$.

A study of rotational spectra is one of the powerful methods that the experimentalist has available to obtain information about molecular structure. The value of B obtained from the pure rotational spectrum can be used to calculate the moment of inertia I, and this quantity in turn gives the internuclear distance r. In this respect, pure rotational spectroscopy supplements x-ray crystallography because it can be used to study the structure of gases, whereas x-ray crystallography is limited to studies of crystalline materials.

EXERCISE 5-6 Calculate the moment of inertia of

(a) HCl^{35} (b) HCl^{37} (c) DCl^{35}

all of which have an equilibrium bond length of 1.275 Å. Using the results of the rigid rotor approximation, calculate the positions of the first three rotational transitions for HCl^{35} and DCl^{35}. Plot these lines on a frequency abscissa.

EXERCISE 5-7 Carbon monoxide absorbs energy in the microwave region of the spectrum at 1.153×10^5 MHz. This absorption can be attributed to the $J = 0$ to $J = 1$ transition. Calculate the internuclear distance and the moment of inertia of CO.

5-3 *INTENSITIES AND SELECTION RULES*

We pointed out in Chapter 4 that the intensity of a spectral line depended on the magnitude of the transition moment and on the relative populations of the two states involved in a transition. Many times it can be shown that, for some classes of transitions, the transition moment $\mathbf{R}^{kn} = 0$. This means that the transition is not allowed, that is, it will have zero intensity. This result, coupled with a statement of the restrictions on the transitions to which it applies, constitutes a "selection rule." Thus, in the case of rotational spectra of most diatomic molecules, $\mathbf{R}^{kn} = 0$ unless $\Delta J \equiv J_k - J_n = \pm 1$ and $\Delta m \equiv m_k - m_n = 0, \pm 1$. Specific examples of calculations of $\mathbf{R}^{kn}$ are given in Exercises 5-8 and 5-9.

EXERCISE 5-8 An electron moves in a one-dimensional box of length a, with unit positive charge at $\frac{1}{2}a$. Show that the transition $\Psi_1 \rightarrow \Psi_2$ is allowed, but the transition $\Psi_1 \rightarrow \Psi_3$ is not. Can you derive a general selection rule for a particle in a one-dimensional box? See Section 3-5 for integral evaluation procedures. The student should note that this problem is artificial, in that, if there is a Coulombic potential, the particle in a box is no longer an appropriate solution to the problem. The point is to illustrate the calculation of a transition moment.

EXERCISE 5-9 Show that the transition $J = 0$ to $J = 2$ is not allowed for rotational transitions in a diatomic molecule. The wave functions are the appropriate spherical harmonics $Y_{J,m}$. Choose $m = 0$. Remember that the dipole moment is a vector that can be written $\mu = \mu_x\mathbf{i} + \mu_y\mathbf{j} + \mu_z\mathbf{k} = \mu_0 (\sin\theta \cos\phi\mathbf{i} + \sin\theta \sin\phi\mathbf{j} + \cos\theta\mathbf{k})$.

Among the allowed transitions just discussed, the relative intensities of the lines in a pure rotational spectrum will depend on the relative populations of the initial state for each transition. These relative populations can be calculated from the Boltzmann distribution law given in Chapter 4. Applied to the rotational states of a diatomic molecule, Eq. 4-26 becomes

$$\frac{N_J}{N_{J=0}} = (2J + 1) \exp\left(-\frac{hcBJ(J + 1)}{kT}\right) \qquad (J = 1, 2, 3, \ldots) \qquad (5\text{-}21)$$

Since Eq. 5-21 contains the degeneracy factor $2J + 1$, the population of rotational states will not continually decrease as their energy increases. Instead, there will be a maximum in a plot of relative population versus quantum number J.

An example of the relative populations of rotational states in HCl is given in Exercise 5-10. If the transition moments were equal for all J (and they are almost independent of J) [3], the intensities of the lines in the pure rotational spectrum of HCl^{35} would vary in a similar manner as the population graph calculated in Exercise 5-10.

EXERCISE 5-10 For HCl^{35} the equilibrium bond length is 1.275 Å. Use the Boltzmann distribution law to calculate the relative populations at room temperature of the states $J = 1, 2, 3, 4, 6$, and 10 at 25°C. Give the state with $J = 0$ a weight of 1.0. Plot the relative populations versus J value. How would a similar diagram look if the calculations were carried out at 500°C?

5-4 THE HARMONIC OSCILLATOR AND VIBRATIONAL SPECTROSCOPY

As a first approximation to the vibrational motion of a diatomic molecule, the molecule will be regarded as a pair of mass points connected by an ideal spring with force constant k. This is called the harmonic oscillator approximation. Before solving the quantum mechanical equation, some of the classical results will be reviewed. Because an ideal spring is one in which the force is proportional to the displacement, and because the potential energy is only a function of the relative distance apart, the Lagrangian function of the internal motion of this system is (see Section 2-3)

$$L = \tfrac{1}{2}\mu\dot{q}^2 - \tfrac{1}{2}kq^2 \tag{5-22}$$

where $q = r - r_e$ is the displacement from the equilibrium internuclear distance. The equation of motion for this one-dimensional harmonic oscillator is

$$\begin{aligned} \frac{d}{dt}(\mu\dot{q}) &= -kq \\ \ddot{q} &= -\frac{k}{\mu}q \end{aligned} \tag{5-23}$$

This is the differential equation for simple harmonic motion which has solutions

$$q = A\cos\left(\frac{k}{\mu}\right)^{\frac{1}{2}} t \tag{5-24}$$

The period of the oscillator τ is the time required for one oscillation and is equal to the reciprocal of the oscillator frequency. If we increase t from some initial value t_0 to the value $t_0 + \tau$, then q must be unchanged. This is true only if the argument of the cosine is increased by 2π. Thus we have

$$q_0 = q_\tau = A \cos\left(\frac{k}{\mu}\right)^{\frac{1}{2}} t_0 = A \cos\left(\frac{k}{\mu}\right)^{\frac{1}{2}} (t_0 + \tau) \tag{5-25}$$

and

$$2\pi = \tau\left(\frac{k}{\mu}\right)^{\frac{1}{2}} \tag{5-26}$$

therefore,

$$\nu = \frac{1}{\tau} = \frac{1}{2\pi}\left(\frac{k}{\mu}\right)^{\frac{1}{2}} \tag{5-27}$$

Thus the frequency of an oscillator is proportional to the square root of the force constant and the reciprocal of the square root of the reduced mass.

To solve the quantum mechanical problem, we first write Hamilton's function, and then transform the appropriate quantum mechanical operator. Thus

$$p_q = \frac{\partial L}{\partial \dot{q}} = \mu \dot{q} \tag{5-28}$$

$$\mathscr{H} = \frac{1}{2\mu} p_q^{\,2} + \frac{1}{2} k q^2 \tag{5-29}$$

and

$$\mathscr{H} = -\frac{\hbar^2}{2\mu}\frac{d^2}{dq^2} + \frac{1}{2} k q^2 \tag{5-30}$$

To find the allowed energies and wave functions, we must solve the eigenvalue equation

$$-\frac{\hbar^2}{2\mu}\frac{d^2\Psi}{dq^2} + \frac{1}{2} k q^2 \Psi = E\Psi \tag{5-31}$$

If Eq. 5-31 is rearranged, it can be written in the form

$$\frac{d^2\Psi}{dq^2} + (\alpha - \beta^2 q^2)\Psi = 0 \tag{5-32}$$

where $\alpha = 2\mu E/\hbar^2$ and $\beta^2 = \mu k/\hbar^2$. To solve Eq. 5-32, the substitution $\xi = (\beta)^{1/2}q$ is the first made. Then $d^2/dq^2 = \beta(d^2/d\xi^2)$ and Eq. 5-32 becomes

$$\frac{d^2\Psi}{d\xi^2} + \left(\frac{\alpha}{\beta} - \xi^2\right)\Psi = 0 \tag{5-33}$$

If Eq. 5-33 is investigated for large values of ξ to see what form $\Psi(\xi)$ takes under these conditions, we obtain the equation

$$\frac{d^2\Psi}{d\xi^2} = \xi^2\Psi \tag{5-34}$$

since $\xi^2 \gg \alpha/\beta$ for large ξ. The solutions to Eq. 5-34 are, therefore, approximately

$$\Psi' = Ae^{\pm\xi^2/2} \tag{5-35}$$

since a factor of ± 1 can be neglected with respect to ξ^2. The solution with the plus sign can be discarded because Ψ' would not have an integrable square under these conditions. This behavior of Ψ at large ξ suggests that a solution to Eq. 5-33 of the form

$$\Psi = u(\xi)e^{-\xi^2/2} \tag{5-36}$$

be tried. If Eq. 5-36 is substituted into Eq. 5-33, it is seen that, if there is to be a solution in the form of Eq. 5-36, then $u(\xi)$ must satisfy the differential equation

$$\frac{d^2u}{d\xi^2} - 2\xi\frac{du}{d\xi} + \left(\frac{\alpha}{\beta} - 1\right)u = 0 \tag{5-37}$$

EXERCISE 5-11 Derive Eq. 5-37 as outlined above. Procedures outlined in Exercise 5-4 can be used for derivative transformations.

Once again, however, Eq. 5-37 is a well-known equation of physics called the Hermite equation, providing that the quantity $\alpha/\beta - 1 = 2v$, where v is an integer. The solutions $u(\xi)$ are called the Hermite polynomials of degree v, and form an orthogonal set of functions [1, 2]. The first few Hermite polynomials are given in Table 5-2.

TABLE 5-2 HERMITE POLYNOMIALS FOR $v = 0$ to $v = 4$

v	$H_v(\xi)$
0	1
1	2ξ
2	$4\xi^2 - 2$
3	$8\xi^3 - 12\xi$
4	$16\xi^4 - 48\xi^2 + 12$

EXERCISE 5-12 By substitution, verify that H_1 and H_3 are solutions of Eq. 5-37.

The corresponding normalized wave functions for the one-dimensional harmonic oscillator are

$$\Psi_v(q) = \left(\frac{(\beta/\pi)^{\frac{1}{2}}}{2^v v!}\right)^{\frac{1}{2}} H_v(\sqrt{\beta}q)e^{-\beta q^2/2} \tag{5-38}$$

From the restriction that $\alpha/\beta - 1 = 2v$ for acceptable solutions of Eq. 5-37 to exist, it is straightforward to show that

$$E_v = \frac{h}{2\pi}\left(\frac{k}{\mu}\right)^{\frac{1}{2}}\left(v + \frac{1}{2}\right) = \left(v + \frac{1}{2}\right)h\nu \tag{5-39}$$

where the energy of the system is restricted to the discrete set of values $\frac{1}{2}, \frac{3}{2}, \frac{5}{2}, \ldots,$ times the energy $h\nu$ associated with the classical frequency of oscillation. It should be noticed that even in its lowest state ($v = 0$), a quantum mechanical oscillator still has energy $\frac{1}{2}h\nu$. This is called the vibrational zero-point energy. It is the zero-point energy difference between C—H and C—D bonds that gives rise to the isotope effects used in the kinetic study of the mechanism of chemical reactions [4].

EXERCISE 5-13 Calculate the ratio of zero-point energies of a C—H and a C—D bond vibration. You may assume that these bond vibrations follow Eq. 5-39, and that the force constants for the C—H and C—D vibrations are the same.

In vibrational or infrared spectroscopy, transitions between allowed vibrational states are observed. Within the context of the harmonic oscillator approximation, the selection rule for vibrational transitions is $\Delta v = \pm 1$. The allowed transitions are then given by

$$\omega_v = \frac{E_{v+1} - E_v}{hc} \tag{5-40}$$

where $\omega_v = v/c$ is the fundamental vibration frequency in wave number units. It should be noted that at this level of approximation, a transition between $v = 0$ and $v = 1$ has the same energy as one between $v = 1$ and $v = 2$ because all of the vibrational levels are equally spaced. It is shown below how this statement must be modified for real oscillators. Since the energy spacing between most vibrational levels is much greater than kT at room temperature, only the lowest state has an appreciable population at this temperature.

EXERCISE 5-14 HCl^{35} absorbs radiation at 2885.9 cm^{-1}. Using the Boltzmann distribution law, calculate the relative populations of the ground and first excited vibrational states at 25°C. You may give the ground state an arbitrary population of 1.000.

We next consider the selection rules for vibrational spectroscopy. The dipole moment of a vibrating diatomic molecule will obviously be a function of the equilibrium internuclear distance. For small displacements, this dipole moment can be expanded in a power series as a function of the displacement coordinate

$$\mu = \mu_0 + \left(\frac{d\mu}{dq}\right)_{q=0} q + \cdots \tag{5-41}$$

The first term in Eq. 5-41 is the permanent dipole moment, the second term involves the change in the dipole moment upon displacement, and higher terms in the expansion are neglected.

The transition moment for a transition $\Psi_0 \rightarrow \Psi_k$ is then

$$\begin{aligned} R_v^{0k} &= \int_{\text{all space}} \Psi_k \left[\mu_0 + \left(\frac{d\mu}{dq}\right)_{q=0} q\right] \Psi_0 \, dq \\ &= \int \Psi_k \mu_0 \Psi_0 \, dq + \int \Psi_k \left(\frac{d\mu}{dq}\right)_{q=0} q \Psi_0 \, dq \end{aligned} \tag{5-42}$$

The first integral in Eq. 5-42 vanishes because μ_0 is a constant, and Ψ_0 and Ψ_k are orthogonal. In order to obtain a nonvanishing transition moment, the second integral must be nonzero. This means that the quantity $(d\mu/dq)_{q=0}$ must be nonzero. The important point to realize is that, for a vibrational transition to be infrared allowed, there must be a change in the dipole moment during a vibrational cycle. Thus homonuclear diatomic molecules will have no infrared spectrum. By using the second integral in Eq. 5-42 and the recursion relations of Hermite polynomials [2], the general selection rule can be derived.

5-5 VIBRATION–ROTATION SPECTROSCOPY

In any real molecule, vibration and rotation are taking place at the same time. The first two vibrational energy levels and the rotational sublevels can be represented schematically as in Figure 5-2. A transition will start at a rotational sublevel of the $v = 0$ vibrational state and end at a rotational sublevel of the $v = 1$ state. The selection rules for these transitions for most

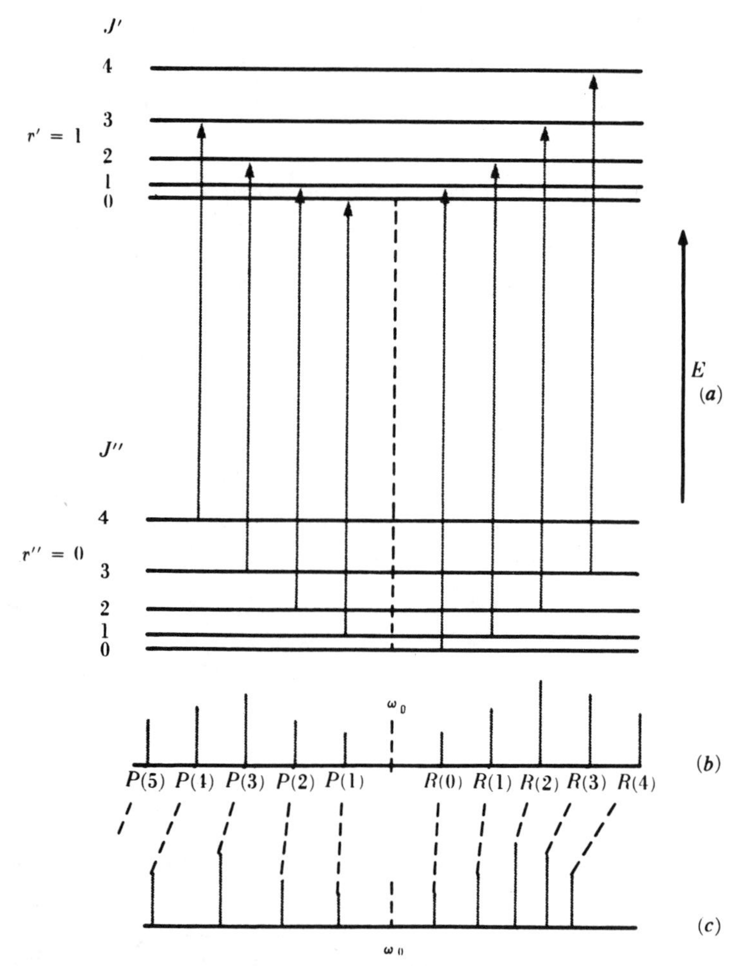

FIGURE 5-2 *Schematic representation of the transitions giving rise to the vibration-rotation spectrum of a diatomic molecule:* (a) *energy level diagram showing the first few P and R branch transitions;* (b) *idealized vibration-rotation spectrum;* (c) *vibration-rotation spectrum that takes vibration-rotation coupling into account. The heights of the lines in* (b) *and* (c) *are to give a general idea of the intensity distribution in rotation-vibration spectra.*

diatomic molecules are a combination of those from rotational and vibrational spectroscopy. Thus for absorption

$$\Delta v = \pm 1 \qquad \Delta J = \pm 1$$

It is convenient to discuss these transitions in two groups. For $\Delta v = 1$, $\Delta J = +1$, and, taking $v = 0$ as the lowest state, we obtain

$$\begin{aligned}\omega_{vR} &= \tfrac{3}{2}\omega_v + B(J'' + 1)(J'' + 2) - \tfrac{1}{2}\omega_v - BJ''(J'' + 1) \\ &= \omega_v + 2B(J'' + 1) \qquad (J'' = 0, 1, 2, 3)\end{aligned} \tag{5-43a}$$

where, as before, J'' is the quantum number of the lowest rotational state. This group of transitions ($\Delta J = +1$) is called the R branch of the spectrum. The lines of the R branch are labeled by the rotational quantum number of the lowest state. This labeling is illustrated in the idealized vibration-rotation spectrum in Figure 5-2b.

Similarly, for the case $\Delta v = 1$, $\Delta J = -1$, it can be shown that

$$\omega_{v,R} = \omega_v - 2BJ'' \qquad (J'' = 1, 2, 3) \tag{5-43b}$$

This group of lines is called the P branch. The P branch lines and their labeling are shown on the left-hand side of the idealized vibration-rotation spectrum in Figure 5-2b. Thus, within the context of the rigid rotor–harmonic oscillator approximation, the vibration-rotation spectrum is made up of two sets of equally spaced lines with spacing $2B$ with a gap between sets, the center of which is the fundamental vibration frequency ω_v.

In a few molecules, transitions corresponding to $\Delta J = 0$ are also allowed, and this group of lines, when present, is called the Q branch. In real molecules, the appearance of the spectrum is modified by vibration rotation coupling (see Section 5-6) and looks like the bottom spectrum in Figure 5-2c.

5-6 MORE EXACT THEORY OF VIBRATION–ROTATION SPECTROSCOPY

It is important at this point to inquire about the validity of the rigid rotor and harmonic oscillator approximations. From physical intuition, it can be argued that there are three ways that the above picture of the vibration-rotation motion of a diatomic molecule should be modified for real molecules. These three modifications are as follows.

1. Correction of the rotational motion for the effects of *centrifugal stretching*.

2. Correction of the vibrational motion for the *anharmonicity* of the vibration.

3. Correction of the rotational constant for changes in the moment of inertia for different vibrational states. This correction is called *vibration-rotation coupling*.

The physical argument for centrifugal stretching is easily visualized for a diatomic molecule. As the molecule rotates faster—that is, is excited to higher rotational states—it will stretch due to the centrifugal force on the end atoms. This will increase the moment of inertia, and consequently decrease the rotational constant B. For an actual molecule, then, the spacing between rotational lines should decrease as the quantum number J increases.

The anharmonicity correction is most easily visualized by a consideration of Figure 5-3. The potential energy function for a harmonic oscillator is a

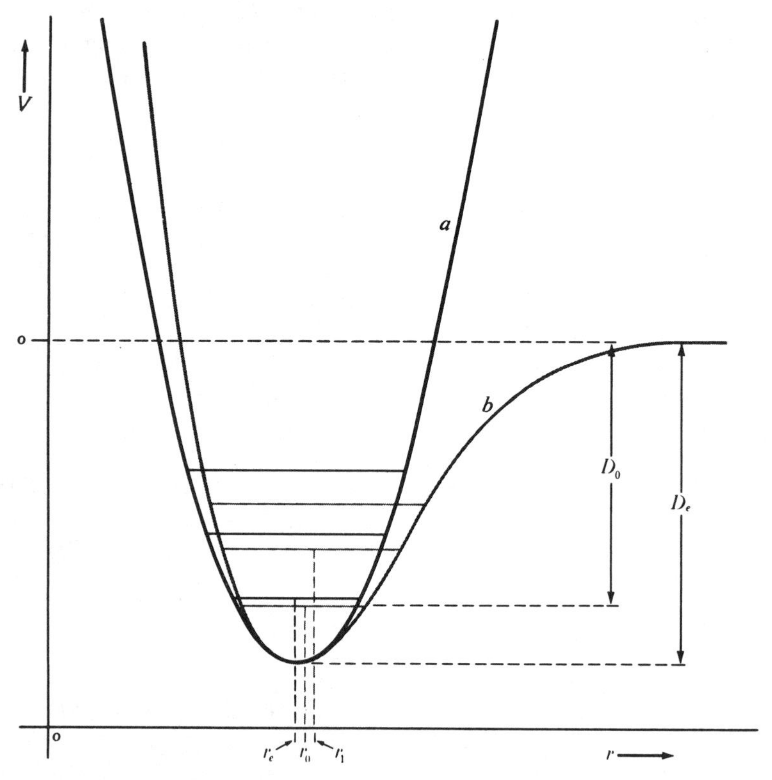

FIGURE 5-3 *Comparison of harmonic oscillator potential* (*a*) *with a more realistic potential function for a diatomic molecule* (*b*). *The first three vibrational energy levels are also indicated for each function. The effect of anharmonicity on the vibrational spacings should be noted as well as the significance of* r_e, r_0, *and* r_1. *It is conventional to take the combined energies of the isolated atoms as the zero of energy for a diatomic molecule. The extrapolated and measured dissociation energies* (D_e, D_0) *are also shown.*

parabola (line a in Figure 5-3). This is immediately seen to be unsatisfactory, for it is known that, given enough vibrational energy, a molecule will dissociate. A parabolic potential energy curve predicts no dissociation. An actual potential energy curve for a diatomic molecule (line b in Figure 5-3) does predict dissociation. (The origin of these potential energy curves will be discussed in Chapter 7.) The deviation of the actual potential energy curve from the parabolic curve is a measure of the anharmonicity of the vibration. It is clear from Figure 5-3 that anharmonicity effects will become more important the higher the vibrational quantum number v.

Vibration-rotation coupling arises from the fact that the average internuclear distance in an excited vibrational state will be larger than in the ground state. Such average internuclear distances are shown for the two lowest energy vibrational states in Figure 5-3. This effect is also due mainly to vibrational anharmonicity. The rotational constant B should, therefore, be a function of the vibrational quantum number v.

These three effects can be treated quantitatively by using perturbation theory or by arguments from classical mechanics. In what follows, a qualitative picture of the effect will be given and the mathematical details will be illustrated in exercises for the interested student.

For centrifugal stretching, the stretching effect as the molecule is excited to higher rotational states will be opposed by the force constant of the bond. The force constant is, in turn, related to the vibrational frequency ω. Furthermore, one would expect this effect to become more important the higher the value of J. The energy levels for a real rotor, then, should include a correction term to the rigid rotor energy, and this correction term would be expected to involve ω, and become more important as J increases. Applying a classical mechanical argument, one obtains

$$E_J = BJ(J+1) - \frac{4B^3}{\omega_v^2} J^2(J+1)^2 \tag{5-44a}$$

$$= BJ(J+1) - DJ^2(J+1)^2 \tag{5-44b}$$

where D is called the centrifugal stretching constant. It should be noted that since $B \ll \omega_v$, D is very small. The second term becomes relatively more important for large J since it depends on higher powers of J.

Inclusion of the centrifugal stretching correction will have the effect that the lines in a pure rotational spectrum will no longer be evenly spaced. For transitions at higher J, the spacing between observed lines will become progressively smaller. Consideration of Eq. 5-44a shows that accurate measurements of the rotational spectrum of a diatomic molecule can also give an estimate of the fundamental vibrational frequency.

To take into account the anharmonicity of vibration, we expand the potential function for a diatomic molecule as a power series; thus

$$V = \frac{1}{2}\left(\frac{\partial^2 V}{\partial q^2}\right)_{q=0} q^2 + \frac{1}{3!}\left(\frac{\partial^3 V}{\partial q^3}\right)_{q=0} q^3 + \cdots \tag{5-45}$$

Such a power series expansion uses the bottom of the potential well as the zero of energy and has no first derivative term because the first derivative must vanish at $q = 0$. The harmonic oscillator approximation only used a potential term in q^2, and therefore the derivative $(\partial^2 V/\partial q^2)_{q=0}$ can be identified with the harmonic oscillator force constant k. Using higher terms in Eq. 5-45 as perturbations gives an anharmonicity correction to the harmonic oscillator energy levels (see Exercise 5-17). Using a more realistic potential function (the Morse function discussed in Section 5-7) for which the Schrödinger equation can be solved exactly, one obtains an expression for the energy of the vth state [6]

$$\omega_v = (v + \tfrac{1}{2})\omega_e - \omega_e x_e(v + \tfrac{1}{2})^2 \tag{5-46}$$

In Eq. 5-46, ω_e is the vibrational energy that the molecule would have in the absence of any anharmonicity. In most molecules, anharmonicity affects even the lowest vibrational levels so that the observed fundamental frequency ω_v will be slightly different from ω_e. Also, we see that the correction term to the vibrational energy involves the square of the vibrational quantum number. This means, of course, that the correction term will become more important at higher v values. This is the expected result.

Anharmonicity has two important effects on the vibrational spectrum of a diatomic molecule. First of all, it modifies the energy levels so that they are no longer evenly spaced. This means that the transition frequency ω_v that one observes will be a function of the quantum number v for the lowest state involved in that transition. Secondly, anharmonicity modifies the vibrational wave functions, and this results in modified selection rules. For a real diatomic molecule, $\Delta v = \pm 2, 3, \cdots$ transitions can be observed, although they are weak. These transitions are called overtones. The $\Delta v = \pm 1$ transition is called the fundamental. From the frequencies of the fundamental and various overtones, the anharmonicity constant $\omega_e x_e$ and the extrapolated vibrational frequency ω_e can be calculated.

To gain a feeling for the magnitude of these effects, the student should work Exercises 5-15 and 5-16.

EXERCISE 5-15 The observed frequencies in cm^{-1} of the fundamental and first three overtones of HCl^{35} are 2885.9, 5668.1, 8347.0, and 10,923.1, respectively. Calculate the extrapolated vibrational frequency ω_e and the anharmonicity constant of HCl^{35}. *Hint:* Derive an expression for the transition energies and use a graphical method to determine ω_e.

EXERCISE 5-16 Some of the rotational transitions for HCl^{35} are given in the following table.

Transition $J \rightarrow J+1$	$\omega_R(\mathrm{cm}^{-1})$
$3 \rightarrow 4$	83.03
$4 \rightarrow 5$	103.62
$5 \rightarrow 6$	124.30
$6 \rightarrow 7$	145.03
$7 \rightarrow 8$	165.51
$8 \rightarrow 9$	185.86
$9 \rightarrow 10$	206.38
$10 \rightarrow 11$	226.50

Derive a general expression for the energy of a rotational transition with the centrifugal stretching constant included. Calculate a value for B and D for HCl from the above data. *Hint:* Use a graphical method similar to that in Exercise 5-15.

EXERCISE 5-17 To calculate the effects of anharmonicity on the harmonic oscillator energy levels, consider a perturbation Hamiltonian of the form $\hat{\mathscr{H}} = aq^3 + bq^4$. This amounts to keeping two additional terms in the expansion 5-45. Transform to the variable ξ and write down the two integrals needed to evaluate the first order correction to the harmonic oscillator energy. Arguing from the symmetry (even or odd upon exchange of q and $-q$) show that the integral involving q^3 equals zero. This can also be done by making use of the recursion relation for Hermite polynomials,

$$H_{v+1}(\xi) - 2\xi H_v(\xi) + 2vH_{v-1}(\xi) = 0$$

and arguing from the orthonormality properties of the Hermite polynomials. Use the relation above to evaluate the integral involving q^4. Show that the correction term involves v^2. Note that only including the q^4 term in first order does not give exactly the v dependence of Eq. 5-46. See p. 160 of [5] for help.

To correct for vibration-rotation coupling, the observed rotational constant B is regarded as a function of the vibrational quantum number v. Thus

$$B_v = B_e - \alpha(v + \tfrac{1}{2}) \tag{5-47}$$

where α is the vibration rotation coupling constant and B_e is the (hypothetical) rotational constant in the absence of vibration. The quantity B_v is the effective value of the rotational constant in the vth vibrational state. In a vibration-rotation spectrum, vibration-rotation coupling causes the lines in the R branch to become more closely spaced and the lines in the P branch to become farther apart as J'' increases. An actual vibration-rotation spectrum of the first overtone band of HCl (Figure 5-4) clearly shows these effects.

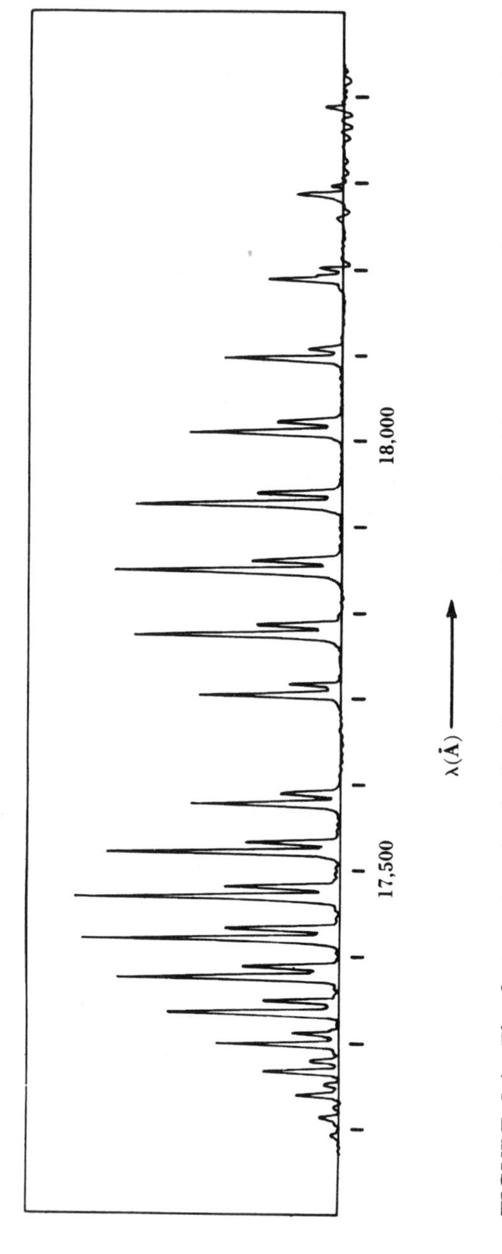

FIGURE 5-4 *The first overtone band of* HCl *taken on a Cary model 14 spectrometer. The strong set of lines is from* HCl^{35}; *the weak set is from* HCl^{37}. *The unequal spacings between lines show the effects of vibration-rotation coupling. Each division on the wavelength scale equals* 100 Å.

Because of vibration-rotation coupling, the expression for the allowed transitions in an actual vibration-rotation spectrum are more complex than indicated by Eq. 5-43. Thus

$$\omega_{v,R} = \omega_v + B'J'(J'+1) - B''J''(J''+1) \tag{5-48}$$

where a different B value must be used for the upper and lower states. Equation 5-48 leads to the following expressions for the energy of the lines in the P and R branches.

$$R \text{ branch } \omega_{v,R} = \omega_v + 2B' + (3B' - B'')J'' + (B' - B'')(J'')^2 \qquad (J'' = 0, 1, \cdots) \tag{5-49a}$$

$$P \text{ branch } \omega_{v,R} = \omega_v - (B' + B'')J'' + (B' - B'')(J'')^2 \qquad (J'' = 1, 2, \cdots) \tag{5-49b}$$

A consideration of Figure 5-2c shows how B'' and B' can be evaluated from the observed spectrum. For example, the difference in energy of $P(2)$ and $R(0)$ gives a spacing $6B''$.

5-7 *THE MORSE FUNCTION, DISSOCIATION ENERGY, AND THE ANHARMONICITY CONSTANT*

An empirical equation for the potential energy of a diatomic molecule, which in many cases is quite accurate, was proposed by P. M. Morse. It is

$$V = D_e(1 - e^{-\beta q})^2 \tag{5-50}$$

where again $q = (r - r_e)$, D_e is the dissociation energy from the potential minimum, and β is a constant which is related to molecular parameters by the relation

$$\beta = \omega_e(2\pi^2 c\mu / D_e h)^{\frac{1}{2}} \tag{5-51}$$

Both D_e and β are in cm^{-1}.

EXERCISE 5-18 Expand the Morse function (Eq. 5-50) in a Maclaurin series about $q = 0$. Using the coefficient of the q^2 term as $\frac{1}{2}k$, relate the force constant k to the parameter β of the Morse function.

Note that for $q = \infty$, $V = D_e$ the dissociation energy. This dissociation energy differs from the measured dissociation energy of a molecule, however.

The quantity D_e is the energy that would be necessary to dissociate the molecule if it could be at the minimum of the potential energy curve. Because of the zero-point energy, this is impossible, and therefore

$$D_e = D_0 + \tfrac{1}{2}h\nu_0 \tag{5-52}$$

where D_0 is the measured dissociation energy in ergs (see Figure 5-3).

There is an approximate relationship between the dissociation energy and the anharmonicity constant that can be derived as follows. The maximum vibration energy a molecule can have is

$$(E_v)_{\max} = D_e \tag{5-53}$$

For large v, the vibrational levels are very closely spaced, and, near dissociation, E_v may be considered a continuous function of v without much error. Converting to wave numbers,

$$\omega_v = (v + \tfrac{1}{2})\omega_e - (v + \tfrac{1}{2})^2\omega_e x_e$$

$$\frac{\partial \omega_v}{\partial v} = \omega_e - 2\left(v + \frac{1}{2}\right)\omega_e x_e \tag{5-54}$$

At dissociation, the vibrational energy must be a maximum. Therefore,

$$\frac{\partial \omega_v}{\partial v} = \omega_e - 2\left(v_{\max} + \frac{1}{2}\right)\omega_e x_e = 0 \tag{5-55}$$

$$v_{\max} = \frac{\omega_e}{2\omega_e x_e} - \frac{1}{2} \tag{5-56}$$

But

$$\omega_{v_{\max}} = D_e = (v_{\max} + \tfrac{1}{2})\omega_e - (v_{\max} + \tfrac{1}{2})^2\omega_e x_e \tag{5-57}$$

where D_e is the dissociation energy in cm^{-1}. Substituting Eq. 5-56 into Eq. 5-57 and rearranging, we obtain

$$D_e = \frac{\omega_e{}^2}{4\omega_e x_e} \tag{5-58}$$

or, in terms of the measured dissociation energy D_0 in cm^{-1},

$$D_0 = \frac{\omega_e{}^2}{4\omega_e x_e} - \frac{1}{2}\omega_e \tag{5-59}$$

Therefore, from the known dissociation energy and fundamental vibrational frequency, the anharmonicity constant can be estimated or vice versa. It should be emphasized that the above treatment is empirical and approximate.

EXERCISE 5-19 Using the Morse function, evaluate the derivative $(d^4V/dq^4)_{q=0}$. Use this value in the results of Exercise 5-17 to evaluate the anharmonicity constant derived from perturbation theory. This should be compared with the quantity derived from the exact solution to the Schrödinger equation using the Morse potential of

$$\omega_e x_e = \frac{\omega_e \beta}{2\pi c}\left(\frac{hc}{8D_e\mu}\right)^{\frac{1}{2}}, \qquad D_e \text{ in cm}^{-1}$$

5-8 VIBRATIONAL SPECTROSCOPY OF COMPLEX MOLECULES

To specify completely the instantaneous state of a molecule containing N atoms, one would have to specify the values of $3N$ coordinates, three coordinates for each atom. It can be said that there is a total of $3N$ degrees of freedom for a molecule containing N atoms. For systems where the potential energy is only a function of the relative positions of the atoms, one can always transform to internal coordinates and separate out the motion of the center of mass. This leaves one with $3N - 3$ internal degrees of freedom and three degrees of freedom for the translational motion of the molecule.

The $3N - 3$ internal degrees of freedom can be further separated into rotational and vibrational degrees of freedom. There will either be two or three rotational degrees of freedom depending on whether the molecule is linear or nonlinear. Thus for a linear molecule there will be $3N - 5$ vibrational degrees of freedom, whereas for a nonlinear molecule there will be $3N - 6$ vibrational degrees of freedom. For a diatomic molecule (which must be linear), $N = 2$, and there is only one vibrational degree of freedom. This is the displacement coordinate $q = (r - r_e)$ used above. For polyatomic molecules, there will be many vibrational degrees of freedom. The detailed treatment of the vibrational motion of these polyatomic molecules is beyond the scope of this book, but the basic approach to the problem will be summarized.

To a fairly high degree of approximation, it is always possible to separate the translational and rotational motions of a molecule from its vibrational motion. This is done by transforming the $3N$ Cartesian coordinates, with respect to some coordinate system fixed in the laboratory, to a new set of coordinates with respect to a coordinate system whose origin is at the center of mass of the molecule. This molecule-fixed coordinate system is moving with the molecule and is rotating in such a way that the $3N - 6$ coordinates in this moving coordinate system have no displacements which correspond to a translation or a rotation of the molecule. The six coordinates which are needed to account for the original $3N$ can be taken as the three coordinates of the center of mass and the three Eulerian angles describing the configuration of the rotating coordinate system. For more details, see [7].

The $3N - 6$ ($3N - 5$ for a linear molecule) coordinates which are left may be thought of as vibrational coordinates. If these coordinates are called $q_1, q_2, \ldots, q_{3N-6}$, then the potential energy of the molecule for small displacements of the coordinates can be expanded in a power series

$$V = \sum_{i=1}^{3N-6} \left(\frac{\partial V}{\partial q_i}\right)_0 q_i + \frac{1}{2} \sum_{i,j=1}^{3N-6} \left(\frac{\partial^2 V}{\partial q_i \, \partial q_j}\right)_0 q_i q_j + \text{higher terms} \tag{5-60}$$

In Eq. 5-60, the potential energy is taken as zero at the equilibrium nuclear configuration, and all of the derivatives are evaluated at that configuration. Futhermore, at the equilibrium configuration, the potential energy of the system is a minimum, and all of the $(\partial V/\partial q_i)_0 = 0$. Neglecting terms higher than 2, the potential energy is then

$$V = \frac{1}{2} \sum_{i,j=1}^{3N-6} f_{ij} q_i q_j \tag{5-61}$$

where

$$f_{ij} = \left(\frac{\partial^2 V}{\partial q_i \, \partial q_j}\right)_0 \tag{5-62}$$

By applying a second coordinate transformation, it is always possible to find a new set of coordinates, called Q_i, that eliminates all of the cross terms in the sum in Eq. 5-61. Using these new coordinates, Eq. 5-61 becomes

$$V = \frac{1}{2} \sum_{i}^{3N-6} F_i Q_i^2 \tag{5-63}$$

In this equation, the potential energy is a sum of terms, each one of which is similar to the harmonic oscillator expression for a one-dimensional oscillator. The F_i are the corresponding force constants. Because the Q_i enable the potential energy to be written in this simple form, they are called normal coordinates, and the modes of vibration which they describe are called normal modes. By deriving the expressions which relate the Q_i to the q_i, these normal modes can be described in terms of various stretching and bending motions of the atoms. As an example, the normal modes for water are shown in Figure 5-5.

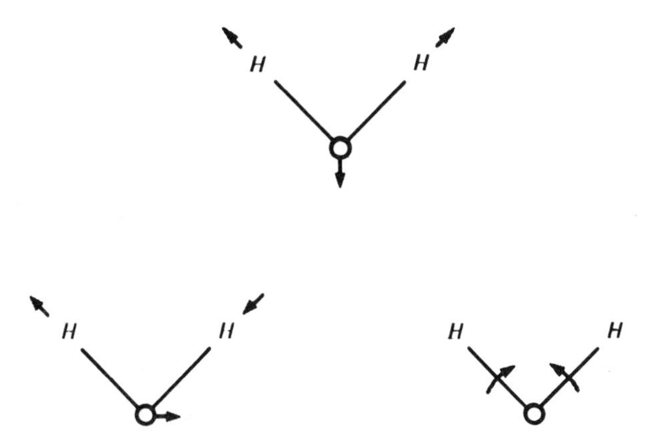

FIGURE 5-5 *Normal modes of the water molecule. The frequencies are* 3652, 3656, *and* 1545 cm^{-1} *for the symmetric stretching, asymmetric stretching, and symmetric bending modes, respectively.*

It should be emphasized that these normal modes are vibrations which involve all of the atoms in a molecule.

A complete normal coordinate analysis is only feasible for small molecules or for larger molecules that have high symmetry. For large asymmetric molecules, the number of vibrational degrees of freedom becomes very large, and an exact analysis of the vibrational motion of such a molecule is impractical, if not impossible. In the infrared spectrum of such molecules, it is usually possible to pick out certain group or bond vibrations, however. These group or bond vibrations occur at approximately the same frequency regardless of what the rest of the molecule is like, and the appearance of many bands in an infrared spectrum can be correlated with the presence of one of these groups or bonds in a molecule. For example, a carbonyl group will always give rise to a band in the infrared between 1,650 and 1,750 cm^{-1}, and, conversely, the presence of this band is the spectrum of an unknown compound can be used to establish the presence of a carbonyl group. An —OH group which is not hydrogen bonded occurs at 3,600 cm^{-1}. If the —OH group is hydrogen bonded, its absorption band occurs at lower energies (3,400–3,500 cm^{-1}). A large amount of data now exists which correlates infrared frequencies with molecular structure, and the interested student may consult other works for more information on this point [8].

5-9 RAMAN SPECTROSCOPY

Infrared spectroscopy gives information about vibrational energies in molecules as long as there is a change in dipole moment during the vibration. As we pointed out in Section 5-4, all homonuclear diatomic molecules do not have dipole moments and, therefore, they will not have an infrared spectrum. In addition, many totally symmetric vibrations in polyatomic molecules do not have a change in dipole moment during the vibration. Consequently, these vibrations will not be infrared active. Many of these vibrations that cannot be studied by infrared can be studied by *Raman spectroscopy*.

In Raman spectroscopy, an intense beam of monochromatic light, usually produced in modern instruments by a laser, is passed through a sample; the light scattered at right angles to the incident beam is analyzed according to frequency. The theoretical development of Raman spectroscopy requires a treatment of light scattering. A detailed treatment would be too long for the present book (the reader is referred to [9] for an excellent simple yet fundamental development), but some of the physical ideas responsible for Raman scattering can be outlined.

When matter interacts with an electric field, electrons in the sample are polarized, giving rise to an induced dipole moment. The *polarizability* of a molecule is the proportionality constant between this induced dipole moment and the electric field (Eq. 5-64).

$$\mu_{\text{ind}} = \alpha E \tag{5-64}$$

Equation 5-64 has been written as a scalar equation for simplicity. In general, μ_{ind} and E are vectors and α is a tensor [10]. If the electric field comes from a light wave, it will oscillate with the frequency of the light wave, ν, and the induced dipole moment will also oscillate with the frequency ν (Eq. 5-65).

$$\mu_{\text{ind}} = \alpha E_0 \cos 2\pi\nu t \tag{5-65}$$

Classical electromagnetic theory tells us, however, that oscillating dipoles emit radiation at the oscillating frequency. Consequently, when a light beam passes through a sample in one direction, the interaction of this light beam with the electrons in the sample causes them to oscillate according to Eq. 5-65, and these oscillating dipoles emit light in other directions. This "scattered" light is at the same frequency as the incident radiation and is called *Rayleigh scattering*.

It is reasonable to expect the polarizability of electrons to depend on the positions of the nuclei and, therefore, to depend on the vibrational coordinate $q = r - r_e$. (A diatomic molecule is assumed here for simplicity.) We can expand α in a Maclaurin series similar to our expansion for the dipole moment (Eq. 5-42).

$$\alpha = \alpha_0 + \left(\frac{d\alpha}{dq}\right)_{q=0} q + \cdots \tag{5-66}$$

Since the molecule is vibrating, q will be changing with the vibrational frequency ν_0, and we can write Eq. 5-66 as

$$\alpha = \alpha_0 + \left(\frac{d\alpha}{dq}\right)_{q=0} A \cos 2\pi\nu_0 t \tag{5-67}$$

where A represents the amplitude of the vibration. Substituting Eq. 5-67 into 5-65, we get

$$\mu_{\text{ind}} = \alpha_0 E_0 \cos 2\pi\nu t + A E_0 \left(\frac{d\alpha}{dq}\right)_{q=0} \cos 2\pi\nu_0 t \cos 2\pi\nu t$$

and using the trigonometric identity

$$\cos\alpha \cos\beta = \tfrac{1}{2}[\cos(\alpha + \beta) + \cos(\alpha - \beta)]$$

we get

$$\mu_{\text{ind}} = \alpha_0 E_0 \cos 2\pi\nu t + \frac{AE_0}{2}\left(\frac{d\alpha}{dq}\right)_{q=0} \times [\cos 2\pi(\nu + \nu_0)t + \cos 2\pi(\nu - \nu_0)t] \tag{5-68}$$

Two important physical ideas appear in Eq. 5-68. The first is that there is not only an induced dipole at the frequency of the incident light (ν) but also an induced dipole at the frequency ν plus and minus the vibrational frequency ν_0. Thus, there will be scattered light not only at frequency ν but also at frequencies $\nu + \nu_0$ and $\nu - \nu_0$. This latter phenomenon is called Raman scattering, and by observing these scattered lines, one can determine the vibrational frequency ν_0. The second physical concept in Eq. 5-68 is that the magnitude of the Raman scattering depends on $(d\alpha/dq)_{q=0}$, the change in polarizability upon vibration. Thus, the selection rules for Raman spectroscopy are different from infrared spectroscopy, and vibrational frequencies from homonuclear diatomic molecules and other infrared inactive vibrations can be determined.

The preceding derivation is a classical one and is valid as long as the frequency of the incident light is much greater than any natural electronic absorption frequencies. When ν is the same as the frequency of an electronic absorption band, one can still get Raman bands, but the intensities of these bands can only be calculated quantum mechanically. Raman spectroscopy in an absorption band is called resonance Raman spectroscopy [11]. Both normal and resonance Raman spectroscopy have been valuable tools in recent biophysical studies.

5-10 SUMMARY

1. Using the rigid rotor approximation, the allowed rotational states of a diatomic molecule were found to have energy $\omega_R = BJ(J+1)$, where J is the rotational quantum number and B is a parameter determined by the molecule.
2. The vibrational states of a diatomic molecule were found to have energy $\omega_v = (v + \frac{1}{2})\omega_0$, where v is the vibrational quantum number.
3. Rotational transitions are only allowed if $\Delta J = \pm 1$, $\Delta m = 0, \pm 1$. Within the context of the harmonic oscillator approximation, vibrational transitions are also allowed if $\Delta v = \pm 1$. These restrictions on allowed transitions are called selection rules.
4. In an infrared spectrum, transitions take place between a number of rotational sublevels in the initial and final vibrational states. This leads to sharp vibration-rotation spectra for light diatomic molecules and vibrational band spectra for heavier diatomic and polyatomic molecules.
5. The harmonic oscillator–rigid rotor energy expression was corrected for centrifugal stretching, vibrational anharmonicity, and vibration-rotation coupling.
6. A brief discussion of vibrations in polyatomic molecules and Raman spectroscopy was given.
7. The student should be familiar with the following terms: rotational degeneracy, average internuclear distance, equilibrium internuclear distance, fundamental frequency, overtones, vibrational zero-point energy, Morse function, extrapolated dissociation energy, measured dissociation energy, normal mode, and dipole moment.

REFERENCES

1. H. Margenau and G. M. Murphy, *The Mathematics of Physics and Chemistry* (D. Van Nostrand, Inc., Princeton, N.J., 1943), pp. 61ff.
2. H. Eyring, J. Walter, and G. E. Kimball, *Quantum Chemistry* (J. Wiley & Sons, Inc., New York, 1944), pp. 52ff.
3. G. Herzberg, *Spectra of Diatomic Molecules* (D. Van Nostrand, Inc., Princeton, N.J., 1950), pp. 125ff.
4. W. C. Gardiner, Jr., *Rates and Mechanisms of Chemical Reactions* (W. A. Benjamin, Inc., New York, 1969).
5. G. M. Barrow, *An Introduction to Molecular Spectroscopy* (McGraw-Hill Book Co., New York, 1962), pp. 59ff.
6. L. Pauling and E. B. Wilson, *Introduction to Quantum Mechanics* (McGraw-Hill Book Co., New York, 1935).
7. E. B. Wilson, Jr., J. C. Decius, and P. C. Cross, *Molecular Vibrations* (McGraw-Hill Book Co., New York, 1955), Chaps. 2 and 11.

8. L. J. Bellamy, *The Infrared Spectra of Complex Molecules* (J. Wiley & Sons., Inc., New York, 1954).
9. A. G. Marshall, *Biophysical Chemistry* (J. Wiley & Sons, Inc., New York, 1978), Chaps. 15 and 17.
10. I. N. Levine, *Molecular Spectroscopy* (J. Wiley & Sons, Inc., New York, 1975), pp. 186ff.
11. D. P. Strommen and K. Nakamoto, *J. Chem. Ed.*, **54**, 474 (1977).

Chapter 6

THE ELECTRONIC STRUCTURE OF ATOMS

THE IDEA of the atom is a construct that lies at the very heart of chemistry. Since the development of the modern atomic theory by John Dalton in the period 1803–1807 the science of chemistry has been concerned with the nature of the atom and the way in which atoms join together to form molecules. It was, therefore, quite natural to turn the new ideas coming out of quantum mechanics toward the structure of the atom to see if a satisfactory model for its experimentally determined properties could be obtained.

It was pointed out in Chapter 3 that the Bohr theory could account for the spectrum of the hydrogen atom, but that it failed to account for the properties of more complex atoms. In this chapter, we will use the postulates given in Chapter 3 to calculate the allowed energies of a hydrogen atom. The structure of more complex atoms will then be discussed in the light of the hydrogen atom results.

6-1 THE HYDROGEN ATOM AND THE HYDROGEN-LIKE IONS

It will be convenient to treat the hydrogen atom and the hydrogen-like ions, He^+, Li^{2+}, etc., in one group because they differ from one another only in their nuclear charge. These atoms and ions have a nucleus of charge $+Ze$ and mass M and one electron of charge $-e$ and mass m. The symbol

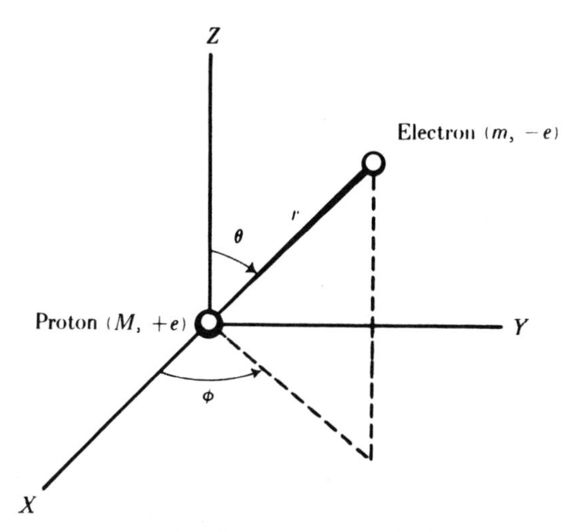

FIGURE 6-1 *Quantities used in the discussion of the hydrogen atom.*

e represents the absolute value of the electronic charge. The coordinates for the hydrogen-atom problem are shown in Figure 6-1.

The potential energy of the system is the attractive energy between the electron and the nucleus. Thus

$$V = -\frac{Ze^2}{r} \tag{6-1}$$

where r is the distance between the nucleus and the electron. Since the potential energy is a function only of the internal coordiates, we can immediately separate the coordinates of the center of mass and only concern ourselves with the internal motion of the system. (Since the proton mass is 1846 times larger than the electron mass, very little error is made if the reduced mass μ in Eq. 6-2 is replaced by the electron mass m. In many discussions of the hydrogen atom the electron mass is used instead of the reduced mass.) The appropriate Hamiltonian operator is

$$\mathscr{H} = -\frac{\hbar^2}{2\mu}\nabla^2 - \frac{Ze^2}{r} \tag{6-2}$$

To find the allowed energies for the hydrogen atom, the eigenvalue equation

$$\mathscr{H}\Psi = E\Psi \tag{6-3}$$

must be solved. Since this is a centrosymmetric problem, it will be most convenient to use spherical polar coordinates. Using the expression for

∇^2 in this coordinate system (Eq. 1-31) one obtains the rather formidable looking equation

$$\frac{1}{r^2}\frac{\partial}{\partial r}\left(r^2\frac{\partial \Psi}{\partial r}\right)+\frac{1}{r^2\sin\theta}\frac{\partial}{\partial\theta}\left(\sin\theta\frac{\partial\Psi}{\partial\theta}\right) + \frac{1}{r^2\sin^2\theta}\frac{\partial^2\Psi}{\partial\phi^2}+\frac{2\mu}{\hbar^2}\left(E+\frac{Ze^2}{r}\right)\Psi=0 \tag{6-4}$$

We will follow the same procedure for solving this equation that was followed for the problem of rotational motion. That is, a solution of the form

$$\Psi = R(r)\Theta(\theta)\Phi(\phi) \tag{6-5}$$

will be sought. After straightforward algebraic manipulation similar to that used in Chapter 5, the variables in Eq. 6-4 can be separated and one obtains three single-variable equations.

$$\frac{1}{\Phi}\frac{d^2\Phi}{d\phi^2} = -m^2 \tag{6-6}$$

$$\frac{1}{\sin\theta}\frac{d}{d\theta}\left(\sin\theta\frac{d\Theta}{d\theta}\right)-\frac{m^2}{\sin^2\theta}\Theta+\lambda\Theta=0 \tag{6-7}$$

$$\frac{1}{r^2}\frac{d}{dr}\left(r^2\frac{dR}{dr}\right)+\left[\frac{2\mu}{\hbar^2}\left(E+\frac{Ze^2}{r}\right)-\frac{\lambda}{r^2}\right]R=0 \tag{6-8}$$

where the constants m and λ are introduced in the separation procedure (see Section 5-1 if you have forgotten the details of the separation of variables argument). Equations 6-6 and 6-7 are exactly the same as the equations obtained in the discussion of rotational motion, and their solutions have already been given in Section 5-1. In the Θ functions for electronic motion, the quantum number l is used instead of J, however. Thus the constant $\lambda = l(l+1)$.

Equation 6-8 can be manipulated to get it in the form of another equation of classical physics, the associated Laguerre equation [1, 2]. For our purposes, we need only know that solutions to Eq. 6-8 that are everywhere finite, single valued, and have integrable squares, exist only for the conditions that

$$-\frac{Z^2\mu e^4}{2\hbar^2E}=n^2 \qquad (n=1,2,3,4,\cdots) \tag{6-9a}$$

and

$$0 \leqq l \leqq n-1 \tag{6-9b}$$

This means that the energy of the electron in a hydrogen atom or hydrogen-like ion is restricted to the values

$$E_n = -\frac{Z^2\mu e^4}{2\hbar^2}\frac{1}{n^2} \tag{6-10}$$

This is the same as the expression obtained from Bohr theory.

EXERCISE 6-1 Use Eq. 6-10 to calculate the ionization potential of a hydrogen atom; of the He^+ ion. The calculation will be obvious if you know n for the initial and final states in each case.

The R functions obtained as solutions of Eq. 6-8 are the associated Laguerre polynomials and are usually called radial wave functions. Their detailed form depends on the values of the quantum numbers n and l. Some of these radial wave functions are given in Table 6-1

TABLE 6-1 SOME NORMALIZED RADIAL WAVE FUNCTIONS FOR THE HYDROGEN ATOM[a]

$$R(1, 0) = 2\left(\frac{Z}{a_0}\right)^{\frac{3}{2}} e^{-\sigma}$$

$$R(2, 0) = \left(\frac{Z}{2a_0}\right)^{\frac{3}{2}}(2-\sigma)e^{-\sigma/2}$$

$$R(2, 1) = 3^{-\frac{1}{2}}\left(\frac{Z}{2a_0}\right)^{\frac{3}{2}}\sigma e^{-\sigma/2} \qquad \sigma = \frac{Zr}{a_0}$$

$$R(3, 0) = \frac{2}{27}\left(\frac{Z}{3a_0}\right)^{\frac{3}{2}}(27 - 18\sigma + 2\sigma^2)e^{-\sigma/3}$$

$$R(3, 1) = \frac{1}{81\sqrt{3}}\left(\frac{2Z}{a_0}\right)^{\frac{3}{2}}(6-\sigma)\sigma e^{-\sigma/3}$$

[a] The numbers in parentheses are the values of n and l. The quantity a_0 is the radius of the first Bohr orbit, 0.529 Å.

EXERCISE 6-2 Show by direct substitution that $R(1, 0)$ and $R(2, 1)$ are solutions of Eq. 6-8.

The total wave function for the hydrogen atom is the product of the suitably normalized radial and angular functions R and $\Theta\Phi$. Thus

$$\Psi(n, l, m) = R(n, l)\Theta(l, m)\Phi(m) = R(n, l)\,Y_{l,m} \tag{6-11}$$

where the explicit dependence on quantum numbers is indicated in parentheses. The $Y_{l,m}$ are the spherical harmonics introduced in Chapter 5. It can be seen from Eq. 6-11 that the allowed states of the hydrogen atom, sometimes called hydrogen-like orbitals, depend on three quantum numbers, n, l, and m. A brief discussion of the significance of each of these quantum numbers will now be given.

The quantum number n is called the principal quantum number. For the hydrogen atom and the hydrogen-like ions, this quantum number determines the energy (for more complex atoms, the energy will also depend on l) and the total number of nodes in the wave function. There will always be $n - 1$ nodes in the total wave function if the node at infinity is neglected.

The quantum number l is called the azimuthal quantum number. It is the quantum number associated with the total angular momentum of the electron. In quantum mechanical language, the functions $Y_{l,m}$ are eigenfunctions of the operator $\hat{L}^2$ with eigenvalue $l(l + 1)\hbar^2$. That is,

$$\hat{L}^2 Y_{l,m} = l(l + 1)\hbar^2 Y_{l,m} \tag{6-12}$$

The number l is restricted to integral values between 0 and $n - 1$, and it gives the number of nodes in the angular part of the wave function.

The quantum number m is called the magnetic quantum number. It is associated with the component of angular momentum along a specific axis in the atom, usually called the Z axis. Since atoms are spherically symmetric, there is no way to define a specific axis unless the atom is placed in an electric or magnetic field. The quantum number m does not have any effect on the properties of the hydrogen atom unless such fields are present. It does determine the degeneracy of a state, however, since there are $2l + 1$ values of m for a state with quantum number l. The number m is restricted to the values $l, l - 1, \ldots, -l + 1, -l$. The functions $\Phi(m)$ are eigenfunctions of the operator $\hat{L}_z$. Thus

$$\hat{L}_z \Phi(m) = m\hbar\Phi(m) \tag{6-13}$$

In the presence of a magnetic field, the states corresponding to different values of m will have different energies. This splitting of states with different m values by a magnetic field is called the Zeeman effect (see Exercise 6-3).

EXERCISE 6-3 It is well known, from electricity and magnetism, that a charge moving in an orbit gives rise to a magnetic field. An electron that has nonzero angular momentum will also have this property. The orbital motion of the electron can be treated as though a magnetic dipole were present in the atom whose z component is given by the relation

$$\hat{\mu}_z = -\beta\hat{L}_z$$

where β is a constant called the Bohr magneton equal to $e\hbar/2mc = 9.2732 \times 10^{-21}$ erg G^{-1}. The energy of interaction of such an orbital magnetic moment with a static magnetic field is

$$E = -\boldsymbol{\mu} \cdot \mathbf{H}$$

Taking the direction of **H** as defining the Z axis, use first order perturbation theory to calculate the energies of the three states $\Psi(2, 1)$.

EXERCISE 6-4 Write down the function for the $1s$ hydrogen orbital. Show that it is normalized. What would the corresponding function be for the He^+ ion?

EXERCISE 6-5 What is the degeneracy of the function $\Psi(4, 3)$? What is the maximum value that the quantum number m can have in this function?

EXERCISE 6-6 Evaluate the integrals $\langle Y_{0,0}|\cos\theta|Y_{0,0}\rangle$ and $\langle Y_{1,0}|\cos\theta|Y_{0,0}\rangle$. Integrals such as these will be important in the treatment of selection rules for the hydrogen atom spectrum.

6-2 THE PHYSICAL SIGNIFICANCE OF HYDROGEN-LIKE ORBITALS

It is important that the physical significance of the hydrogen-atom solutions be understood because, as we see later, these solutions form the building blocks for the solution to more complex problems. The significance of these solutions arises from Postulate I, which states that the function $\Psi^*\Psi\ d\tau$ represents a probability. We will first discuss the radial functions $R(n, l)$ in this context.

There are two ways to discuss the radial functions. The first way makes use of a plot of the function R^2 versus r/a_0, the distance from the nucleus in atomic units. Such a plot is shown for several radial functions in Figure 6-2. This quantity $R^2\ dr$ only has significance for a fixed value of the angular variables θ and ϕ. Fixing θ and ϕ determines a line beginning at the origin. The function $R^2\ dr$ then is proportional to the probability of finding the electron in a small unit of length dr at different locations along this line. Figure 6-2 shows that, for s orbitals, the electron has a nonvanishing probability of being at the nucleus, whereas for all other orbitals the value of R^2 at the nucleus is zero. This fact will be important when electron magnetic resonance spectroscopy is discussed in Chapter 9.

A second way to discuss the significance of the radial wave functions is to integrate over the angular variables and plot the resulting function F. Thus we wish to find

$$F = \int_0^{\pi} \int_0^{2\pi} R^2 r^2 \sin\theta\ dr\ d\theta\ d\phi = 4\pi r^2 R^2\ dr \qquad (6\text{-}14)$$

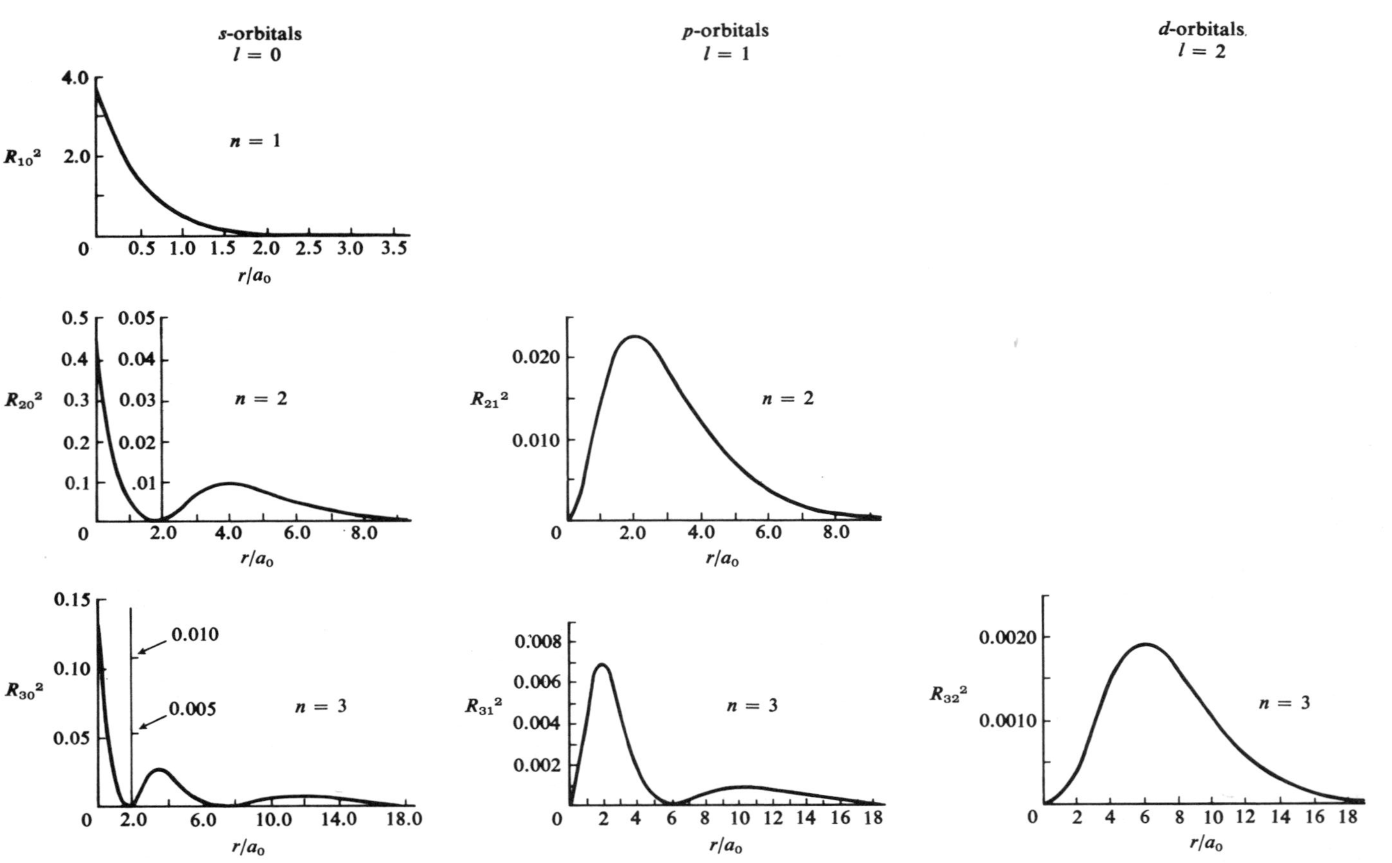

FIGURE 6-2 *Plots of the radial probability density* R^2 *as a function of the distance r from the nucleus for several atomic orbitals.* R^2 *is in units of* a_0^{-3}. *Distances are in units of* a_0. *Note the change in density scale for the different orbitals; also,* R^2 *is nonzero*

The function $4\pi r^2 R^2$ is called the radial distribution function and gives the probability of finding the electron in a spherical shell of thickness dr at a distance r away from the nucleus. Plots of the radial distribution function for several orbitals are shown in Figure 6-3, and a good discussion of the significance of this function can be found in [3].

Some other features of the hydrogenic wave functions which should be noted from Figures 6-2 and 6-3 are the following:

1. There are no nodes in the radial part of the $2p$ function. Since there must be one node in the total wave function, it must be in the angular part of the function. Similar arguments should be considered about the nodes in the other functions.

2. An electron in a $2s$ orbital has a higher probability of being close to the nucleus than an electron in a $2p$ orbital. We will use this fact to rationalize the difference in energy between $2s$ and $2p$ orbitals in atoms with more than one electron.

Several exercises will be used to illustrate some features of the radial wave functions.

EXERCISE 6-7 Calculate the most probable value of r for an electron in a hydrogen $1s$ orbital.

EXERCISE 6-8 For an electron in a $1s$ orbital of a hydrogen atom, calculate $\langle r \rangle$, and $\langle r^2 \rangle^{\frac{1}{2}}$ (root mean square value of r), and $\langle 1/r \rangle$.

EXERCISE 6-9 The operator corresponding to the potential energy of an electron in a hydrogen-like ion with nuclear charge Z is $V = -Ze^2/r$. Calculate the expectation value of the potential energy $\langle V \rangle$ for an electron in the lowest energy state of this hydrogen-like ion.

EXERCISE 6-10 Use the total energy of a hydrogen-like ion (see Eq. 6-10) and the results of Exercise 6-9 to calculate the expectation value of the kinetic energy $\langle T \rangle$ of a hydrogen-like ion with charge Z. How is $\langle T \rangle$ related to $\langle V \rangle$? This is a specific example of a general theorem, called the virial theorem, that will be discussed in more detail in Chapter 7 [note $(\mu e^4/\hbar^2) = e^2/a_0$].

EXERCISE 6-11 Calculate the probability of finding an electron in a $1s$ orbital outside of the first Bohr orbit (a_0).

EXERCISE 6-12 Calculate the most probable value of r for an electron in a $1s$ orbital of the He^+ ion. Compare this result with that from Exercise 6-7.

A problem of representation arises when discussing the angular dependence of atomic orbitals. The student is probably familiar with the geometrical pictures (such as those in Figure 6-4) in elementary textbooks. These pictures can be called "charge cloud" drawings. From these pictures, it is easy

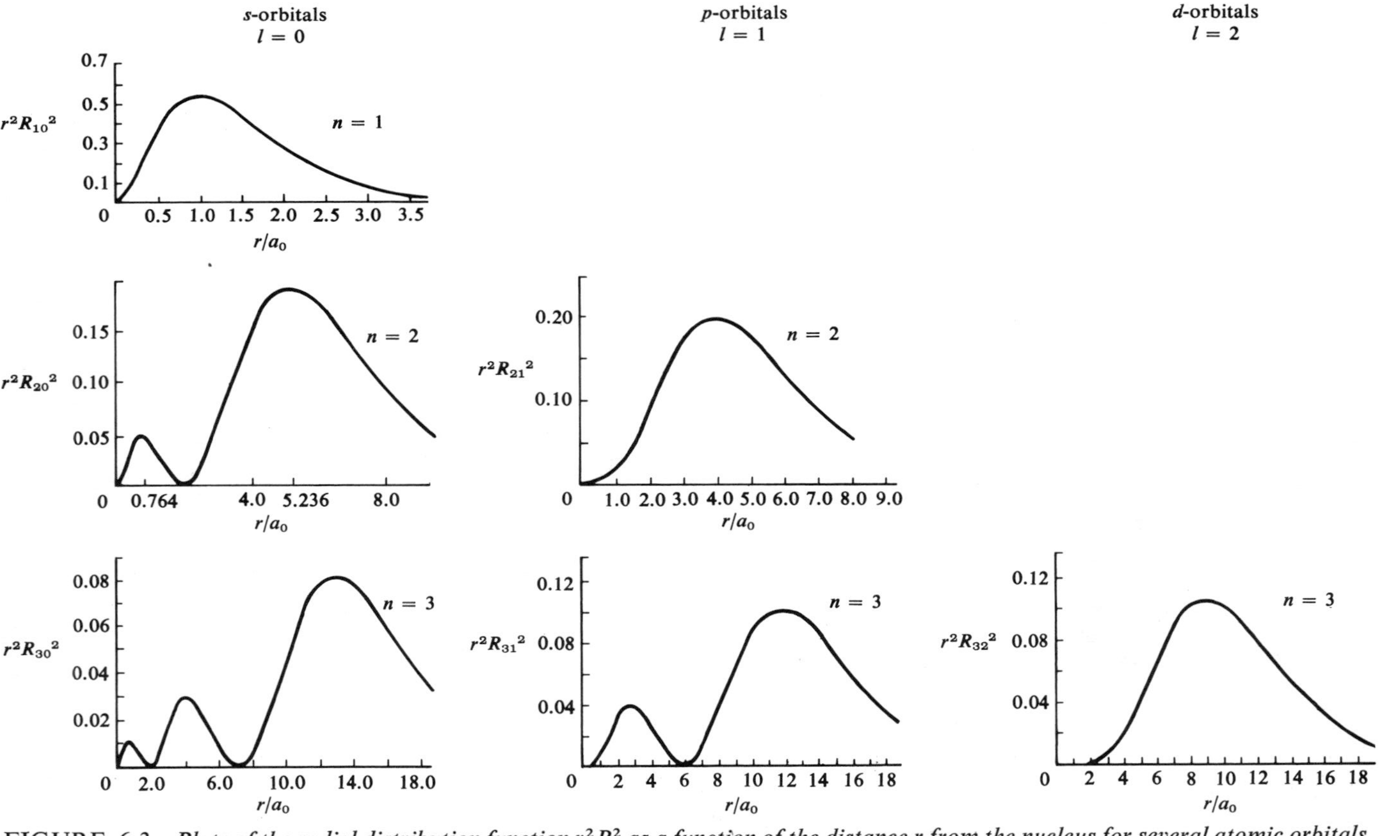

FIGURE 6-3 *Plots of the radial distribution function r^2R^2 as a function of the distance r from the nucleus for several atomic orbitals. Note that the probability of finding a 2s electron within one Bohr radius of the nucleus is greater than the probability of finding a 2p electron in the same region. This shows why the 1s electrons are more effective in screening 2p electrons from the nuclear charge than in screening 2s electrons. The result is that 2s and 2p orbitals will no longer have the same energy in atoms with more than one*

to jump to the conclusion that the electron is a smeared-out charge with the shape shown. This conclusion is misleading, however, for the electron is not a smeared-out charge. A more rigorous interpretation of these pictures can be made if one regards the outline of the charge cloud as a line on a contour map. The significance of the charge cloud drawings is that, if an experimentalist could make a large number of measurements of the position of the electron, 90% (or some other chosen fraction) of them would fall inside the contour line. Also, the probability of finding an electron inside this surface is not uniform. This probability will vary along any line starting at the origin as the function $R^2\,dr$ varies.

In describing the angular functions, one has the option of either depicting the function $\Theta\Phi(l, m)$ or the function $[\Theta\Phi(l, m)]^2$. The difference (besides slight differences in shape) is that the former function will have different signs in different regions of space whereas the latter function will be everywhere positive. We will choose to depict $\Theta\Phi(l, m)$ in most cases because the sign of the wave function is useful in discussing symmetry and in the evaluation of certain integrals which will occur in later sections. The student should keep in mind that the sign of the wave function has no physical significance. It is the quantity $[\Theta\Phi(l, m)]^2$ that is related to a probability, and this function is everywhere positive.

With these remarks in mind, some description of the angular dependence of hydrogen-like orbitals will be given. The function $\Theta\Phi(0, 0)$ is simply a

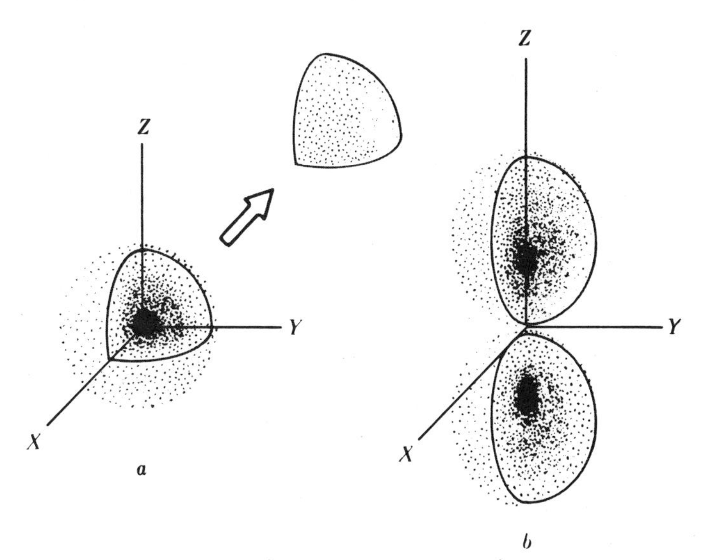

FIGURE 6-4 *Angular dependence of hydrogen-like orbitals.* (*a*) 1*s orbital;* (*b*) 2*p orbital. The contour surfaces shown represent a volume wherein the probability of finding the electron will be* 0.9.

constant (see Table 5-1). The function has the same value regardless of the value of θ and ϕ and, therefore, the electron distribution is spherically symmetrical as shown in Figure 6-4a. The plus sign indicates that the function $\Theta\Phi(0, 0)$ is everywhere positive.

A diagram representing a p_z orbital is shown in Figure 6-4b. This function is

$$\Theta\Phi(1, 0) = \left(\frac{3}{4\pi}\right)^{\frac{1}{2}} \cos\theta \tag{6-15}$$

The orbital characterized by this function will have its maximum value at $\theta = 0$ and $\theta = 180°$, and will vanish at $\theta = 90°$. The function is positive on one side of the XY plane and negative on the other side. Whenever an orbital has this property, the orbital is said to be antisymmetric with respect to reflection in the XY plane.

EXERCISE 6-13 Plot the function $\Theta\Phi(1, 0)$ in the ZX plane. Let θ go from 0 along X to $\frac{1}{2}\pi$ along Z to π along $-X$, etc. Place a point corresponding to the value of $\Theta\Phi(1, 0)$ in the ZX plane at each value of θ. About four points should be taken in each quadrant. What is the relation of the plot you have just made to Figure 6-4b?

EXERCISE 6-14 What is the value of the function $\Theta\Phi(1, 0)$ at a point ($x = 1$, $z = 1$, $y = 2$) in a Cartesian coordinate system? (Remember that $\cos\theta = z/r$ and that r is the magnitude of the vector $\mathbf{r} = x\mathbf{i} + y\mathbf{j} + z\mathbf{k}$). What is its value at ($x = 1$, $z = -1$, $y = -2$). An antisymmetric function, rigorously defined, is one which changes sign when $x \to -x$, $y \to -y$, and $z \to -z$. Note that $\Theta\Phi(1, 0)$ is antisymmetric.

Making a drawing of the functions $\Theta\Phi(1, 1)$ and $\Theta\Phi(1, -1)$ presents a problem because both functions have an imaginary part. These functions are

$$\Theta\Phi(1, \pm 1) = \left(\frac{3}{8\pi}\right)^{\frac{1}{2}} \sin\theta e^{\pm i\phi}$$

The functions $|\Theta\Phi(1, 1)|^2$ and $|\Theta\Phi(1, -1)|^2$ are real, however, and are

$$|\Theta\Phi(1, 1)|^2 = |\Theta\Phi(1, -1)|^2 = \frac{3}{8\pi} \sin^2\theta \tag{6-16}$$

A consideration of Eq. 6-16 shows that both functions have the same spatial distribution, and that the spatial distribution has a maximum in the XY plane ($\theta = 90°$), and is zero along the Z axis ($\theta = 0$). A picture representing this electron distribution is shown in Figure 6-5. In the interpretations of the

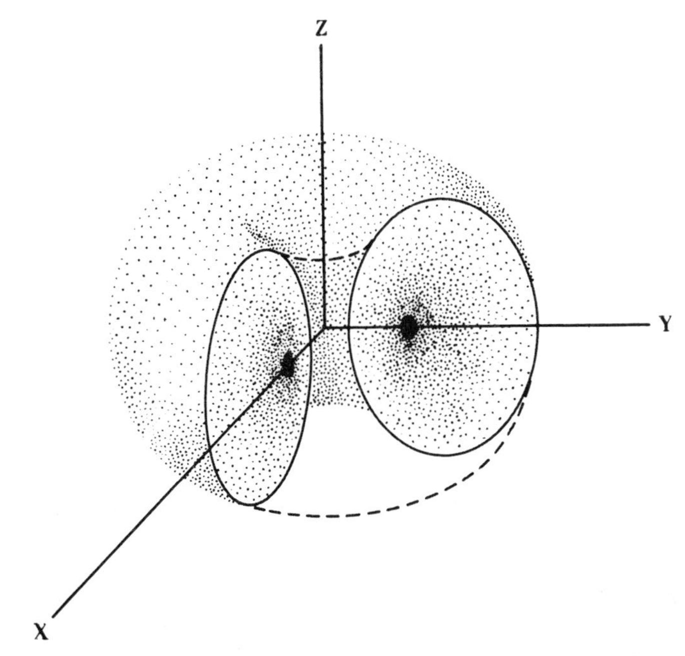

FIGURE 6-5 *The angular dependence of the functions* $|\Theta\Phi(1, \pm 1)|^2$. *Note that both functions have the same probability density. They are different because the electron in* $\Theta\Phi(1, 1)$ *can be thought of as moving counterclockwise about the Z axis whereas an electron in* $\Theta\Phi(1, -1)$ *is moving clockwise about the Z axis.*

physical significance of these functions, it must be kept in mind that the functions $\Theta\Phi(1, \pm 1)$ are eigenfunctions of $\hat{L}_z$ with eigenvalues $\pm 1\hbar$ respectively (see Exercise 5-3). In $\Theta\Phi(1, +1)$, therefore, the electron can be thought of as moving counterclockwise about the Z axis, whereas in $\Theta\Phi(1, -1)$ the electron is moving clockwise.

It is a general property of degenerate wave functions, as $\Theta\Phi(1, \pm 1)$ are in the absence of fields, that any linear combination of the members of the degenerate set is also an acceptable wave function and has the same energy as the original functions. We can, therefore, construct the two new functions

$$\psi_1 = \frac{1}{2}[\Theta\Phi(1, 1) + \Theta\Phi(1, -1)] = A \sin\theta \cos\phi \tag{6-17a}$$

$$\psi_2 = \frac{1}{2i}[\Theta\Phi(1, 1) - \Theta\Phi(1, -1)] = A \sin\theta \sin\phi \tag{6-17b}$$

where A is the appropriate normalizing factor. These are the functions called p_x and p_y orbitals. They have the same shape as the p_z orbital shown in

Figure 6-4b, except that the maximum in the electron density is along the X and Y axes, respectively. These $2p$ orbitals are shown schematically in Figure 6-6. It should be noted that, in this representation, all three p orbitals have a nodal plane perpendicular to the axis of symmetry of the electron distribution.

EXERCISE 6-15 Show that Eqs. 6-17 are true.

EXERCISE 6-16 Evaluate the normalizing constant A in Eqs. 6-17.

EXERCISE 6-17 Plot the function ψ_1 in the XY plane. Note that $\sin\theta$ is always *one* in the XY plane and that ϕ goes from 0 to 2π. How does this plot compare with the one obtained in Exercise 6-13?

EXERCISE 6-18 Calculate an expression for the probability distribution

$$P = [\Theta\Phi(1, 0)]^2 + [\Theta\Phi(1, 1)]^2 + [\Theta\Phi(1, -1)]^2$$

What does this probability distribution look like?

As pointed out above, d orbitals have $l = 2$, five possible values of m, and two nodes in the angular part of the wave function. The directional properties of d orbitals are very important in inorganic chemistry because the unique chemistry of the transition metal ions depends upon d orbital properties.

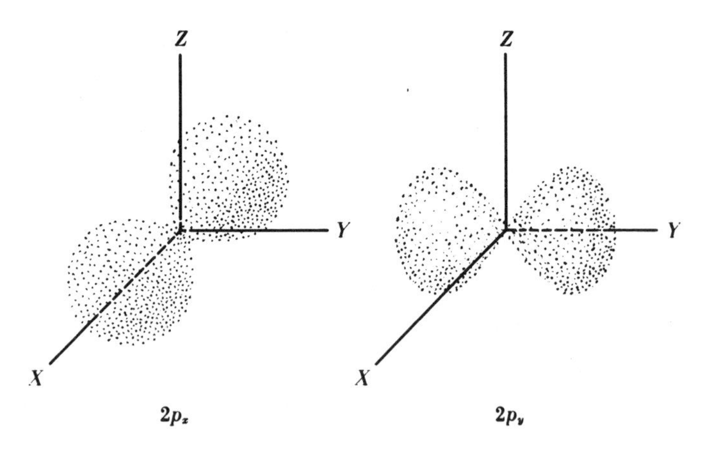

FIGURE 6-6 *Probability contour drawings for the $2p_x$ and $2p_y$ orbitals. The relative electron density inside the surfaces goes as the radial distribution function for a $2p$ orbital.*

Similar procedures can be used for representing d orbitals as those used for p orbitals above. Consider first the three functions

$$|\Theta\Phi(2, \pm 2)|^2 = \frac{1}{2\pi}\frac{15}{16}\sin^4\theta \tag{6-18a}$$

$$|\Theta\Phi(2, \pm 1)|^2 = \frac{1}{2\pi}\frac{15}{4}\sin^2\theta\cos^2\theta \tag{6-18b}$$

$$\Theta\Phi(2, 0) = \frac{1}{\sqrt{2\pi}}\left(\frac{5}{8}\right)^{\frac{1}{2}}(3\cos^2\theta - 1) \tag{6-18c}$$

Plots of the functions 6-18a and 6-18b are shown in Figure 6-7 and show the electron distribution in the $d_{\pm 2}$ and $d_{\pm 1}$ orbitals. Figure 6-8 shows a plot of the d_0 (also called d_{z^2}) orbital (Eq. 6-18c) as well as plots of the appropriate linear combinations of the $d_{\pm 2}$ and $d_{\pm 1}$ orbitals to give the spatially oriented orbitals d_{xz}, d_{xy}, d_{yz}, and $d_{x^2-y^2}$. The descriptions of the five d orbitals shown in Figure 6-8 are the ones usually used in discussing the structure and chemistry of the transition metal ions. The angular functions appropriate to these orbitals are

$$d_{xz} = A_{xz}\sin\theta\cos\theta\cos\phi \tag{6-19a}$$

$$d_{yz} = A_{yz}\sin\theta\cos\theta\sin\phi \tag{6-19b}$$

$$d_{x^2-y^2} = A_{x^2-y^2}\sin^2\theta\cos 2\phi \tag{6-19c}$$

$$d_{xy} = A_{xy}\sin^2\theta\sin 2\phi \tag{6-19d}$$

where the A's are appropriate normalizing constants.

EXERCISE 6-19 Using the same procedure as was outlined above for p orbitals, derive expressions for each of the d orbitals in Eq. 6-19 in terms of the angular momentum eigenfunctions (Eqs. 6-18a and 6-18b). Write down the normalizing constants from your results without evaluating any integrals.

It should be emphasized again that for a set of degenerate orbitals, *any* linear combination of the members of the set is an acceptable orbital also. It is required, however, that any *set* of degenerate orbitals (called a representation) possess members that are orthogonal to one another. Thus the set of orbitals described by Eq. 6-18 would be acceptable, as would the set shown in Figure 6-8. The five orbitals d_0, d_{+1}, d_{xz}, $d_{x^2-y^2}$, and d_{xy}, would not be acceptable since d_{+1} and d_{xz} are not orthogonal to one another.

EXERCISE 6-20 Show that the orbitals d_{+1} and d_{xz} are not orthogonal.

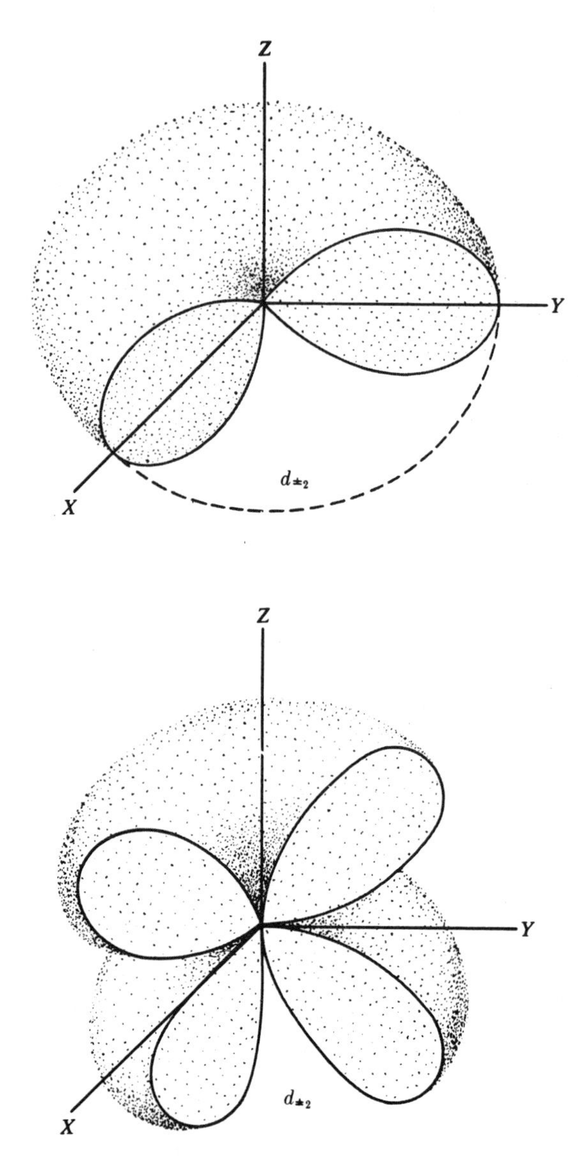

FIGURE 6-7 *The angular dependence of the functions* $|\Theta\Phi(2, \pm 2)|^2$ *and* $|\Theta\Phi(2, \pm 1)|^2$. *The distribution in the function* $|\Theta\Phi(2, \pm 2)|^2$ *is similar to that shown in Figure 6-5 except that it is more elongated.*

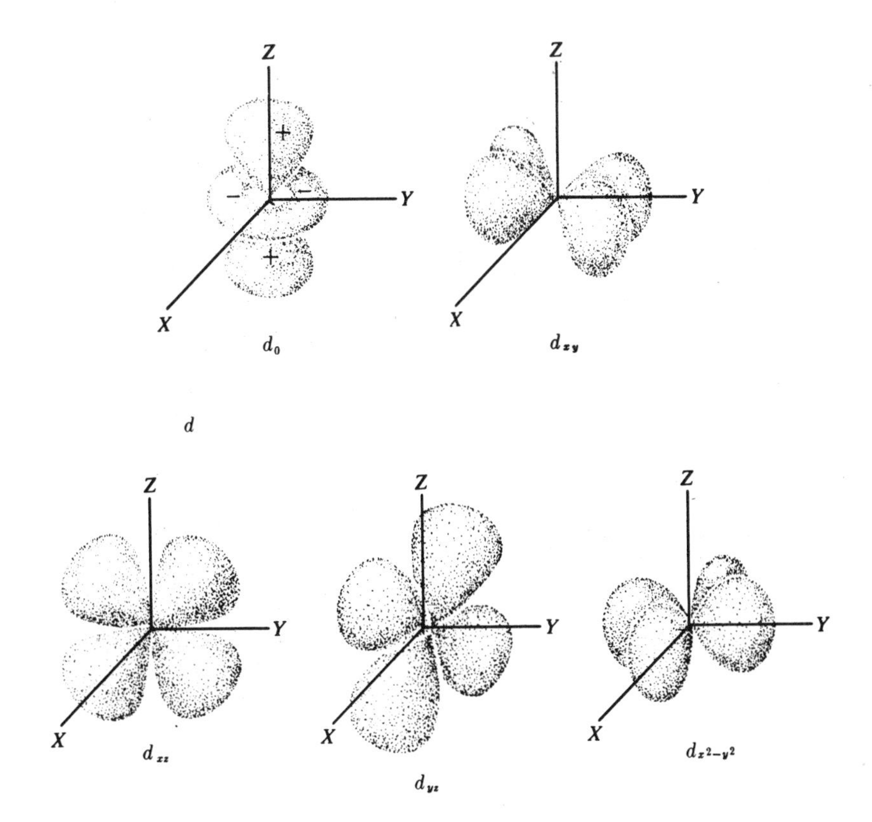

FIGURE 6-8 *The representation of the five d orbitals usually used to discuss the structure and chemistry of transition metal ions.*

6-3 THE SPECTRUM OF ATOMIC HYDROGEN: SELECTION RULES

The formula giving the energy of the observed transitions in the hydrogen atom has already been given in Chapter 3. These transitions are now interpreted as transitions between the allowed states of the hydrogen atom, $\Psi(n, l, m)$. We have not said anything about selection rules, however. It is shown below that these depend on the angular part of the functions Ψ.

The selection rules derived below hold true for all atoms that have only one electron outside of a closed shell—the alkali metals, for example. The selection rules also hold for more complex atoms provided they are interpreted in terms of the eigenvalues with respect to the operator for total electronic angular momentum of the atom. The calculation of these angular momentum eigenvalues is discussed in Section 6-10.

The instantaneous dipole moment of a hydrogen atom is $-e\mathbf{r}$, where $\mathbf{r}$ is the vector from the nucleus to the electron. The transition moment is then

$$\mathbf{R} = -\langle\Psi(n', l', m') \,|\, e\mathbf{r} \,|\, \Psi(n'', l'', m'')\rangle \tag{6-20}$$

where the primes and double primes have their usual significance. To evaluate this integral, each component must be evaluated separately. Thus Eq. 6-20 becomes the three equations

$$R_x = e\int_0^\infty\int_0^\pi\int_0^{2\pi} \Psi^*(n', l', m')(r \sin\theta \cos\phi) \times \Psi(n'', l'', m'')r^2 \sin\theta \, dr \, d\theta \, d\phi \tag{6-21a}$$

$$R_y = e\int_0^\infty\int_0^\pi\int_0^{2\pi} (\Psi^*)'(\Psi)''r \sin\theta \sin\phi r^2 \sin\theta \, dr \, d\theta \, d\phi \tag{6-21b}$$

$$R_z = e\int_0^\infty\int_0^\pi\int_0^{2\pi} (\Psi^*)'(\Psi)''r \cos\theta \, r^2 \sin\theta \, dr \, d\theta \, d\phi \tag{6-21c}$$

These integrals may be evaluated for general Legendre polynomials to obtain the selection rules [1, 2]. See Exercise 6-6 for an example of the calculation of the angular contribution to R_z.

$$\begin{aligned} &\Delta l = \pm 1 \\ &\Delta m = 0, \pm 1 \\ &\Delta n \quad \text{no restriction} \end{aligned} \tag{6-22}$$

We will not go through this mathematical procedure; rather, several exercises will be given to illustrate these selection rules.

EXERCISE 6-21 Show that, for a hydrogen atom, the transition $\Psi_{1s} \to \Psi_{2s}$ is not allowed.

EXERCISE 6-22 Show that the transition $\Psi_{1s} \to \Psi_{3p}$ is allowed.

6-4 *ATOMIC UNITS*

To save writing down a lot of constants, it is customary to use atomic units (a.u.) in atomic and molecular calculations. These units are as follow.

Unit of length: a_0 (radius of first Bohr orbit) $= \hbar^2/me^2 = 0.529$ Å

Unit of energy: H(Hartree) $= e^2/a_0 = 27.2$ eV

In terms of a.u., the energy of a hydrogen atom in its ground state is -13.6 eV or

$$E = -\frac{me^4}{2\hbar^2} = -\frac{1}{2}\frac{e^2}{a_0} = -\frac{1}{2}\,\mathrm{H} \tag{6-23}$$

The Hamiltonian operator in atomic units can be derived from the corresponding operator in cgs units by changing $-(\hbar^2/2m)\nabla^2$ to $-\frac{1}{2}\nabla^2$, and by changing the potential energy terms from Ze^2/r to Z/r. To see how this comes about, consider the Hamiltonian for the hydrogen atom

$$\hat{\mathscr{H}}(\text{cgs}) = -\frac{\hbar^2}{2m}\nabla^2 - \frac{Ze^2}{r} \tag{6-24}$$

We now let $r(\text{cm}) = a_0 r'(\text{a.u.})$, and take advantage of the fact that

$$\frac{\partial^2}{\partial r^2} = \frac{1}{a_0{}^2}\frac{\partial^2}{\partial (r')^2} \tag{6-25}$$

Making these substitutions, the Hamiltonian in cgs units is converted to the Hamiltonian in a.u. multiplied by e^2/a_0.

$$\begin{aligned} \hat{\mathscr{H}}(\text{cgs}) &= \frac{\hbar^2}{2m}\frac{1}{a_0{}^2}\nabla^2(\text{a.u.}) - \frac{Ze^2}{a_0 r'} \\ &= \left(\frac{1}{2}\nabla^2 - \frac{Z}{r'}\right)\frac{e^2}{a_0} = \hat{\mathscr{H}}(\text{a.u.})\left(\frac{e^2}{a_0}\right) \end{aligned} \tag{6-26}$$

The a.u. of angular momentum is $\hbar$. This means that the quantum numbers l and m are angular momenta in a.u. For example, the function $e^{-2i\phi}$ is an eigenfunction of $\hat{L}_z$ with eigenvalue -2.

6-5 *THE HELIUM ATOM AND THE VARIATIONAL PRINCIPLE*

Now that the quantum mechanical problem of the hydrogen atom has been solved, the next step is to find the allowed energy states and wave functions for more complex atoms. The simplest of these is the helium atom. The helium atom contains a nucleus with charge $+2$ and two electrons which, for convenience, will be designated 1 and 2 (see Figure 6-9). Using our common procedure, we can immediately write down the appropriate Hamiltonian operator, which, in a.u., is

$$\hat{\mathscr{H}} = -\frac{1}{2}[\nabla^2(1) + \nabla^2(2)] - \frac{2}{r_{N1}} - \frac{2}{r_{N2}} + \frac{1}{r_{12}} \tag{6-27}$$

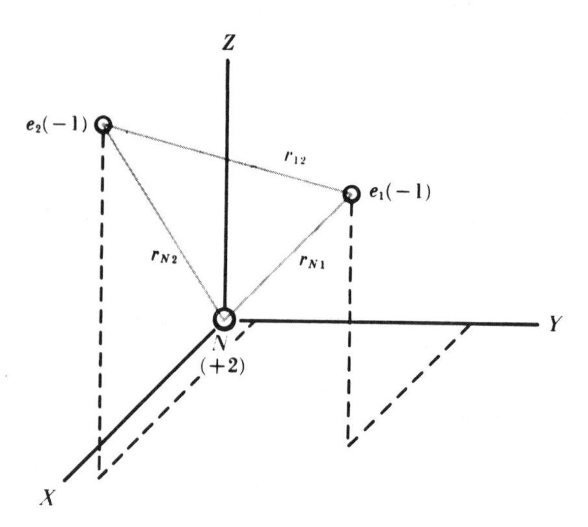

FIGURE 6-9 *Quantities used in the discussion of the helium atom. The quantities in parentheses are the charges on the particles.*

The term in brackets contains the kinetic energy operators for electrons 1 and 2, respectively, and the remainder is the potential energy for the system. Note that there are three terms in the potential energy. These correspond to the nuclear attraction for electron 1, the nuclear attraction for electron 2, and the mutual repulsion between the two electrons, respectively.

We now have to solve the eigenvalue equation

$$\mathscr{H}\Psi = E\Psi \tag{6-28}$$

for $\mathscr{H}$ given in Eq. 6-27, and the problem is completed. Unfortunately, the task is frustrated at this point because Eq. 6-28 is now insoluble in closed analytical form. That is, there are no known analytic functions Ψ which, when operated on by the Hamiltonian operator of Eq. 6-27, will give a number times the function back again. The failure to obtain solutions to Eq. 6-28 is due to the presence of the $1/r_{12}$ term in the Hamiltonian. This term is called the electron repulsion term, and much present research in quantum chemistry is devoted to trying to find a satisfactory way to take this term into account.

Two-electron quantum mechanical problems, and in general many-electron problems, are qualitatively different from one-electron problems. Since the electrons interact with one another, the wave function for the atom will depend on the coordinates of both electrons. We will see in Section 6-8 that electrons are indistinguishable. This fact means that it will be impossible to talk rigorously about the properties of a single electron in a many-electron atom or molecule. What we determine as closely as our approximation methods and our computer time will allow is the *energy* of a particular *state* of the whole atom or molecule and the *many-electron wave function.* We emphasize this point here because one of the most powerful approximation schemes that we have available involves writing the many-electron wave function in terms of *one-electron* functions similar to the hydrogen atom solutions. These one-electron functions that are used to construct the many-electron wave function are called *atomic orbitals.*

To see how this approximation arises, let us investigate what the solutions to the helium atom would look like if the $1/r_{12}$ term were left out of the Hamiltonian. The Hamiltonian in Eq. 6-27 then becomes

$$\hat{\mathscr{H}}_0 = \frac{1}{2}(\nabla_1{}^2 + \nabla_2{}^2) - \frac{Z}{r_{N1}} - \frac{Z}{r_{N2}} \tag{6-29}$$

which the student will note can be written in the form

$$\hat{\mathscr{H}}_0 = \hat{h}_0(1) + \hat{h}_0(2)$$

where $\hat{h}_0(1)$ depends only on the coordinates of electron 1, and $\hat{h}_0(2)$ depends only on the coordinates of electron 2. Recalling our previous discussion at the end of Chapter 3 about the separation of variables, we know that a solution to the equation

$$\hat{\mathscr{H}}_0 \Psi_0 = E_0 \Psi_0 \tag{6-30}$$

can be found using

$$\Psi = \phi_0(1)\phi_0(2) \tag{6-31}$$

We know further that the total energy E is the sum of the two "one-electron" energies ε_1 and ε_2. Substituting Eq. 6-31 into Eq. 6-30 and separating variables, we obtain two identical equations

$$\hat{h}_0(1)\phi_0(1) = -\frac{1}{2}\nabla_1{}^2\phi_0(1) - \frac{Z}{r_{N1}}\phi_0(1)$$

$$= \varepsilon_1{}^0\phi_0(1) \tag{6-32a}$$

$$\hat{h}_0(2)\phi_0(2) = -\frac{1}{2}\nabla_2{}^2\phi_0(2) - \frac{Z}{r_{N2}}\phi_0(2)$$

$$= \varepsilon_2{}^0\phi_0(2) \tag{6-32b}$$

These are the same as the eigenvalue equations for a hydrogen atom with nuclear charge Z, however, and the solutions to these are known. Thus we can immediately write

$$\phi_0(1) = R(n, l)\Theta(l, m)\Phi(m)(1)$$

$$\phi_0(2) = R(n, l)\Theta(l, m)\Phi(m)(2)$$

and the energies in a.u. are

$$\varepsilon_1 = -\frac{Z^2}{2}\left(\frac{1}{n_1{}^2}\right)$$

$$\varepsilon_2 = -\frac{Z^2}{2}\left(\frac{1}{n_2{}^2}\right) \tag{6-33}$$

$$E_0 = -\frac{Z^2}{2}\left(\frac{1}{n_1{}^2} + \frac{1}{n_2{}^2}\right)$$

At this level of approximation, the energy of the helium atom in its ground, or lowest energy, state ($n_1 = n_2 = 1$) is

$$E_0 = 2Z^2E_{\text{H}} \tag{6-34}$$

where E_{H} is the energy of a hydrogen atom in its ground state, $-\frac{1}{2}$H or -13.6 eV. Notice also that the two-electron wave function is now written as a product of two one-electron functions.

For the helium atom there are two ways to compare the calculated results with experimental results. One way is to compare values of the total binding energy E_{He}. The other is to compare values of the first ionization potential, $(\text{IP})_{\text{He}}$. The first ionization potential is equal to the energy change for the reaction

$$\text{He} \rightarrow \text{He}^+ + e^-$$

and is, therefore,

$$(\text{IP})_{\text{He}} = E_{\text{He}^+} - E_{\text{He}} \tag{6-35}$$

Since the energy of He^+ can be calculated exactly (it is a hydrogen-like ion), the accuracy of the calculated ionization potential reflects the accuracy of the calculated energy of the helium atom. The energy of He^+ is

$$E_{\text{He}^+} = 4E_{\text{H}} = -2\text{ H} = -54.4\text{ eV} \tag{6-36}$$

The experimental values of E_{He} and IP are -2.905 H (-78.98 eV) and 0.904 H (24.6 eV), respectively.

These experimental values will now be compared with values calculated using different levels of approximation.

The most naive calculation uses the approximate wave function, Eq. 6-31, with $Z = 2$, the actual nuclear charge, and the approximate Hamiltonian, Eq. 6-29, to calculate the expectation value of the energy. Thus, using Eq. 6-34,

$$E_{He} = 2(4)E_H = -4\text{H} = -109 \text{ eV} \tag{6-37a}$$

and

$$(\text{IP})_{He} = -(-4\text{H} + 2\text{H}) = 2\text{H} = 54.4 \text{ eV} \tag{6-37b}$$

It can be seen that the energy is in error by 38%, the ionization potential by over 100%. This calculation is not very satisfying.

We should not be surprised that the answer we have just calculated is not very good because we assumed that the electrons do not interact at all (by leaving the $1/r_{12}$ term out of the calculation). We can get better numerical agreement by using the state function Ψ_0 of Eq. 6-31 and the exact Hamiltonian (Eq. 6-27) to calculate the energy. The student may verify that this leads to

$$E_{He} = 2Z^2 E_H + \left(\Psi_0 \left| \frac{1}{r_{12}} \right| \Psi_0\right) \tag{6-38}$$

The second term in Eq. 6-38 may be thought of as a first order perturbation correction to the zero order energy, Eq. 6-37a. The integral is algebraically difficult to evaluate, and we will simply give its value. For details the student is referred to [2]. Its value is

$$\left(\Psi_0 \left| \frac{1}{r_{12}} \right| \Psi_0\right) = \frac{5}{8} Z \tag{6-39}$$

where Z is the nuclear charge used in the functions $\phi_0(1)$ and $\phi_0(2)$. Using $Z = 2$, we then obtain

$$E_{He} = -4\text{H} + \tfrac{5}{4}\text{H} = \tfrac{11}{4} = -74.8 \text{ eV}$$

$$(\text{IP})_{He} = -(-74.8 + 54.4) = 20.4 \text{ eV}$$

These are much more satisfactory answers, being off by only 5% and 17%, respectively. It can be seen that the ionization potential is a much more sensitive criterion of the "goodness" of our calculation than is the total binding energy.

EXERCISE 6-23 Calculate the expectation values of $\langle V \rangle$ and $\langle T \rangle$ for helium at this level of approximation. The virial theorem requires that $\langle T \rangle = -\frac{1}{2}\langle V \rangle$. Is the virial theorem satisfied? (See Exercise 6-10 and Section 7-4 for more about the virial theorem.)

To improve the calculation using the simple product function (Eq. 6-31), we must correct the orbitals ϕ_0 for the fact that each electron moves in a potential that is the sum of the nuclear attraction and the screening effect of the other electron. We want to introduce an effective nuclear charge Z' in ϕ_0 so that it will have the form

$$\phi_0 = Ae^{-Z'r/a_0} \tag{6-40}$$

Our problem is to find the value of Z' that gives the *best* value of the energy. To solve this problem, we need to introduce the second major approximation method of quantum mechanics—the variation method.

The variation method rests on a result that can be proved from classical physics.

This result, which we will give without proof, is called the variation principle. In the form that we use, it states the following:

Given any approximate wave function satisfying the boundary conditions of the problem, the expectation value of the energy calculated from this function will always be higher than the true energy of the ground state.

This principle suggests a procedure for solving quantum mechanical problems. The procedure is to guess several functions, called trial functions, calculate the expectation value of the energy for each one, choose the one with the lowest energy, and conclude that this is the best function that can be obtained from the original guesses.

Scientists like to use a more systematic procedure than the one just described. One way to be more systematic is to start with a trial function containing parameters. The expectation value of the energy is then calculated and minimized with respect to these parameters. In this way, a larger number of guesses can be made with a single function. The resulting wave function is then the best one available with the particular parametric form chosen.

In our treatment of the helium atom, the arbitrary parameter is Z' and we wish to calculate E as a function of Z' and then find the value of Z' that minimizes E.

To calculate E as a function of Z', it is convenient to rewrite the Hamiltonian (Eq. 6-27) in the following form:

$$\begin{aligned}\mathscr{H} &= -\tfrac{1}{2}\nabla_1^2 - \tfrac{1}{2}\nabla_2^2 - \frac{Z'}{r_1} - \frac{Z'}{r_2} - \frac{Z}{r_1} + \frac{Z'}{r_1} - \frac{Z}{r_2} + \frac{Z'}{r_2} + \frac{1}{r_{12}} \\ &= \hat{h}_0'(1) + \hat{h}_0'(2) - (Z - Z')\frac{1}{r_1} - (Z - Z')\frac{1}{r_2} + \frac{1}{r_{12}}\end{aligned} \tag{6-41}$$

where $\hat{h}_0' = -\frac{1}{2}\nabla^2 - Z'/r$. This can be done by adding and subtracting Z'/r_1 and Z'/r_2 to Eq. 6-27 and regrouping terms. Using the fact that the ϕ_0 of Eq. 6-40 is an eigenfunction of $\hat{h}_0'$ and the results for $\langle 1/r \rangle$ and $\langle 1/r_{12} \rangle$ given previously, we can write

$$E_{\text{He}} = (Z')^2 - 2ZZ' + \tfrac{5}{8}Z' \tag{6-42}$$

EXERCISE 6-24 Using the ϕ_0 functions in Eq. 6-40, show that the integral $\langle \Psi | h_0'(1) + h_0'(2) | \Psi_0 \rangle$ equals $-(Z')^2$. Use the normalization condition to break down the two-electron integral above into two one-electron integrals.

EXERCISE 6-25 Show that the integral $\langle 1/r_1 \rangle = \langle 1/r_2 \rangle = Z'$ when evaluated using the ϕ_0 of Eq. 6-40. Use the normalization condition to reduce two-electron integrals to one-electron integrals.

EXERCISE 6-26 Use the results of Exercises 6-24 and 6-25 with the fact that $\langle 1/r_{12} \rangle = \frac{5}{8}Z'$ for the functions (Eq. 6-40) to derive Eq. 6-42.

EXERCISE 6-27 Calculate $\langle V \rangle$ and $\langle T \rangle$ for the above variational treatment of helium. Does $\langle T \rangle = -(1/2)\langle V \rangle$? Compare this result with the answer to Exercise 6-23.

It is of interest to ask the question: What is the energy of a single electron, moving with a probability distribution given by Eq. 6-40 in the field of the helium nucleus, and a second electron with an identical charge distribution? This energy, which we will call the orbital energy of the electron ε_1, is given by the relation

$$\varepsilon_1 = \tfrac{1}{2}(Z')^2 - ZZ' + \tfrac{5}{8}Z' \tag{6-43}$$

where the first term represents the kinetic energy of the electron, the second term represents the potential energy of attraction between the nucleus and the electron distribution, and the third term represents the potential energy of repulsion between the two electron distributions. The student may verify this by evaluating the kinetic energy explicitly and by considering the form of the potential energy integrals. We mention Eq. 6-43 because it is important to note that the total energy of the atom is *not* equal to the sum of the orbital energies of the two electrons. In fact, $E = 2\varepsilon_1 - \langle \phi_0\phi_0 | 1/r_{12} | \phi_0\phi_0 \rangle$. We will have occasion to return to this point in our later discussion of many-electron atoms.

We now apply the variational principle and determine the value of Z' that minimizes the energy in Eq. 6-42. To do this we calculate

$$\frac{dE}{dZ'} = 2Z' - 2Z + \tfrac{5}{8} = 0$$

$$Z' = Z - \tfrac{5}{16}$$

Using this result in Eq. 6-42 gives

$$E_{\mathrm{He}} = (Z - \tfrac{5}{16})^2 - 2Z(Z - \tfrac{5}{16}) + \tfrac{5}{8}(Z - \tfrac{5}{16})$$

which, for $Z = 2$, gives

$$E_{\mathrm{He}} = -(\tfrac{27}{16})^2 = -77.45 \text{ eV}$$

and

$$(\mathrm{IP})_{\mathrm{He}} = 23.05 \text{ eV}$$

These values differ from the experimental ones by only 2% and 6%, respectively. This final calculation is the best that can be done using the independent electron approximation and a single exponential function with the form of Eq. 6-40.

It is interesting to note that, at this level of approximation, each electron is screened from the full charge of +2 by $\frac{5}{16}$. This latter number is a measure of the screening effect that the average distribution of one electron has upon the nuclear charge "seen" by the other electron.

EXERCISE 6-28 Calculate the energy ε_1 for the optimum value of Z' in helium. Compare the magnitude of this energy with the first ionization energy of helium. The orbital energies defined by a generalization of Eq. 6-43 are approximations to experimental ionization energies. This result is a specific case of a theorem called Koopman's theorem [9].

The calculation for helium can be further improved by putting more variational flexibility into our one-electron orbitals. Thus, we could write ϕ_0 as a sum of orbitals such as

$$\phi_0 = N(c_1 e^{-\alpha r} + c_2 r e^{-\beta r} + \cdots)$$
(N is a normalizing constant)

and use both the coefficients $c_1, c_2, \ldots$ and the screening constants $\alpha, \beta, \ldots$ as variational parameters. This procedure ultimately converges to a value of the energy that is still higher than the correct energy. This limiting value of the energy is called the Hartree–Fock limit, and it is the best value of the energy that can be calculated with a wave function that only takes electron repulsion into account in an average way. The wave function corresponding to this energy is called a *self-consistent field* (SCF) wave function.

The difference between the Hartree–Fock energy and the true energy of the atomic or molecular system is called the *correlation energy*. The correlation energy arises because electrons stay out of one another's way, not only in an average sense but also in an instantaneous sense. Thus, the calculation for helium can be improved by including the interelectron distance explicitly in the wave function. In such a calculation, all one-electron features such as atomic orbitals disappear, and one ends up with a good value of the energy and a table of numbers that give the probability of finding the electrons in some region of space. In the best of these calculations, the binding energy can be calculated to greater accuracy than it can be measured (see [4]).

6-6 ORBITAL ANGULAR MOMENTUM

The solutions to the rigid rotor problem and to the hydrogen atom problem both involved functions, the spherical harmonics $Y_{l,m}r$, that were eigenfunctions of the angular momentum operators $\hat{L}^2$ and $\hat{L}_z$. These are just two examples of the important role that angular momentum plays in quantum mechanics. In this section, we discuss the subject of orbital angular momentum in some detail. This treatment serves a dual purpose in that it gives the student some familiarity with quantum mechanical operator algebra, and it lays groundwork for the discussion of electron spin in Section 6-7.

Many of the properties of angular momentum operators have been introduced in previous chapters, usually in the form of exercises. The results that should be familiar are the following:

$$\mathbf{L} = \mathbf{r} \times \mathbf{P} \tag{6-44a}$$

$$\mathbf{L} = L_x \mathbf{i} + L_y \mathbf{j} + L_z \mathbf{k} \tag{6-44b}$$

$$\hat{L}_x = y\hat{p}_z - z\hat{p}_y = -i\hbar\left(y\frac{\partial}{\partial z} - z\frac{\partial}{\partial y}\right) \tag{6-44c}$$

$$\hat{L}_y = z\hat{p}_x - x\hat{p}_z = -i\hbar\left(z\frac{\partial}{\partial x} - x\frac{\partial}{\partial z}\right) \tag{6-44d}$$

$$\hat{L}_z = x\hat{p}_y - y\hat{p}_x = -i\hbar\left(x\frac{\partial}{\partial y} - y\frac{\partial}{\partial x}\right) \tag{6-44e}$$

$$\hat{L}^2 = \hat{L} \cdot \hat{L} = \hat{L}_x^2 + \hat{L}_y^2 + \hat{L}_z^2 \tag{6-44f}$$

(See Exercises 1-6 and 3-10.)

The student should also be acquainted with the appropriate expressions in polar coordinates. These are

$$\hat{L}_x = -i\hbar\left(-\sin\phi\frac{\partial}{\partial\theta} - \cot\theta\cos\phi\frac{\partial}{\partial\phi}\right) \tag{6-45a}$$

$$\hat{L}_y = -i\hbar\left(\cos\phi\frac{\partial}{\partial\theta} - \cot\theta\sin\phi\frac{\partial}{\partial\phi}\right) \tag{6-45b}$$

$$\hat{L}_z = -i\hbar\frac{\partial}{\partial\phi} \tag{6-45c}$$

$$\hat{L}^2 = -\hbar^2\left(\frac{1}{\sin\theta}\frac{\partial}{\partial\theta}\sin\theta\frac{\partial}{\partial\theta} + \frac{1}{\sin^2\theta}\frac{\partial}{\partial\phi^2}\right) \tag{6-45d}$$

EXERCISE 6-29 Derive Eqs. 6-45. This may be done by using the relations

$$r^2 = x^2 + y^2 + z^2$$

$$\cos\theta = \frac{z}{(x^2 + y^2 + z^2)^{\frac{1}{2}}}$$

$$\tan\phi = \frac{y}{x}$$

and the result from calculus that, for example,

$$\begin{aligned}\hat{L}_x &= -i\hbar\left(y\frac{\partial}{\partial z} - z\frac{\partial}{\partial y}\right)\\ &= -i\hbar\left[r\sin\theta\sin\phi\left(\frac{\partial r}{\partial z}\frac{\partial}{\partial r} + \frac{\partial\theta}{\partial z}\frac{\partial}{\partial\theta} + \frac{\partial\phi}{\partial z}\frac{\partial}{\partial\phi}\right)\right.\\ &\qquad\left. - r\cos\theta\left(\frac{\partial r}{\partial y}\frac{\partial}{\partial r} + \frac{\partial\theta}{\partial y}\frac{\partial}{\partial\theta} + \frac{\partial\phi}{\partial y}\frac{\partial}{\partial\phi}\right)\right]\end{aligned}$$

It was pointed out in Chapter 3 that the commutation properties of operators play an important role in quantum mechanics. The student will recall that there is a relationship between the uncertainty principle and the value of the commutator of two operators. If two operators commute, it is possible to measure precisely and simultaneously the properties corresponding to those two operators. If the two operators do not commute, there exists an uncertainty relation between the two corresponding properties. For this reason, we wish to explore the commutation relationships between angular momentum operators.

It can be shown that $\hat{L}^2$ commutes with any component of angular momentum; that is,

$$[\hat{L}^2, \hat{L}_x] = [\hat{L}^2, \hat{L}_y] = [\hat{L}^2, \hat{L}_z] = 0 \tag{6-46}$$

No two components commute with each other, however. In fact,

$$[\hat{L}_x, \hat{L}_y] = \hat{L}_x \hat{L}_y - \hat{L}_y \hat{L}_x = i\hbar \hat{L}_z \tag{6-47a}$$

$$[\hat{L}_y, \hat{L}_z] = i\hbar \hat{L}_x \tag{6-47b}$$

$$[\hat{L}_z, \hat{L}_x] = i\hbar \hat{L}_y \tag{6-47c}$$

EXERCISE 6-30 Prove that $[\hat{L}^2, \hat{L}_z] = 0$ and that $[\hat{L}_x, \hat{L}_y] = i\hbar L_z$. The proof is most convenient in Cartesian coordinates.

The physical significance of these commutation relations for angular momentum is that it is only possible to measure simultaneously and precisely the square of the angular momentum and *one* component. That is, it is impossible to know exactly at a given time more than one component of angular momentum. By convention, the measurable component is usually taken to be the z component, and the Z axis is usually defined by an electric or a magnetic field. For more information about the use of magnetic fields to study electronic states, the student is referred to Chapter 9.

Two other operators of interest are the so-called raising and lowering, shift or ladder operators. These operators, called $\hat{L}_+$ and $\hat{L}_-$, are defined as

$$\hat{L}_+ = \hat{L}_x + i\hat{L}_y \tag{6-48a}$$

$$\hat{L}_- = \hat{L}_x - i\hat{L}_y \tag{6-48b}$$

These operators have an important effect on the spherical harmonics, $Y_{l,m}$. They will also be important in problems involving electron and nuclear spin. The operators $\hat{L}_+$ and $\hat{L}_-$ have the property that

$$\hat{L}_+ Y_{l,m} = \hbar[l(l+1) - m(m+1)]^{\frac{1}{2}} \, Y_{l,m+1} \tag{6-49}$$

$$\hat{L}_- Y_{l,m} = \hbar[l(l+1) - m(m-1)]^{\frac{1}{2}} \, Y_{l,m-1} \tag{6-50}$$

The easiest way to see the effect of these operators is to substitute some of the functions $Y_{l,m}$ into Eqs. 6-49 and 6-50. It can be seen that the effect of $\hat{L}_+$ or $\hat{L}_-$ is to either raise or lower the eigenvalue with respect to $\hat{L}_z$, keeping the eigenvalue with respect to $\hat{L}^2$ the same. If the eigenvalue of $Y_{l,m}$ with respect to $\hat{L}_z$ already has its maximum value, $m = l$, then $\hat{L}_+$ annihilates the function. That is,

$$\hat{L}_+ Y_{l,l} = 0$$

These properties can best be illustrated by some specific examples.

EXERCISE 6-31 Show that $\hat{L}_+ Y_{1,-1} = +\hbar\sqrt{2}\, Y_{1,0}$, and that $\hat{L}_+ Y_{1,1} = 0$. Likewise, show that $\hat{L}_- Y_{1,0} = \hbar\sqrt{2}\, Y_{1,-1}$, and that $\hat{L}_- Y_{1,-1} = 0$. That is, show that

$$(\hat{L}_x + i\hat{L}_y)(\tfrac{3}{8}\pi)^{\frac{1}{2}} \sin\theta\, e^{-i\phi} = +\hbar\sqrt{2}(\tfrac{3}{4}\pi)^{\frac{1}{2}} \cos\theta\, e^{0}$$

One of the most useful properties of the ladder operators is that they permit the expression for $\hat{L}^2$ to be written in a form which makes computation much simpler. From Eq. 6-44f,

$$\hat{L}^2 = \hat{L}_x^{\,2} + \hat{L}_y^{\,2} + \hat{L}_z^{\,2} \tag{6-51}$$

but

$$\begin{aligned} \hat{L}_+\hat{L}_- &= (\hat{L}_x + i\hat{L}_y)(\hat{L}_x - i\hat{L}_y) = \hat{L}_x^{\,2} - i\hat{L}_x\hat{L}_y + i\hat{L}_y\hat{L}_x + \hat{L}_y^{\,2} \\ &= \hat{L}_x^{\,2} + \hat{L}_y^{\,2} - i(\hat{L}_x\hat{L}_y - \hat{L}_y\hat{L}_x) \end{aligned} \tag{6-52}$$

(Note that we have been careful to maintain the order of operations.)

The term $(\hat{L}_x\hat{L}_y - \hat{L}_y\hat{L}_x)$ is the commutator of $\hat{L}_x$ and $\hat{L}_y$, however, and is given by Eq. 6-47a as $i\hbar\hat{L}_z$. Therefore,

$$\hat{L}_+\hat{L}_- = \hat{L}_x^{\,2} + \hat{L}_y^{\,2} + \hbar\hat{L}_z \tag{6-53}$$

and

$$\hat{L}_x^{\,2} + \hat{L}_y^{\,2} = \hat{L}_+\hat{L}_- - \hbar\hat{L}_z \tag{6-54}$$

Therefore, we can write the operator $\hat{L}^2$ in the form

$$\hat{L}^2 = \hat{L}_+\hat{L}_- + \hat{L}_z^{\,2} - \hbar\hat{L}_z \tag{6-55}$$

Likewise, it can also be shown that

$$\hat{L}^2 = \hat{L}_-\hat{L}_+ + \hat{L}_z^{\,2} + \hbar\hat{L}_z \tag{6-56}$$

EXERCISE 6-32 Derive Eq. 6-56 making use only of definitions and commutative relations.

All of the above relations have been for a system with only one electron. For systems with more than one electron, the angular momentum must be calculated vectorially. For example, if a two-electron system is considered we can write

$$\hat{\mathbf{L}}_T = \hat{\mathbf{L}}(1) + \hat{\mathbf{L}}(2) \tag{6-57}$$

where $\hat{\mathbf{L}}_T$ is the total angular momentum of the system, and $\hat{\mathbf{L}}(1)$ and $\hat{\mathbf{L}}(2)$ are the angular momenta of electrons 1 and 2, respectively. The square of the total angular momentum is, therefore,

$$\hat{L}_T^2 = \hat{\mathbf{L}}_T \cdot \hat{\mathbf{L}}_T = \hat{L}^2(1) + \hat{L}^2(2) + 2\hat{\mathbf{L}}(1) \cdot \hat{\mathbf{L}}(2) \tag{6-58}$$

For systems with more than one electron, calculations using $\hat{L}_T^2$ are most conveniently performed using the expressions analogous to Eqs. 6-55 and 6-56. Thus, for the two-electron case,

$$\hat{L}_\alpha = \hat{L}_\alpha(1) + \hat{L}_\alpha(2) \qquad (\alpha = x, y, z) \tag{6-59}$$

and

$$\begin{aligned} \hat{L}_+ \hat{L}_- &= [\hat{L}_x(1) + \hat{L}_x(2)]^2 + [\hat{L}_y(1) + \hat{L}_y(2)]^2 \\ &\quad + i\{[\hat{L}_y(1) + \hat{L}_y(2)][\hat{L}_x(1) + \hat{L}_x(2)] \\ &\quad - [(\hat{L}_x(1) + \hat{L}_x(2))(\hat{L}_y(1) + \hat{L}_y(2))\} \end{aligned} \tag{6-60}$$

After some straightforward algebra in which one must be careful to keep the order of operations straight, Eq. 6-60 becomes

$$\hat{L}_+ \hat{L}_- = \hat{L}_x^2 + \hat{L}_y^2 + \hbar\hat{L}_z \tag{6-61}$$

Substitution of Eq. 6-61 into Eq. 6-58 gives

$$\hat{L}_T^2 = \hat{L}_+ \hat{L}_- + \hat{L}_z^2 - \hbar\hat{L}_z \tag{6-62}$$

identical with Eq. 6-55. It is important to remember, however, that the operators in Eq. 6-62 are operators for *total* angular momentum; for example, $\hat{L}_+ = \hat{L}_+(1) + \hat{L}_+(2)$, $\hat{L}_z = \hat{L}_z(1) + \hat{L}_z(2)$, and so on. Thus Eq. 6-62 holds regardless of the number of particles.

EXERCISE 6-33 Show that the relation $\hat{L}_T^2 = \hat{L}_- \hat{L}_+ + \hat{L}_z^2 + \hbar\hat{L}_z$ is also identical with Eq. 6-58.

6-7 ELECTRON SPIN

Shortly before quantum mechanics was developed, S. Goudsmit and G. Uhlenbeck postulated that an electron had an intrinsic angular momentum which they called "spin angular momentum." This new property, many times just called "spin," has some unusual characteristics. In part, these characteristics seem unusual because *there is no classical analog of spin.* Electron spin does arise when the motion of the electron is treated relativistically, but that subject is beyond the scope of this book.

The experimental evidence supporting the spin hypothesis is quite large. Some examples of things which can be explained by the concept of spin are the following:

1. Stern–Gerlach experiments.
2. Electron and nuclear magnetic resonance spectroscopy.
3. The degeneracy of the excited states of atoms and molecules.
4. The anomalous Zeeman effect.
5. The fine structure splittings of atomic spectra.

To discuss each of these experiments is beyond the scope of this book, and good discussions are already available [4]. Electron and nuclear spin resonance are discussed in Chapter 9.

Since there is no classical analog of spin, we cannot follow our standard procedure, starting with writing down the classical expression, for finding the appropriate quantum mechanical operators, commutation rules, and so on. We therefore introduce spin by a series of postulates. These postulates are justified by the fact that they give the right answers in experiments such as those listed above. The treatment that follows is essentially that of [4]. The postulates are the following.

Postulate I. The operators for spin angular momentum commute and combine in the same way as those for ordinary angular momentum.

We therefore have operators $\hat{S}^2$, $\hat{S}_x$, $\hat{S}_y$, $\hat{S}_z$, $\hat{S}_+$, $\hat{S}_-$, which are exactly analogous to the operators $\hat{L}^2$, $\hat{L}_x$, $\hat{L}_y$, $\hat{L}_z$, $\hat{L}_+$, and $\hat{L}_-$ discussed in Section 6-6.

Postulate II. For a single electron there are only two simultaneous eigenfunctions of $\hat{S}^2$ and $\hat{S}_z$. These are called α and β and have eigenvalue equations

$$\begin{aligned} \hat{S}_z\alpha &= \tfrac{1}{2}\hbar\alpha, \qquad \hat{S}^2\alpha = \tfrac{1}{2}(\tfrac{1}{2}+1)\hbar^2\alpha \\ \hat{S}_z\beta &= -\tfrac{1}{2}\hbar\beta, \qquad \hat{S}^2\beta = \tfrac{1}{2}(\tfrac{1}{2}+1)\hbar^2\beta \end{aligned} \tag{6-63}$$

Spin, like orbital angular momentum, can be expressed in a.u. of $\hbar$. The functions α and β are then said to be eigenfunctions of $\hat{S}_z$ with eigenvalue $\frac{1}{2}$ and $-\frac{1}{2}$, respectively. The functions α and β are taken to be normalized.

That is, the integrals of α^2 and β^2 over all *spin* space equal 1. They are also orthogonal because of Theorem II in Chapter 3.

Postulate III. The spinning electron acts like a magnet, the magnetic dipole moment of which is

$$\hat{\boldsymbol{\mu}} = -g_0 \beta_m \hat{\mathbf{S}} \tag{6-64}$$

The quantities g_0 and β_m are the spectroscopic splitting factor and the Bohr magneton, respectively. These quantities have the values 2.0023 and 9.2732×10^{-21} erg G^{-1}. The value of g_0 was first obtained as an empirical constant but was later derived from the relativistic theory of the electron by Dirac. The minus sign indicates that the direction of the dipole moment vector is antiparallel to the spin vector. From Eq. 6-64 it can easily be seen that the operator for the z component of the magnetic moment is

$$\hat{\mu}_z = -g_0 \beta \hat{S}_z \tag{6-65}$$

Since the spin operators affect only "spin coordinates," they commute with all operators that are a function only of space coordinates. The operators $\hat{S}^2$ and $\hat{S}_z$ will therefore commute with the operators $\mathscr{H}$, $\hat{L}^2$, and $\hat{L}_z$ as long as $\mathscr{H}$ contains no spin terms. We can choose our atomic wave functions, then, to be simultaneous eigenfunctions of all five of these operators. It is customary, therefore, to characterize atomic wave functions by four quantum numbers, n, l, m, and $\hat{S}_z$. A single electron can thus be thought of as moving in a four-dimensional space described by three space coordinates and a spin coordinate.

EXERCISE 6-34 Using the definitions of $\hat{S}_+$ and $\hat{S}_-$, calculate the results of operating on the functions α and β with the operators $\hat{S}_x$ and $\hat{S}_y$. Can you construct two functions from α and β which are eigenfunctions of $\hat{S}_x$?

EXERCISE 6-35 For more than one electron, the operators for the *total* spin angular momentum can be constructed vectorially. See Section 6-6 for details. Thus, for two electrons, $\hat{\mathbf{S}}_T = \hat{\mathbf{S}}_1 + \hat{\mathbf{S}}_2$, $\hat{S}_z = \hat{S}_z(1) + \hat{S}_z(2)$, and so on. For the two-electron case, write the operators $\hat{S}_+$ and $\hat{S}_-$ and then write $\hat{\mathbf{S}}_T{}^2$ in terms of these and $\hat{\mathbf{S}}_z$. Show that the functions

$$\alpha(1)\alpha(2)$$
$$\tfrac{1}{2}\sqrt{2}[\alpha(1)\beta(2) + \beta(1)\alpha(2)]$$
$$\beta(1)\beta(2)$$

are simultaneous eigenfunctions of $\hat{S}_T{}^2$ and $\hat{S}_z$. What are the eigenvalues? These spin functions are the spin functions for a *triplet* state.

6-8 IDENTICAL PARTICLES AND THE PAULI PRINCIPLE

We now wish to consider some aspects of systems containing two or more identical particles. By identical particles, we mean that the particles possess exactly the same properties, and they cannot, therefore, be distinguished from one another by physical measurements of their properties.

In classical mechanical systems there is no fundamental difference between a system in which particles are identical or nonidentical. For example, the laws of classical mechanics hold just as well for a "super" handball game in which two identical handballs are being used simultaneously as for a similar game in which one handball and one tennis ball are being used. The fundamental point is that, in a handball game, one of the identical balls can be identified by its previous history, for example, the one that was just hit by the server. Such is not the case in quantum mechanics, however. Here our information is more limited in that we can measure only the probability of finding "a handball" in a particular region, and we do not have any information about which handball it is. Therefore, in quantum mechanics all of the conclusions that we draw about systems containing identical particles must be independent of the labeling on the particles. For example, consider the case of a pair of noninteracting particles in a one-dimensional box. The Hamiltonian operator for the system will be

$$\mathscr{H} = -\frac{\hbar^2}{2m}[\nabla^2(1) + \nabla^2(2)] \tag{6-66}$$

and it is clear that the Hamiltonian is symmetric in the two particles. That is, interchanging the labels (1) and (2) leaves the Hamiltonian in Eq. 6-66 unchanged. In the eigenvalue equation using the Hamiltonian of Eq. 6-66, the variables for particles (1) and (2) are separable and the solutions are

$$\psi_1 = A \sin \frac{n_1 \pi x}{a} \tag{6-67a}$$

$$\psi_2 = A \sin \frac{n_2 \pi x}{a} \tag{6-67b}$$

Suppose we consider the case where one particle is in the lowest energy orbital and the second is in the next to lowest energy orbital. An acceptable mathematical solution to the two-particle eigenvalue equation would then be

$$\Psi = \psi_1(1)\psi_2(2) \tag{6-68}$$

where the subscripts 1 and 2 refer to the values of the quantum numbers n_1 and n_2. Such a wave function is physically unacceptable, however, because it implies that we can separately locate particles (1) and (2). That is, it unequivocally states that particle (1) is in state ψ_1 and particle (2) is in state ψ_2. Such an assertion is impossible to verify experimentally. The only thing that can be experimentally verified is that there *is* a particle in state ψ_1 and a particle in state ψ_2.

It is a straightforward procedure to construct wave functions which will take this indistinguishability into account. Since the Hamiltonian operator of Eq. 6-66 is symmetric in the two electrons, an equally good solution to the eigenvalue equation is

$$\Psi = \psi_2(1)\psi_1(2) \tag{6-69}$$

Also, any linear combinations of Eqs. 6-68 and 6-69 will be an equally good solution. Two of these linear combinations will be of special interest. They are

$$\Psi_+ = \tfrac{1}{2}\sqrt{2}[\psi_1(1)\psi_2(2) + \psi_2(1)\psi_1(2)] \tag{6-70}$$

$$\Psi_- = \tfrac{1}{2}\sqrt{2}[\psi_1(1)\psi_2(2) - \psi_2(1)\psi_1(2)] \tag{6-71}$$

Equation 6-69 is a symmetrical linear combination. That is, it remains unchanged when the labels on particles (1) and (2) are interchanged. Equation 6-71 is an antisymmetrical linear combination. It changes sign when the labels on particles (1) and (2) are interchanged. In operator language, we can introduce a permutation operator $\hat{P}(1, 2)$ which tells one to interchange the coordinates on particles (1) and (2) in the function that follows. Equations 6-70 and 6-71 will both be eigenfunctions of the operator $\hat{P}(1, 2)$ with eigenvalues $+1$ and -1, respectively. The important thing about Eqs. 6-70 and 6-71 is that they are consistent with the fact that we can tell that there *is* a particle in state ψ_1 and state ψ_2, but we cannot tell which particle it is.

EXERCISE 6-36 The factor $\frac{1}{2}\sqrt{2}$ in Eqs. 6-70 and 6-71 is a normalizing constant. Given ψ_1 and ψ_2 are normalized functions for a particle in a one-dimensional box, show that the functions Ψ_+ and Ψ_- are normalized.

It will be a general property of wave functions for systems containing many identical particles that they must be properly symmetrized to take the indistinguishability of the identical particles into account. This can be taken to be an additional postulate to those given previously. To complete the postulate, the particular kind of symmetrized wave function to use must be stated. This choice is determined by experimental observation. The Pauli exclusion principle, which originally was given for electrons, but which applies to all particles with half-integral spin (that is, protons, ^{13}C nuclei, and so on), is one form of this postulate. It states the following.

All acceptable wave functions for half-integral spin particles must be antisymmetric upon permutation of the coordinates of any two particles. Such particles are called Fermi particles or fermions. For particles with integral spin, the wave function must be symmetric with respect to permutation of the coordinates of any two particles. Such particles are called Bose–Einstein particles or bosons.

The symmetry properties of acceptable wave functions play an important role in the development of statistical mechanics. The student is referred to [6] for a good elementary discussion of this point.

To illustrate the application of the Pauli principle to a specific case, let us write down all of the possible functions of the form of Eq. 6-31 that are approximate solutions for the ground state of the helium atom. These solutions will also include the electron spin coordinate. The four possible functions are

$$\begin{aligned}
\Psi_1 &= \phi_{1s}(1)\phi_{1s}(2)\alpha(1)\alpha(2) \\
\Psi_2 &= \phi_{1s}(1)\phi_{1s}(2)\alpha(1)\beta(2) \\
\Psi_3 &= \phi_{1s}(1)\phi_{1s}(2)\beta(1)\alpha(2) \\
\Psi_4 &= \phi_{1s}(1)\phi_{1s}(2)\beta(1)\beta(2)
\end{aligned} \tag{6-72}$$

All four of these functions are correct solutions to the eigenvalue Eq. 6-30, but none of the four satisfies the Pauli principle. This can be seen by actually operating on the four functions with the permutation operator $\hat{P}(1, 2)$. The functions Ψ_1 and Ψ_4 are eigenfunctions of $\hat{P}(1, 2)$ with eigenvalue $+1$, and Ψ_2 and Ψ_3 are not eigenfunctions of $\hat{P}(1, 2)$. In fact,

$$\begin{aligned}
\hat{P}(1, 2)\Psi_2 &= \Psi_3 \\
\hat{P}(1, 2)\Psi_3 &= \Psi_2
\end{aligned} \tag{6-73}$$

In order to preserve the indistinguishability of the electrons, we must therefore take linear combinations of Ψ_2 and Ψ_3. The appropriate ones are those similar to Eqs. 6-70 and 6-71, and are

$$\Psi_+ = \tfrac{1}{2}\sqrt{2}\{\phi_{1s}(1)\phi_{1s}(2)[\alpha(1)\beta(2) + \beta(1)\alpha(2)]\} \tag{6-74}$$

$$\Psi_- = \tfrac{1}{2}\sqrt{2}\{\phi_{1s}(1)\phi_{1s}(2)[\alpha(1)\beta(2) - \beta(1)\alpha(2)]\} \tag{6-75}$$

with Ψ_+ having the eigenvalue $+1$ with respect to $\hat{P}(1, 2)$, and Ψ_- having the eigenvalue -1. We therefore see that the only acceptable ground state wave function for the helium atom at this level of approximation is given by Eq. 6-75.

If one were to calculate the eigenvalue with respect to the total spin angular momentum operator $\hat{S}_T^2$ for the function Ψ_- in Eq. 6-75, this eigenvalue would be zero. The state characterized by Eq. 6-75 would therefore have a spin degeneracy of $(2\hat{S}_T + 1) = 1$. The spin degeneracy, in turn, determines the multiplicity of a state. For $\hat{S}_T = 0$, the multiplicity of the state is *one* and the state is called a singlet state. For the case that $\hat{S}_T = 1$, $(2\hat{S}_T + 1)$ equals *three*, and the state is called a *triplet* state (see Exercises 6-35 and 6-38).

EXERCISE 6-37 The function 6-75 can be written as the product of a space part and a spin part. Thus $\Psi_- = \phi_{1s}(1)\phi_{1s}(2)\{\frac{1}{2}\sqrt{2}[\alpha(1)\beta(2) - \beta(1)\alpha(2)]\}$. Show that the spin part of this function is an eigenfunction of $\hat{S}_T^2$ with eigenvalue 0.

EXERCISE 6-38 An approximate product wave function for an excited state of a helium atom is

$$\Psi^E = \phi_{1s}(1)\phi_{2s}(2)$$

Construct *two* excited states wave functions that obey the Pauli principle. How do the space and spin parts of these wave functions differ?

6-9 INDEPENDENT ELECTRON THEORY OF COMPLEX ATOMS

Following the same procedure as that used for the helium atom, we can immediately write down the Hamiltonian operator for any atom. This operator is

$$\mathscr{H} = -\frac{1}{2}\sum_i \nabla_i^2 - \sum_i \frac{Z_N}{r_{iN}} + \sum_{i<j}\sum \frac{1}{r_{ij}} \qquad (6\text{-}76)$$

The first term is the sum of the kinetic energy operators for the electrons. The second term is the attractive energy of the electrons for the nucleus. The third term is the electron repulsion energy. Of course, the eigenvalue equation, using Eq. 6-76 for the operator, is impossible to solve exactly in terms of analytical functions.

To find approximate solutions for many-electron atoms, we will follow the procedure introduced at the end of our discussion of the helium atom. At the lowest level of approximation, we will use a trial function of the form

$$\Psi = \prod_{i=1}^{n} \phi_i(i) \qquad (6\text{-}77)$$

where the ϕ_i are atomic orbitals, including spin, and where there is one atomic orbital for each electron. These atomic orbitals may contain effective nuclear charges as variational parameters, and the values of these parameters can then be chosen to minimize the energy.

At this point, the meaning of the word *orbital* needs to be reemphasized because the concept of an orbital plays an important part in the remainder of the text. An orbital is a one-electron wave function. That is, it is a solution to an eigenvalue equation in which the Hamiltonian depends on only the coordinates of a single electron. Orbitals, and the energies of electrons occupying them, have physical significance only for systems that contain a single electron such as H, He^+, and the hydrogen molecule ion H_2^+ (discussed in Chapter 7). In more complex systems, the *states* of the atom or molecule are often expressed as products or sums of products of orbitals, but it must be kept in mind that such descriptions are convenient approximations in which a many-electron wave function is described by a product of one-electron wave functions. In the best calculation that has been done on the helium atom, all traces of atomic orbitals have disappeared. The orbital concept has great value for the chemist, however, in spite of the fact that numerical results calculated using wave functions based on a simple orbital approximation are, in general, poor. The fact that the orbital concept correctly rationalizes many qualitative observations connected with atoms and the periodicity of these elements is an outstanding achievement, and its importance should not be minimized. In addition, the orbital approximation serves as a starting point for more sophisticated treatments of atomic structure. We will have more to say about these more sophisticated treatments later.

A single product function such as Eq. 6-77 does not obey the Pauli principle. In order to construct a function that does satisfy the Pauli principle, we must take a linear combination of product functions so that the resulting function will change sign upon permutation of the coordinates of any two electrons.

Instead of discussing this procedure in general, let us consider the specific case of the lithium atom, a three-electron problem. Including spin, the lowest energy independent electron orbitals and their orbital energies are

χ_i	ε_i	Alternate notation	
$\chi_1 = \phi(1, 0, 0)\alpha$	ε_{1s}	1	
$\chi_2 = \phi(1, 0, 0)\beta$	ε_{1s}	$\bar{1}$	(a bar over the number indicates β spin)
$\chi_3 = \phi(2, 0, 0)\alpha$	ε_{2s}	2	
$\chi_4 = \phi(2, 0, 0)\beta$	ε_{2s}	$\bar{2}$	

(6-78)

where $\phi(1, 0, 0)$ is a $1s$ type atomic orbital and $\phi(2, 0, 0)$ is a $2s$ type.

We now must place the three electrons in these orbitals to form product functions and then find a linear combination of the product functions that is a simultaneous eigenfunction of $\hat{P}(1, 2)$, $\hat{P}(1, 3)$, and $\hat{P}(2, 3)$ with eigenvalues -1. We first investigate what happens when all three electrons are placed into one of the lowest energy orbitals, χ_1 or χ_2. The student can easily verify that such a product function is symmetric with respect to the interchange of the coordinates of any two electrons. Hence all such configurations may be ruled out. A similar argument can be used to rule out configurations with *two* electrons in any one orbital. We are thus left with the fact that the lowest energy state for the lithium atom must have one electron in each of χ_1, χ_2, and χ_3. (One electron can also be placed in χ_1, χ_2, and χ_4 to give a state with the same energy in the absence of external fields.) The final antisymmetric wave function must, therefore, be a linear combination of product functions involving these three independent electron orbitals.

There is an elegant way to construct such a wave function which was introduced by J. C. Slater. This method makes use of the property of determinants that requires that a determinant change sign upon interchange of any two rows or columns. We thus label the columns of a determinant with the electron number and the rows with the orbital number, and the determinant automatically gives an antisymmetric wave function. Thus for lithium we have

$$\begin{matrix} & 1 & 2 & 3 \leftarrow \text{electron number} \\ \Psi = \begin{matrix} 1 \\ 2 \\ 3 \end{matrix} & \left| \begin{matrix} \chi_1(1) \\ \chi_2(1) \\ \chi_3(1) \end{matrix} \right. & \begin{matrix} \chi_1(2) \\ \chi_2(2) \\ \chi_3(2) \end{matrix} & \left. \begin{matrix} \chi_1(3) \\ \chi_3(3) \\ \chi_2(3) \end{matrix} \right| \dfrac{1}{(3!)^{\frac{1}{2}}} \\ \uparrow & & & \\ \text{orbital number} & & & \end{matrix} \tag{6-79}$$

The $1/(3!)^{\frac{1}{2}}$ is a normalizing factor because an $N \times N$ determinant will have $N!$ terms. Such determinants are called Slater determinants and are usually abbreviated by only writing the diagonal term $|\chi_1(1)\ \chi_2(2)\ \chi_3(3)|$.

EXERCISE 6-39 Evaluate the determinant given in Eq. 6-79 and show that it is an eigenfunction of $\hat{P}(1, 3)$ with eigenvalue -1.

It is of interest to use the properties of Slater determinants to rule out the product functions that would result if we tried to place two or more electrons into one of the orbitals, say, χ_1. If the Slater determinant is written down for such a case, it can immediately be seen that two or more rows will be identical. Such a determinant must vanish (see Section 1-2). This result allows us to make a more useful statement of the Pauli principle for problems involving the independent electron approximation. Since, if two electrons are in the same orbital, they must have the same set of four quantum numbers (three space and one spin), the Pauli principle can be stated in the following form. *No two electrons can have the same set of four quantum numbers.*

Such a statement of the Pauli principle is a logical consequence of the more general statement given in Section 6-7 and can be used whenever an independent electron model is employed.

We now wish to continue our treatment of the lithium atom. To simplify working out the details, the alternate notation of Eqs. 6-78 will be used. Thus, for an orbital product we write

$$\phi(1, 0, 0)\alpha(1)\phi(1, 0, 0)\beta(2)\phi(2, 0, 0)\alpha(3) = \phi_{1s}\alpha(1)\phi_{1s}\beta(2)\phi_{2s}\alpha(3) = 1\bar{1}2$$

where the order of the symbols indicates the electron labels and a bar indicates β spin. With this notation, the Slater determinant for the ground state of a lithium atom becomes

$$\psi_G = \frac{1}{(3!)^{\frac{1}{2}}}\begin{vmatrix} 1 & 1 & 1 \\ \bar{1} & \bar{1} & \bar{1} \\ 2 & 2 & 2 \end{vmatrix} \tag{6-80}$$

The expansion of Eq. 6-80 yields (compare with your answer to Exercise 6-34)

$$\psi_G = \frac{1}{\sqrt{6}}(1\bar{1}2 - 12\bar{1} - \bar{1}12 + 21\bar{1} + \bar{1}21 - 2\bar{1}1) \tag{6-81}$$

The Hamiltonian for lithium is obtained from Eq. 6-76 by letting the index i go from one to three.

$$\begin{aligned} \hat{\mathscr{H}} = &-\frac{1}{2}\nabla_1^2 - \frac{Z}{r_1} - \frac{1}{2}\nabla_2^2 - \frac{Z}{r_2} - \frac{1}{2}\nabla_3^2 - \frac{Z}{r_3} \\ &+ \frac{1}{r_{12}} + \frac{1}{r_{13}} + \frac{1}{r_{23}} \end{aligned} \tag{6-82}$$

It can be seen that this Hamiltonian is in two parts, a one-electron part (the first six terms) and a two-electron part (the last three terms). To simplify the rotation, let us write this Hamiltonian as

$$\hat{\mathscr{H}} = \hat{\mathscr{H}}^{(0)} + \hat{\mathscr{H}}^{(1)}$$

where $\mathscr{H}^{(0)}$ contains all of the one-electron terms and $\hat{\mathscr{H}}^{(1)}$ contains all of the two-electron terms.

We now wish to evaluate the energy of the lithium atom.

$$\begin{aligned} E = \langle\Psi_a|\hat{\mathscr{H}}|\Psi_G\rangle &= \langle\Psi_G|\hat{\mathscr{H}}^{(0)}|\Psi_G\rangle + \langle\Psi_a|\hat{\mathscr{H}}^{(1)}|\Psi_a\rangle \\ &- \quad E^{(0)} \quad + \quad E^{(1)} \end{aligned} \tag{6-83}$$

Attacking the $E^{(0)}$ term first, we write

$$\begin{aligned} E^{(0)} = \langle \Psi_a | \mathscr{H}^0 | \Psi_a \rangle = \tfrac{1}{6}\{ & \langle 1\bar{1}2 | \mathscr{H}^{(0)} | 1\bar{1}2 \rangle \\ & - \langle 1\bar{1}2 | \mathscr{H}^0 | 12\bar{1} \rangle - \langle 1\bar{1}2 | \mathscr{H}^0 | \bar{1}12 \rangle + \cdots \\ & + \langle 2\bar{1}1 | \mathscr{H}^{(0)} | 2\bar{1}1 \rangle \} \end{aligned} \tag{6-84}$$

The student should be able to see, by comparing Eq. 6-81 with Eq. 6-84, that there will be 36 terms in the braces of Eq. 6-84. We now divide these terms into types to see if we can bring some order into the evaluation of Eq. 6-84. We consider first integrals in which the orbital product is *identical* on both sides of the operator. There will be six such integrals in Eq. 6-84. One of these integrals is $\langle 1\bar{1}2 | \mathscr{H}^{(0)} | 1\bar{1}2 \rangle$. Since $\mathscr{H}^{(0)}$ does not contain electron spin in any of its terms, we can immediately integrate over the spin variable of electrons 1, 2, and 3. Thus,

$$\begin{aligned} \langle 1\bar{1}2 | \mathscr{H}^{(0)} | 1\bar{1}2 \rangle &= \langle \phi_{1s}\phi_{1s}\phi_{2s} | \mathscr{H}^{(0)} | \phi_{1s}\phi_{1s}\phi_{2s} \rangle \cdot \langle \alpha | \alpha \rangle \langle \beta | \beta \rangle \langle \alpha | \alpha \rangle \\ &= \langle \phi_{1s}\phi_{1s}\phi_{2s} | \mathscr{H}^{(0)} | \phi_{1s}\phi_{1s}\phi_{2s} \rangle \cdot 1 \end{aligned} \tag{6-85}$$

since the spin functions are normalized. Note that the product of integrals $\langle \alpha | \alpha \rangle \langle \beta | \beta \rangle \langle \alpha | \alpha \rangle$ means

$$\int \alpha^2 \, d\tau_s(1) \int \beta^2 \, d\tau_s(2) \int \alpha^2 \, d\tau_s(3) = 1 \cdot 1 \cdot 1 = 1$$

To go one step further, note that $\mathscr{H}^{(0)}$ can be written as a sum of 3 one-electron terms. This should be obvious from the way the terms have been grouped in Eq. 6-82. Thus, we can write

$$\langle \phi_{1s}\phi_{1s}\phi_{2s} | \hat{\mathscr{H}}^{(0)} | \phi_{1s}\phi_{1s}\phi_{1s} \rangle = \langle \phi_{1s}\phi_{1s}\phi_{2s} | \hat{h}(1) + \hat{h}(2) + \hat{h}(3) | \phi_{1s}\phi_{1s}\phi_{2s} \rangle \tag{6-86}$$

where the $\hat{h}(i)$ are one-electron operators. Expanding the last term in Eq. 6-86, we obtain

$$\begin{aligned} & \langle \phi_{1s}\phi_{1s}\phi_{2s} | \hat{h}(1) + \hat{h}(2) + \hat{h}(3) | \phi_{1s}\phi_{1s}\phi_{2s} \rangle \\ &= \langle \phi_{1s} | \hat{h}(1) | \phi_{1s} \rangle \langle \phi_{1s}\phi_{2s} | \phi_{1s}\phi_{2s} \rangle + \langle \phi_{1s} | \hat{h}(2) | \phi_{1s} \rangle \langle \phi_{1s}\phi_{2s} | \phi_{1s}\phi_{2s} \rangle \\ & \quad + \langle \phi_{2s} | \hat{h}(3) | \phi_{2s} \rangle \langle \phi_{1s}\phi_{1s} | \phi_{1s}\phi_{1s} \rangle \\ &= \langle \phi_{1s} | \hat{h}(1) | \phi_{1s} \rangle \cdot 1 + \langle \phi_{1s} | \hat{h}(i) | \phi_{1s} \rangle \cdot 1 + \langle \phi_{2s} | \hat{h}(3) | \phi_{2s} \rangle \cdot 1 \\ &= \quad \varepsilon_{1s}{}^0 \quad + \quad \varepsilon_{1s}{}^0 \quad + \quad \varepsilon_{2s}{}^0 \\ &= \quad 2\varepsilon_{1s}{}^0 + \varepsilon_{2s}{}^0 \end{aligned} \tag{6-87}$$

where $\varepsilon_{1s}{}^0$ and $\varepsilon_{2s}{}^0$ are one-electron energies to be discussed in more detail later. The second integral in each of the terms in Eq. 6-87 equals one because of the normalization condition.

We mentioned above that there would be six integrals of the type of Eq. 6-85. Analysis of each of these integrals shows that they all give the same result as Eq. 6-87. Thus, the contribution of all six of these integrals to E^0 is $6(2\varepsilon_{1s} + \varepsilon_{2s})$.

EXERCISE 6-40 Using the arguments developed above, show that the integral $\langle \bar{1}21|\mathscr{H}^{(0)}|\bar{1}21\rangle = 2\varepsilon_{1s} + \varepsilon_{2s}$.

We now consider the 30 remaining integrals in which the orbital product is not the same on both sides of the operator. We find that these fall into two classes: those that differ in spin and those that differ in space. As an example of an integral where the spin products differ, we consider $\langle 1\bar{1}2|\hat{\mathscr{H}}^{(0)}|\bar{1}12\rangle$. This integral, and all others like it, can be shown to vanish because of the orthogonality of the spin functions α and β. Thus,

$$\langle 1\bar{1}2|\hat{\mathscr{H}}^{(0)}|\bar{1}12\rangle = \langle \phi_{1s}\phi_{1s}\phi_{2s}|\hat{\mathscr{H}}^0|\phi_{1s}\phi_{1s}\phi_{2s}\rangle\langle\alpha|\beta\rangle\langle\beta|\alpha\rangle\langle\alpha|\alpha\rangle = 0$$

EXERCISE 6-41 Consider Eq. 6-81 and determine how many integrals of the type we have just evaluated will be in Eq. 6-84. List them using the shorthand nomenclature.

The second class of integrals are those in which the spin products are the same, but the space products on the two sides of the operator are different. These integrals also vanish. To see this, consider the integral $\langle \bar{1}12|\hat{\mathscr{H}}^0|\bar{1}21\rangle$. Expanding this integral gives

$$\begin{aligned}
\langle \bar{1}12|\hat{\mathscr{H}}^0|\bar{1}21\rangle &= \langle \phi_{1s}\phi_{1s}\phi_{2s}|\hat{\mathscr{H}}^0|\phi_{1s}\phi_{2s}\phi_{1s}\rangle\langle\beta|\beta\rangle\langle\alpha|\alpha\rangle\langle\alpha|\alpha\rangle \\
&= \langle \phi_{1s}\phi_{1s}\phi_{2s}|\hat{h}(1) + \hat{h}(2) + \hat{h}(3)|\phi_{1s}\phi_{2s}\phi_{1s}\rangle\cdot 1\cdot 1\cdot 1 \\
&= \langle \phi_{1s}|\hat{h}(1)|\phi_{1s}\rangle\langle\phi_{1s}\phi_{2s}|\phi_{2s}\phi_{1s}\rangle \\
&\quad + \langle \phi_{1s}|\hat{h}(2)|\phi_{2s}\rangle\langle\phi_{1s}\phi_{2s}|\phi_{1s}\phi_{1s}\rangle \\
&\quad + \langle \phi_{2s}|\hat{h}(3)|\phi_{1s}\rangle\langle\phi_{1s}\phi_{1s}|\phi_{1s}\phi_{2s}\rangle
\end{aligned} \tag{6-88}$$

There are two considerations involved in the evaluation of the integrals in Eq. 6-88. The first is that the basis functions ϕ_{1s} and ϕ_{2s} are chosen to be orthonormal. This choice means that $\langle\phi_\mu|\phi_\nu\rangle = \delta_{\mu\nu}$. The second is that the basis functions are chosen to be eigenfunctions of the one-electron operator $\hat{h}(i)$. Thus, all integrals of the type $\langle\phi_\mu|h(i)|\phi_\nu\rangle$ vanish unless $\mu = \nu$. Applying these considerations to each term in Eq. 6-88 shows that each term will vanish.

EXERCISE 6-42 Write the integral $\langle 12\bar{1}|\hat{\mathscr{H}}^0|21\bar{1}\rangle$ and show which integrals vanish because of the orthonormality condition.

The $E^{(0)}$ term in Eq. 6-83 is, therefore, equal to

$$E^{(0)} = \tfrac{1}{6}\{6(2\varepsilon_{1s}{}^0 + \varepsilon_{2s}{}^0)\} = 2\varepsilon_{1s}{}^0 + \varepsilon_{2s}{}^0 \tag{6-89}$$

The student should note that the answer is identical to what would have been obtained if the simple product wave function $1\bar{1}2$ had been used for Ψ_G. This general result holds for all one-electron operators. One can calculate the value of the property for such operators by using a simple product rather than a Slater determinant.

We now apply the same kind of reasoning to the evaluation of the $E^{(1)}$ term. Here we find that the Slater determinant does make a difference. Since $\hat{\mathscr{H}}^{(1)}$ doesn't contain spin terms, all integrals where the order of spin functions is different vanish as they did with the one-electron terms. That leaves us with six integrals of the type $\langle 1\bar{1}2|\hat{\mathscr{H}}^{(1)}|1\bar{1}2\rangle$ where the order of the space parts is identical on both sides of the operator and six integrals of the type $\langle 1\bar{1}2|\hat{\mathscr{H}}^{(1)}|2\bar{1}1\rangle$ when the order of the space parts is different. Let us consider the identical order type first. Expanding and keeping track of the order of functions that indicates the electron number, we obtain

$$\begin{aligned}
\langle 1\bar{1}2|\hat{\mathscr{H}}^{(1)}|1\bar{1}2\rangle &= \left\langle \phi_{1s}\phi_{1s}\phi_{2s}\left|\frac{1}{r_{12}} + \frac{1}{r_{13}} + \frac{1}{r_{23}}\right|\phi_{1s}\phi_{1s}\phi_{2s}\right\rangle \cdot 1 \\
&= \left\langle \phi_{1s}\phi_{1s}\left|\frac{1}{r_{12}}\right|\phi_{1s}\phi_{1s}\right\rangle \underbrace{\langle\phi_{2s}|\phi_{2s}\rangle}_{1} \\
&\quad + \left\langle \phi_{1s}\phi_{2s}\left|\frac{1}{r_{13}}\right|\phi_{1s}\phi_{2s}\right\rangle \underbrace{\langle\phi_{1s}|\phi_{1s}\rangle}_{1} \\
&\quad + \left\langle \phi_{1s}\phi_{2s}\left|\frac{1}{r_{23}}\right|\phi_{1s}\phi_{2s}\right\rangle \underbrace{\langle\phi_{1s}|\phi_{1s}\rangle}_{1} \\
&= J_{1s,1s} + 2J_{1s,2s}
\end{aligned} \tag{6-90}$$

where the symbol J_{ij} has been introduced to indicate the integral $\langle\phi_i\phi_j|(1/r_{ij})|\phi_i\phi_j\rangle$. Evaluation of each of the other integrals of this type gives an identical result.

Expansion of one of the integrals where the order of space functions is different gives, for example,

$$\begin{aligned}
\langle 1\bar{1}2|\mathscr{H}^{(1)}|2\bar{1}1\rangle &= \left\langle \phi_{1s}\phi_{1s}\phi_{2s}\left|\frac{1}{r_{12}}+\frac{1}{r_{13}}+\frac{1}{r_{23}}\right|\phi_{2s}\phi_{1s}\phi_{1s}\right\rangle \\
&= \left\langle \phi_{1s}\phi_{1s}\left|\frac{1}{r_{12}}\right|\phi_{2s}\phi_{1s}\right\rangle \underbrace{\langle \phi_{2s}|\phi_{1s}\rangle}_{0} \\
&\quad + \left\langle \phi_{1s}\phi_{2s}\left|\frac{1}{r_{13}}\right|\phi_{2s}\phi_{1s}\right\rangle \underbrace{\langle \phi_{1s}|\phi_{1s}\rangle}_{1} \\
&\quad + \underbrace{\langle \phi_{1s}|\phi_{2s}\rangle}_{0} \left\langle \phi_{1s}\phi_{2s}\left|\frac{1}{r_{23}}\right|\phi_{1s}\phi_{1s}\right\rangle \\
&= K_{ij}
\end{aligned} \tag{6-91}$$

where the symbol K_{ij}, called an exchange integral, indicates the integral $\langle \phi_i\phi_j|(1/r_{ij})|\phi_j\phi_i\rangle$. The remaining integrals give the same result. Notice that the labels on the two orbitals are exchanged in the two halves of the integral. A consideration of the signs of each term in Eq. 6-81 shows that all of the integrals of this latter type will have a negative sign in front of them in the expansion of $E^{(1)}$. Thus, the complete expression for $E^{(1)}$ is

$$\begin{aligned}
E^{(1)} &= \tfrac{1}{6}\{6(J_{1s,1s} + 2J_{1s,2s}) - 6K_{1s,2s}\} \\
&= J_{1s,1s} + 2J_{1s,2s} - K_{1s,2s}
\end{aligned} \tag{6-92}$$

and the total energy of the lithium atom is

$$E = 2\varepsilon_{1s}{}^{0} + \varepsilon_{2s}{}^{0} + J_{1s,1s} + 2J_{1s,2s} - K_{1s,2s} \tag{6-93}$$

To go further in the calculation of the energy, we need to choose the detailed form of the basis functions and evaluate the integrals. The one-electron integrals are fairly simple to evaluate, but the two-electron integrals are algebraically complicated. Students interested in learning how to evaluate these integrals are referred, for example, to [7]. A tabulation of atomic and molecular integrals can be found in [8].

If we choose hydrogen-like $1s$ and $2s$ orbitals with orbital exponents equal to the actual nuclear charge ($Z = 3$), we obtain in Hartrees

$$\varepsilon_{1s} = -\frac{Z^2}{2} = -\frac{9}{2}; \qquad \varepsilon_{2s} = -\frac{1}{2}\left(\frac{Z^2}{4}\right) = -\frac{9}{8}$$

$$J_{1s,1s} = \frac{5}{8}Z = \frac{15}{8}; \qquad J_{1s,2s} = \frac{17}{81}Z = \frac{17}{27}$$

$$K_{1s,2s} = \frac{16}{729}Z = \frac{16}{243}$$

The total energy in atomic units is thus

$$E = 2\left(\frac{-9}{2}\right) - \frac{9}{8} + \frac{15}{8} + 2\left(\frac{17}{27}\right) - \frac{16}{243}$$

$$= -7.0566\text{H} = -192.01 \text{ eV} \tag{6-94}$$

This number should be compared to the true ground state energy of -203.5 eV, an error of 5.6%.

It was pointed out in the discussion of the helium atom that the ionization energy was a much more sensitive measure of the goodness of a quantum calculation than the total energy. It is instructive to calculate the first ionization energy of Li using the same approximations in the Li^+ calculation that we used above in the Li calculation. This calculation is done in Exercise 6-43. If the student works this exercise correctly, the answer will be negative.

EXERCISE 6-43 Using the same approximations as above ($Z = 3$, $J_{1s,1s} = 5Z/8$), calculate the energy of Li^+. Use this calculated energy and the energy of Li calculated above to calculate the first ionization energy of Li.

At this level of approximation, Li^+ is more stable than Li, a result that is strikingly contrary to reality.

Introducing an effective nuclear charge into the $1s$ and $2s$ orbitals can be expected to improve our calculation. Unfortunately, putting different scale factors into the $1s$ and $2s$ orbitals removes the orthogonality of these orbitals, and the integral evaluations that were simplified using this condition become algebraically much more involved. The problem can still be solved without a computer, however, and the reader is referred to Pilar's book [9] for a discussion of this procedure. When the optimum scale factors that minimize the energy are found, one obtains

$$\begin{aligned} E &= -7.414 \text{ a.u.} = -201.7 \text{ eV} \\ Z'_{1s} &= 2.694 \qquad Z'_{2s} = 1.534 \end{aligned} \tag{6-95}$$

Note that the total energy is now in error by only 0.86%.

Since we have already solved the helium atom problem (and, therefore, Li^+) at this level of approximation, it is instructive to calculate the first ionization energy for this "better" calculation.

EXERCISE 6-44 Calculate the energy of Li^+ when the optimum scale factor is used in the 1*s* orbital. (Use Eq. 6-42 for Li^+ to find this optimum scale factor. It is not necessarily the same as for Li.) Use this energy to calculate the first ionization energy of Li at this level of approximation.

Exercise 6-44 shows that the ionization energy is now positive (as it should be) and is in fairly good agreement with the experimental value of 5.37 eV.

As we pointed out in the discussion of helium, further improvement in energy calculations for atoms involves expanding the variational flexibility of the basis orbitals by writing each orbital as a linear combination of several atomic orbitals, each with its own coefficient and orbital exponent, or by writing the ground state wave function as a sum of Slater determinants, each of which corresponds to an excited configuration of the atom. A procedure utilizing the first approach leads to the Hartree–Fock energy when the basis set is made suitably large. This calculation is the best that can be done with a wave function consisting of a single Slater determinant. Such a calculation takes average electron repulsion into account, but it does not include instantaneous electron correlation.

The second approach is a method of adding electron correlation without explicitly introducing the interelectron distance. This approach, which uses a linear combination of Slater determinants, is called *configuration interaction*. Such calculations begin to account for electron correlation, but they also greatly increase the computational complexity and, therefore, also increase computer costs. A good survey of these procedures is given by Clementi [10].

6-10 THE AUFBAU PRINCIPLE AND THE PERIODIC TABLE

One of the tests of quantum theory is to see if it can give a rationale to the repetition of physical and chemical properties that is the basis for the periodic table. We wish to answer such questions as: Why do we get an alkali metal after 2, 8, 8, and then 18 elements? A related question is: Why do the transition elements enter where they do? Students who recall their general chemistry course will remember that these facts were rationalized by an orbital energy diagram such as the one in Figure 6-10. These orbitals, in order of increasing energy, are then filled with two electrons each to satisfy the Pauli principle. We now discuss the basis for, and the adequacy of, the diagram in Figure 6-10.

The student should recall that when the hydrogen atom problem was solved, the energy of the atomic orbitals depended on only the quantum number n.

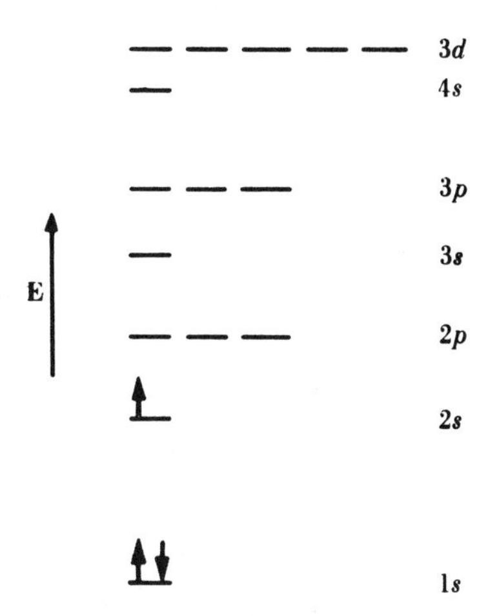

FIGURE 6-10 *Schematic representation of the orbital energies in complex atoms. Electrons are placed in these orbitals to correspond to the arrangement in the lithium atom. Note the threefold degeneracy of the p orbitals and the fivefold degeneracy of the d orbitals.*

This means that in a hydrogen atom, the $2s$ and $2p$ orbitals are degenerate. Similarly, the $3s$, $3p$, and $3d$ orbitals all have the same energy. In Figure 6-10, however, the energy depends not only on n but also on l. The $2s$ and $2p$ orbitals no longer have identical energies, and neither do the $3s$, $3p$, and $3d$ orbitals. How does quantum theory account for this change?

Table 6-2 gives the energies of the ground state of helium and lithium and one excited state of lithium calculated using the independent electron model developed in the preceding section.

Let us use the results of the preceding section to compare the energies of the ground and excited states of lithium given in Table 6-2. We know from the experimental study of the spectrum of the lithium atom that the energy difference between the two states ^2P and ^2S is 1.849 eV or 0.068 Hartree. Thus, the electron configuration which, in the independent electron model, has an electron in a $2p$ orbital has a higher energy than the configuration with the

TABLE 6-2 INDEPENDENT ELECTRON ENERGIES FOR THE GROUND STATES OF He AND Li AND AN EXCITED STATE OF Li. THE SYMBOLS ARE GIVEN FOR LATER REFERENCE

Atom	*(State)*	*Symbol*	*Wave function*	*Calculated energy*
He	(ground)	^1S	$\lvert 1s(1)\overline{1s}(2)\rvert$	$2\varepsilon_{1s}{}^0 + J_{1s,1s}$
Li	(ground)	^2S	$\lvert 1s(1)\overline{1s}(2)2s(3)\rvert$	$2\varepsilon_{1s}{}^0 + \varepsilon_{2s}{}^0 + J_{1s,1s} + 2J_{1s,2s} - K_{1s,2s}$
Li	(excited)	^2P	$\lvert 1s(1)\overline{1s}(2)2p(3)\rvert$	$2\varepsilon_{1s}{}^0 + \varepsilon_{2p}{}^0 + J_{1s,1s} + 2J_{1s,2p} - K_{1s-2p}$

electron in a $2s$ orbital. To rationalize this fact using a simple orbital picture, we draw an orbital energy diagram like Figure 6-10 and say that *the energy of a 2p orbital is greater than the energy of a 2s orbital.* The previous statement is italicized because the $2s$ and $2p$ orbitals arise from the approximation method used to solve the many-electron atomic problem and, in fact, they do not represent the quantum states of individual electrons.

Looking at our theoretical expression for the energies of the $2s$ and $2p$ states of lithium, we can see that the energy difference between these two states is given by

$$\Delta E(^2S \rightarrow {}^2P) = \varepsilon_{2p}{}^0 - \varepsilon_{2s}{}^0 + 2(J_{1s,2p} - J_{1s,2s}) - (K_{1s,2p} - K_{1s,2s}) \tag{6-96}$$

Even though the orbital picture of the atom is theoretically incorrect, it is still valuable because of the intuitive grasp that it gives about many atomic properties. For example, Figure 6-11 plots the probability of finding an electron (r^2R^2) as a function of distance away from the nucleus of the optimized orbitals ϕ_{1s}, ϕ_{2s}, and ϕ_{2p} for a lithium atom. We can interpret the orbital exponent in the $2s$ orbital as an *effective nuclear charge* in which the electron in the $2s$ orbital is "screened" from the nuclear charge by the electrons in the $1s$ orbital. Figure 6-11 shows that the probability of finding a $2s$ electron outside of the $1s$ electron shell is much larger than it is inside that shell. In addition, Figure 6-11 also shows that an electron in a $2p$ orbital will be more effectively screened by the $1s$ electron than an electron in a $2s$ orbital. This comes about because the $2s$ orbital has a probability "bump" inside the $1s$ electron distribution that the $2p$ orbital does not have. Many books speak of the "penetration" of the $2s$ electron being better than that of a $2p$ electron.

We can use Figure 6-11 to give qualitative arguments about the magnitudes of various terms in Eq. 6-96. First of all, we can argue that $\varepsilon_{2p}{}^0 > \varepsilon_{1s}{}^0$ since more screening means a smaller orbital exponent and ε^0 becomes less negative as the orbital exponent becomes smaller. Secondly, since the probability of finding a $2p$ electron in the $1s$ electron region is smaller than the probability of finding a $2s$ electron in that region, $J_{1s,2p}$ and $K_{1s,2p}$ will be smaller than $J_{1s,2s}$ and $K_{1s,2s}$. Thus, a number of terms in Eq. 6-96 go into making the ^2P state higher in energy than the ^2S state. It should be clear to the student that there are a lot of detailed considerations buried in the statement: *In many-electron atoms, an electron in a 2p orbital has a higher energy than an electron in a 2s orbital.*

Always remembering that these buried considerations exist, we can continue to use Figure 6-10 to write ground state electron configurations of atoms. Thus, for beryllium and boron, we write

$$(\text{Be})\ \Psi_G = |1s(1)\overline{1s}(2)2s(3)\overline{2s}(4)|$$
$$(\text{B})\ \Psi_G = |1s(1)\overline{1s}(2)2s(3)\overline{2s}(4)2p(5)|$$

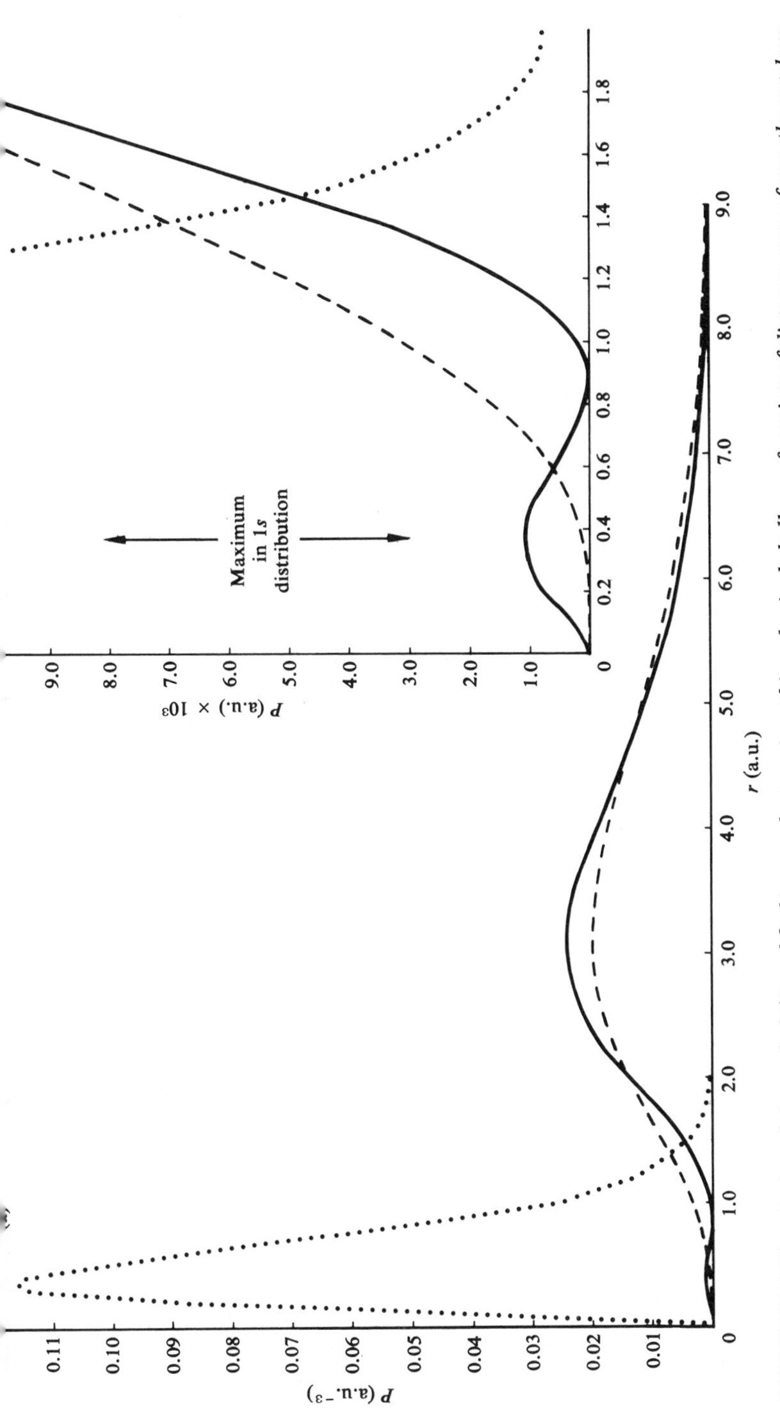

FIGURE 6-11 *A plot of the probability of finding an electron in a thin spherical shell as a function of distance away from the nucleus for the* 1s, 2s, *and* 2p *orbitals in an optimized lithium atom wave function. In (a) is shown the fact that most of the electron density for a* 2s *and* 2p *orbital lies outside of the* 1s *distribution. In (b) is shown an expanded view of the small r region showing that the hump in the* 2s *electron distribution occurs at the maximum in the* 1s *distribution. This plot explains why a* 2s *electron is less screened from the nuclear charge and why, for example,* $J_{1s,2s}$ *is greater than* $J_{1s,2p}$. (· · · = 1s, —— = 2s, - - - = 2p).

For carbon, a complication arises because there are a number of different ways that two electrons can be arranged in the three degenerate p orbitals without violating the Pauli principle, and there is nothing in Figure 6-10 that allows us to decide which of these configurations will have the lowest energy. We can write all of the possible single determinantal wave functions for the p^2 configuration and label them according to their values of S_z and L_z. Table 6-3 does just this. The values of S_z and L_z for these wave functions are instructive because they allow us to designate atomic states according to their values of the spin and orbital angular momentum. The total angular momentum quantum number for the state is indicated by the capital letters S, P, D, F, etc. for values of $L_T = 0, 1, 2, 3, \ldots$. The total spin quantum number is indicated by a left superscript that gives the spin degeneracy or the multiplicity $(2S_T + 1)$ of the state. Thus, if $S_T = 0$, the multiplicity equals 1 and the state is a singlet. If $S_T = 1$, $2S_T + 1 = 3$ and the state is a triplet. The student should now be able to rationalize the symbols given for the states in Table 6-2.

We can now draw some qualitative correlations between some of the determinants of Table 6-2 and certain electron arrangements in Figure 6-10. Thus,

TABLE 6-3 POSSIBLE SINGLE DETERMINANTAL WAVE FUNCTIONS FOR THE CONFIGURATION $2P^2$

i	Ψ_i	L_z	S_z
1	$\lvert 1s\overline{1s}2s\overline{2s}2p_1\overline{2p_1}\rvert$	2	0
2	$\lvert 1s\overline{1s}2s\overline{2s}2p_12p_0\rvert$	1	1
3	$\lvert 1s\overline{1s}2s\overline{2s}2p_12p_0\rvert$	1	0
4	$\lvert 1s\overline{1s}2s\overline{2s}\overline{2p_1}2p_0\rvert$	1	0
5	$\lvert 1s\overline{1s}2s\overline{2s}\overline{2p_1}\overline{2p_0}\rvert$	1	−1
6	$\lvert 1s\overline{1s}2s\overline{2s}2p_12p_{-1}\rvert$	0	1
7	$\lvert 1s\overline{1s}2s\overline{2s}2p_1\overline{2p_{-1}}\rvert$	0	0
8	$\lvert 1s\overline{1s}2s\overline{2s}2p_1\overline{2p_{-1}}\rvert$	0	0
9	$\lvert 1s\overline{1s}2s\overline{2s}\overline{2p_1}\overline{2p_{-1}}\rvert$	0	−1
10	$\lvert 1s\overline{1s}2s\overline{2s}2p_0\overline{2p_0}\rvert$	0	0
11	$\lvert 1s\overline{1s}2s\overline{2s}2p_02p_{-1}\rvert$	−1	1
12	$\lvert 1s\overline{1s}2s\overline{2s}2p_02p_{-1}\rvert$	−1	0
13	$\lvert 1s\overline{1s}2s\overline{2s}\overline{2p_0}2p_{-1}\rvert$	−1	0
14	$\lvert 1s\overline{1s}2s\overline{2s}\overline{2p_0}\overline{2p_{-1}}\rvert$	−1	−1
15	$\lvert 1s\overline{1s}2s\overline{2s}2p_{-1}\overline{2p_{-1}}\rvert$	−2	0

Note: The subscripts on the $2p$ orbitals indicate the eigenvalue of that orbital with respect to the operator $\hat{L}_z$.

Ψ_1, Ψ_{10}, and Ψ_{15} each contain two electrons in one of the $2p$ orbitals with their spins paired. These configurations must clearly be associated with singlet ($S_T = 0$) states. Ψ_2, Ψ_5, Ψ_6, Ψ_9, Ψ_{11}, and Ψ_{14} clearly have one electron in each of two p orbitals with spins parallel. These configurations must be associated with triplet states ($S_T = 1$). In fact, by looking at the maximum value of L_z, we can conclude that there must be a state with $L_T = 2$ and, therefore, a D state. Since all spins are paired in Ψ_1 and Ψ_{15}, this state must be a ^1D state. The maximum value of L_z associated with any of the functions Ψ_2, Ψ_5, Ψ_6, Ψ_9, and Ψ_{11} is 1 and, therefore, there must be a ^3P state. A ^1D state is fivefold degenerate corresponding to the five possible values of L_z, and a ^3P state is ninefold degenerate (a spin degeneracy of three coupled with an orbital degeneracy of three). This accounts for 14 of the 15 determinants of Table 6-3. Since the only kind of state with a degeneracy of 1 is a ^1S state, this state must be the remaining possible one for a carbon atom.

The preceding argument demonstrates that the electron configuration $2p^2$ gives rise to three states, ^1S, ^1D, and ^3P, but what is the ordering of the energy of these states? To determine this, we refer to two empirical statements known as Hund's rules. Hund's first rule is

Other things being equal, the state of highest multiplicity will be the most stable.

Hund's second rule is

Among levels having the same electronic configuration and the same multiplicity, the most stable level is the one with the largest angular momentum.

Thus, for a carbon atom, the group of ^3P states will have the lowest energy because it has the highest multiplicity (3). Among the ^1D and ^1S states, the ^1D will be the most stable because it has the largest total angular momentum.

The physical basis for Hund's first rule may be seen by considering the properties of two-electron triplet and singlet wave functions. From advanced quantum mechanical considerations, it can be shown that the probability of finding two electrons with the *same* spin at the same point in space is *zero*. No such restriction applies to electrons with opposite spins, however. These statements are another form of the Pauli principle. Since, in a triplet state, the electrons have the same spin, they cannot be allowed to come into the same region of space if the Pauli principle is to be obeyed. This keeping of the electrons apart in a triplet wave function is accomplished by virtue of the fact that the space part of the function must be antisymmetric. (In Exercise 6-35 it was shown that the spin functions for a triplet state were symmetric. The space part of a function for a triplet state must therefore be antisymmetric in order for the total wave function to be antisymmetric.) Since the electrons are kept apart in this function, the electron repulsion energy is smaller than if the electrons were allowed to come close together.

In a singlet wave function, the electrons have opposite spins and there is nothing in the Pauli principle which keeps these electrons from being in the same region of space. This is consistent with the fact that the space part of a two-electron singlet function is always symmetric with respect to exchange of the two electrons. Since the electrons can come close to one another, the electron repulsion energy between them will be relatively large. As a consequence of this difference in the space parts of the triplet and singlet wave functions, the triplet state for a given configuration will tend to have a lower energy than the singlet state for that configuration. See Exercise 6-45 for an illustration of this point.

EXERCISE 6-45 Let ψ_1 and ψ_2 be the functions for a particle in a one-dimensional box with $n = 1$ and $n = 2$. If one electron were in each of these orbitals, the space part of the triplet and singlet wave functions would be

$$\Psi_A = \tfrac{1}{2}\sqrt{2}[\psi_1\psi_2 - \psi_2\psi_1] \text{ and } \Psi_S = \tfrac{1}{2}\sqrt{2}[\psi_1\psi_2 + \psi_2\psi_1]$$

respectively. Suppose that particle 1 is in a small element of length dx at $x = 0.250a$ and particle 2 is in a small element of length dx at $x = 0.255a$. The quantity a is the length of the box. Show that Ψ_A has a very small value under these conditions while Ψ_S can be large. What happens to Ψ_A if both electrons are at $x = 0.250a$? This problem shows how an antisymmetric space function keeps the electrons apart.

EXERCISE 6-46 The wave functions for two *excited states* of a helium atom are

$$\Psi(^3S) = \frac{1}{\sqrt{2}}(\phi_{1s}(1)\phi_{2s}(2) - \phi_{2s}(1)\phi_{1s}(2))\frac{1}{\sqrt{2}}\begin{pmatrix}\alpha\alpha \\ \alpha\beta + \beta\alpha \\ \beta\beta\end{pmatrix}$$

$$\Psi(^1s) = \frac{1}{\sqrt{2}}(\phi_{1s}(1)\phi_{2s}(2) + \phi_{2s}(1)\phi_{1s}(2))\frac{1}{\sqrt{2}}(\alpha\beta - \beta\alpha)$$

Choose one of the components of the triplet wave function and use the Hamiltonian for helium to calculate the energy of ^3S state in terms of ε_i, J_{ij}, and K_{ij}. Proceeding in the same manner, calculate the energy of the ^1S state. Calculate

$$\Delta E(\text{singlet} - \text{triplet}) = E(^1S) - E(^3S)$$

Since all K_{ij} are positive, the triplet state has a lower energy than the singlet state for this excited state of helium.

The second rule can be rationalized in the following way. If the electrons in an atom have a large total angular momentum, it is an indication that the electrons tend to move around the nucleus in the same direction. Therefore, they are better able to keep out of each other's way than if they were going in opposite or random directions. W. Kauzmann's example [4] is a good one to illustrate this point. Think about people emptying a football stadium by walking around the track. The "repulsion interaction" will be much smaller if they are all going in the same direction around the track than if they are going in random directions. It should be pointed out that Hund's rules apply strictly only to the ground state of an atom. For excited states they might not give the correct result.

Similar arguments can be used to predict that the ground states of a nitrogen, oxygen, and fluorine atom should be a ^4S, ^3P, and ^2P, respectively. All of these predictions are confirmed by experiment.

When we get to neon, all of the $n = 2$ orbitals are filled, and we have an element with properties very similar to helium. The fundamental idea behind the periodic table is that the chemical properties of atoms are determined by their electron configurations and, therefore, atoms with similar electron configurations will have similar properties. Using the arrangement of orbitals given in Figure 6-10, the third row of the periodic table should also contain eight elements as the $3s$ and $3p$ orbitals are filled with electrons.

The simple orbital picture of Figure 6-10 becomes much more complicated to rationalize when we get to elements of the fourth row of the periodic table. Experimentally, we know that potassium and calcium are an alkali and an alkaline earth metal and, therefore, must have electron configurations Ar$4s$ and Ar$4s^2$, respectively (Ar stands for the argon core $1s^2 2s^2 2p^6 3s^2 3p^6$). Next comes a group of ten elements that make up the first series of transition metals. This series involves filling the five $3d$ orbitals. It is important at this point to remember the experimental fact that the ^2S state of potassium has a lower energy than the ^2D state. A treatment of orbital energies similar to that used to derive Eq. 6-43 shows that ε_{4s} is always greater than ε_{3d}. This fact is not inconsistent with the fact that the ^2S state for potassium has a lower energy than the ^2D state because the energy of the atomic state depends on a number of coulomb and exchange integrals in addition to the orbital energy. For an excellent discussion of this point, the student is referred to [11].

The important conclusion to be drawn from this section is that quantum theory does give a satisfactory account of the periodic table. In addition, using the computational tools discussed in this chapter, we can compute sets of atomic orbitals for atoms up to atomic number 86 and use these orbitals as the starting point for more sophisticated atomic wave functions, for atomic properties other than energy, and for molecular calculations.

6-11 *SUMMARY*

1. The energy levels and wave functions for the hydrogen atom and the hydrogen-like ions were found. The wave functions were found to depend on three quantum numbers, n, l, and m, and the significance of these quantum numbers was discussed.

2. The selection rules for transitions between hydrogen-like orbitals were derived. They are $\Delta l = \pm 1$, $\Delta m = 0, \pm 1$, Δn—no restriction.

3. Atomic units were introduced. Using these units, distances are measured in Bohr radii and energy in Hartrees. (One Hartree = 27.2 eV.)

4. The electronic structure of the helium atom was discussed. It was found that exact analytical solutions to the helium problem could not be obtained

because of the electron repulsion term in the Hamiltonian. The variational method for finding approximate solutions to quantum mechanical problems was introduced and applied to a calculation of the ground state energy of helium.

5. The commutation properties of angular momentum operators were discussed, and ways of treating angular momentum were introduced.

6. The existence of electron spin angular momentum was postulated. The operators for spin angular momentum combine and commute in the same way as the corresponding ones for orbital angular momentum.

7. The restrictions placed on wave functions for systems containing many identical particles were discussed. For electrons and other spin $\frac{1}{2}$ particles, acceptable wave functions must be antisymmetric upon interchange of the coordinates of any two particles. For integral or zero spin particles, acceptable wave functions must be symmetric upon interchange of the coordinates of any two particles.

8. The independent electron theory of complex atoms was developed, and it was shown that an antisymmetrized wave function for such atoms could be written in the form of a Slater determinant. A detailed discussion of the ground state energy of a lithium atom was given.

9. The quantum mechanical rationalization of the periodic table was presented and a critical discussion of the usual atomic orbital energy diagram was given.

10. The student should be familiar with the following terms: principal quantum number, azimuthal quantum number, magnetic quantum number, radial distribution function, node, atomic orbital, raising and lowering operators, antisymmetric wave function, permutation operator, Slater determinant, coulomb and exchange integrals, multiplicity, aufbau principle, Hund's first and second rules, electron correlation, and singlet and triplet states.

REFERENCES

1. H. Margenau and G. M. Murphy, *The Mathematics of Physics and Chemistry* (D. Van Nostrand, Inc., Princeton, N.J., 1943), Vol. 1, pp. 77ff.
2. H. Eyring, J. Walter, and G. E. Kimball, *Quantum Chemistry* (J. Wiley & Sons, Inc., New York, 1949), pp. 63ff.
3. D. J. Marks, *J. Chem. Ed.* **45**, 637 (1968).
4. W. Kauzmann, *Quantum Chemistry* (Academic Press Inc., New York, 1957).
5. J. M. Anderson, *Introduction to Quantum Chemistry* (W. A. Benjamin, Inc., New York, 1969).
6. R. E. Dickerson, *Molecular Thermodynamics* (W. A. Benjamin, Inc., New York, 1969).

7. I. N. Levine, *Quantum Chemistry* (Allyn and Bacon, Boston, 1970), Vol. 1, pp. 221, 222.
8. J. Miller, J. M. Gerhausen, and F. M. Matson, *Quantum Chemistry Integrals and Tables* (University of Texas Press, Austin, 1959).
9. F. L. Pilar, *Elementary Quantum Chemistry* (McGraw-Hill, New York, 1968), pp. 333–336.
10. E. Clementi, *IBM J. Res. Develop.* **9**, 2 (1965).
11. F. L. Pilar, *J. Chem. Ed.* **55**, 2 (1978).

Chapter 7

MOLECULES AND THE CHEMICAL BOND

ALTHOUGH ATOMS are regarded as the most fundamental species in chemistry, it is molecules that occupy the central position in modern chemical research. In Chapter 6, we saw that quantum mechanics can account exactly for the energy levels of the hydrogen atom. In addition, simple quantum theory can be applied to more complicated atoms to give correct qualitative results for many atomic properties, including a rationalization of the modern periodic table. With the expenditure of additional computational labor, made relatively easy by modern electronic computers, good quantitative calculations of atomic properties can be obtained even for quite heavy atoms.

In this chapter, we see if quantum mechanics can give us answers to some important questions about molecules. Examples of the kind of questions we hope to answer are the following.

1. Why do molecules form at all? Why, for example, do two halogen atoms combine to form a halogen molecule while two inert gas atoms do not? An alternate way to phrase this same question is: What is a chemical bond?

2. Why do molecules have the formulas that they do? Can a quantum mechanical description be provided for the law of multiple proportions?

3. Why do molecules form with their characteristic stereochemical arrangement?

4. Can we calculate molecular properties such as energies, transition moments, and dipole moments in agreement with experiment?

It seems almost too basic to point out in answer to the first question that molecules will form whenever the energy of a molecule is less than the sum of the energies of its constituent atoms. This fundamental fact, however, indicates the strategy for molecular quantum calculations. This strategy is to explore the energy of a multiatom system as a function of the distances between the atoms. A minimum in the "potential surface" resulting from such calculations with a depth considerably greater than kT at room temperature corresponds to a stable molecule. No minimum with this depth means that molecule formation will not take place. The student should keep in mind that although much progress has been made in calculating the details of molecular structure, many problems still remain to be solved. We hope that this chapter will induce some students to contribute additional insights to this area of research.

7-1 THE HYDROGEN MOLECULE ION, H_2^+

The simplest molecule is one containing two protons and one electron. This molecule, H_2^+, is called the hydrogen molecule ion. It was discovered by J. J. Thomson in cathode rays, and it has been found to have an equilibrium internuclear distance of 1.060 Å and a binding energy of 2.791 eV. Although H_2^+ is not a common chemical molecule, the quantum mechanical solutions to its structure are important for two reasons:

1. Approximation methods for molecular calculations can be compared to the exact solution and their adequacy evaluated.
2. Solutions to the H_2^+ problem give a set of molecular orbitals that can be used as the starting point for calculations on complex molecules in the same way as hydrogen atom solutions were used as a starting point for calculations on complex atoms.

The nomenclature used in discussing H_2^+ is shown in Figure 7-1. The quantities r_A and r_B are the electronic distances from nuclei A and B, respectively, and R_{AB} is the internuclear distance. The Hamiltonian operator in atomic units is

$$\mathscr{H} = -\frac{1}{2}\nabla^2 - \frac{m}{M_A}\frac{1}{2}\nabla_A{}^2 - \frac{m}{M_B}\frac{1}{2}\nabla_B{}^2 - \frac{1}{r_A} - \frac{1}{r_B} + \frac{1}{R_{AB}} \tag{7-1}$$

where the terms represent, respectively, the kinetic energies of the electron and nuclei A and B, the potential energy of attraction between the electron and the two nuclei, and the nuclear repulsion energy. The allowed energies and wave functions are found by the solution to the eigenvalue equation

$$\mathscr{H}\Psi_i = E_i\Psi_i \tag{7-2}$$

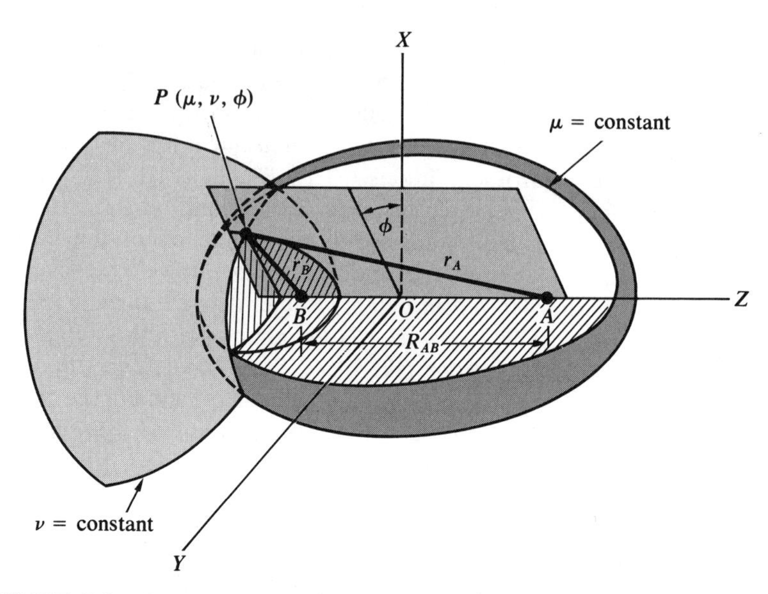

FIGURE 7-1 *Quantities used in the discussion of* H_2^+. *One nucleus is at point A, the other at point B. The electron is located at the point P(μ, ν, φ).*

The student will immediately notice that the problem of H_2^+ is a three-body problem like that of the helium atom. The three-body problem is insoluble in its most general form, but, fortunately, a very good approximation that renders Eq. 7-2 soluble to any degree of accuracy required can be made. This approximation is called the Born–Oppenheimer approximation. It states that the motions of the electrons in a molecule are so rapid that, in studying the electronic properties of molecules, the nuclei may be regarded as fixed. This means that the electronic energy can be found at a set of fixed internuclear distances and then plotted as a function of R_{AB}. When this procedure is followed for H_2^+, a plot such as that shown in Figure 7-2 is obtained. A minimum in the curve corresponds to the formation of a stable molecule, and the depth of the minimum is the dissociation or binding energy.

The student should note that the binding energy is the same as the energy D_e discussed in Chapter 5. Thus, the binding energy is equal to the measured dissociation energy D_0, plus the zero-point energy.

Curves such as that of Figure 7-2 can have a double interpretation. First, they give the electronic energy as a function of internuclear distance. Second, they give the potential energy of the nuclei as a function of the displacement from the equilibrium internuclear distance. Thus, it is the change in electronic energy, upon increasing or decreasing the internuclear distance, that gives rise to the vibrational force constant discussed in Chapter 5.

Mathematically, the Born–Oppenheimer approximation allows us to drop the

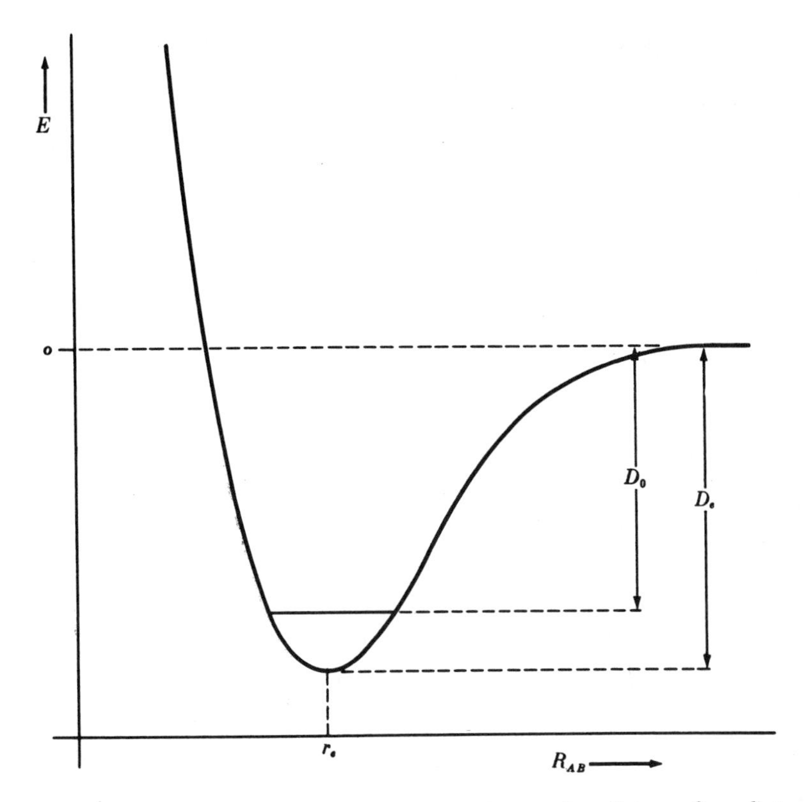

FIGURE 7-2 *A plot of electronic energy versus internuclear distance for a diatomic molecule showing the equilibrium internuclear distance r_e, the extrapolated dissociation energy D_e, and the experimentally measured dissociation energy D_0. The quantity $D_e = D_0 + \frac{1}{2}h\nu_0$, where $\frac{1}{2}h\nu_0$ is the zero-point vibrational energy. D_e is also called the binding energy.*

nuclear kinetic energy terms from the Hamiltonian in Eq. 7-1. We assume, in effect, that the nuclei are fixed. The problem of H_2^+ is then exactly soluble because it is reduced to a one-electron problem.

If elliptical coordinates are used, it is possible to separate variables and find a solution to Eq. 7-2 of the form

$$\Psi = U(\mu)V(v)\Phi(\phi) \tag{7-3}$$

where the three functions U, V, and Φ are a function of only the variables μ, v, and ϕ, respectively. The exact solutions to Eq. 7-3, while not overly difficult, are algebraically tedious. The function $V(v)$ is a linear expansion of Legendre polynomials with coefficients that depend on the energy, angular momentum, and internuclear distance. The $U(\mu)$ part of the wave function also contains a power series expansion in μ with coefficients that depend on the same three parameters. Tables of these coefficients can be found in [1].

The Φ part of Eq. 7-3 reflects the fact that the Hamiltonian given in Eq. 7-1 possesses rotational symmetry about the bond axis and leads to the eigenvalue equation

$$\frac{d^2\phi}{d\phi^2} = -m^2\phi \tag{7-4}$$

which has the familiar solutions

$$\Phi = Ae^{im\phi} \tag{7-5}$$

where $m = 0, \pm 1, \pm 2, \ldots$. These solutions have a physical interpretation based on the fact that they are eigenfunctions of the operator $\hat{L}_z$, where now the Z axis is defined by the direction of the AB bond.

In the case of H_2^+, the energy will depend on m except that states for $\pm m$ will have the same energy. Thus, considering only angular degeneracy, we have a singly degenerate level for $m = 0$, doubly degenerate levels for $m = \pm 1, \pm 2$, and so on. This discussion of the Φ functions for H_2^+ has been given to introduce a nomenclature to describe the states of diatomic molecules. This nomenclature is similar to the s, p, d, f nomenclature used in atomic orbitals except that Greek letters are used. Thus, for various values of m, the orbitals are labeled as follows.

m	*Orbital designation*
0	σ
± 1	π
± 2	δ

It may not always be convenient to use orbital functions of the form of Eq. 7-5; it is similar to the situation encountered with the $\Theta\Phi(1, \pm 1)$ functions for atoms. Thus, orbitals analogous to the p_x and p_y orbitals for atoms can be constructed by taking appropriate linear combinations of the $\pm m$ functions in Eq. 7-5. The same orbital designations given above are used for these linear combinations.

It is instructive to compare a plot of the exact Ψ^2 for the ground state of H_2^+ with the sum of Ψ^2 for two hydrogen atoms located at the position of the nuclei. Such a plot is shown in Figure 7-3. This plot illustrates in a general way what happens to the electron distribution when two atoms form a molecule. Two features should be noted:

1. The probability of finding an electron close to the nucleus increases when molecules form.

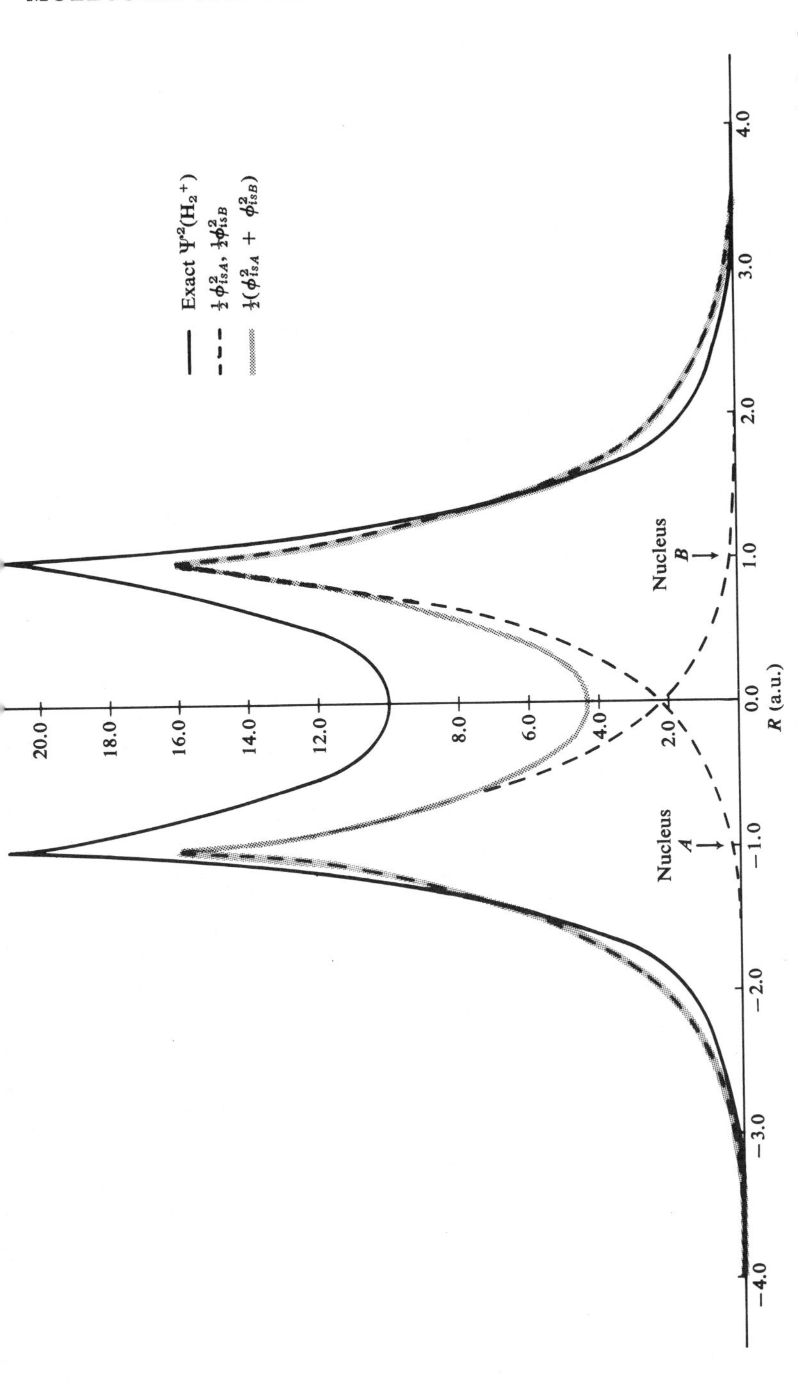

FIGURE 7-3 *Comparison of the electron density* (Ψ^2) *for finding an electron along the* H_2^+ *bond axis. The solid line is calculated from the exact solution for the* H_2^+ *problem. The dashed line corresponds to a suitably normalized electron distribution expected for two unperturbed hydrogen atoms located at the internuclear distance for* H_2^+. *The gray line represents the sum of these two atomic distributions.*

2. The probability of finding an electron in the bonding region between the two nuclei is greatly increased when molecules form.

Both of these increases come at the expense of electron density in the outer wings where the atomic probabilities are higher.

We now explore some of the approximate solutions for H_2^+ and compare these solutions with the exact energy and wave functions. We will then use what we learn to discuss calculations on more complicated diatomic molecules where the exact solution cannot be obtained in analytical form.

7-2 THE USE OF LINEAR VARIATION FUNCTIONS; THE LCAO METHOD

In Chapter 6, we introduced the variation principle that encouraged a strategy of writing trial approximate wave functions containing parameters and then minimizing the energy with respect to those parameters. The variation principle tells us that the energy obtained from such a calculation can never be less than the true energy. A convenient trial function for molecules is a linear variation function of the form

$$\psi_i = \sum_{\mu} C_{i\mu}\chi_{\mu} \tag{7-6}$$

Since molecules are made up of atoms, it is an intuitively plausible first step to assume that an electron distribution in a molecule can be approximated by a sum of atomic electron distributions. This is the physical basis for the molecular orbital or linear combination of atomic orbitals–molecular orbital (LCAO–MO) method of solving quantum mechanical problems. In this method, the ψ_i in Eq. 7-6 is a molecular orbital and the χ_μ are atomic orbitals centered on the appropriate atoms. The coefficients $C_{i\mu}$ are the parameters that will be chosen so as to minimize the energy. The orbitals χ_μ are called *basis* orbitals or basis functions.

We now illustrate a variational calculation for H_2^+. In this calculation, we assume that the molecular orbital Ψ will be some linear combination of hydrogen $1s$ atomic orbitals. Thus, we write

$$\Psi = C_A\chi_A + C_B\chi_B \tag{7-7}$$

where χ_A and χ_B are $1s$ orbitals localized on atoms A and B, respectively. The expectation value of the energy is

$$E \equiv \langle E\rangle = \frac{(\Psi|\mathscr{H}|\Psi)}{(\Psi|\Psi)}$$

$$= \frac{C_A^2(\chi_A|\mathscr{H}|\chi_A) + 2C_AC_B(\chi_A|\mathscr{H}|\chi_B) + C_B^2(\chi_B|\mathscr{H}|\chi_B)}{C_A^2(\chi_A|\chi_A) + 2C_AC_B(\chi_A|\chi_B) + C_B^2(\chi_B|\chi_B)} \tag{7-8}$$

We take advantage of the fact that χ_A and χ_B can be taken to be normalized, and we also introduce a shorthand notation as follows

$$(\chi_\mu | \mathscr{H} | \chi_\nu) \equiv H_{\mu\nu} \tag{7-9a}$$

$$(\chi_\mu | \chi_\nu) \equiv S_{\mu\nu} \tag{7-9b}$$

Using the notation of Eqs. 7-9 and the normalization condition that $(\chi_\mu|\chi_\mu) = 1$, Eq. 7-8 becomes

$$E = \frac{C_A^2 H_{AA} + 2C_A C_B H_{AB} + C_B^2 H_{BB}}{C_A^2 + 2C_A C_B S_{AB} + C_B^2} \tag{7-10}$$

We now wish to find the values of C_A and C_B that make E a minimum. To do this, we must solve the equations

$$\left(\frac{\partial E}{\partial C_A}\right)_{C_B} = 0 \qquad \left(\frac{\partial E}{\partial C_B}\right)_{C_A} = 0 \tag{7-11}$$

With some straightforward algebra, Eq. 7-11 becomes

$$\begin{aligned} C_A(H_{AA} - E) + C_B(H_{AB} - S_{AB}E) &= 0 \\ C_A(H_{AB} - S_{AB}E) + C_B(H_{BB} - E) &= 0 \end{aligned} \tag{7-12}$$

EXERCISE 7-1 Using Eqs. 7-10 and 7-11, derive Eqs. 7-12.

These equations are called *secular equations*. To solve them for C_A and C_B, we make use of the fact that, for a set of simultaneous linear equations without constant terms to have a nontrivial solution (that is, a solution other than $C_A = C_B = 0$), the determinant of the coefficients must vanish. Thus, for Eq. 7-12 to have a solution, it must be true that

$$\begin{vmatrix} H_{AA} - E & H_{AB} - SE \\ H_{BA} - SE & H_{BB} - E \end{vmatrix} = 0 \tag{7-13}$$

where S has been used for S_{AB}. The determinant in Eq. 7-13 is called a *secular determinant*. Since, in the present problem, χ_A and χ_B are identical and Eq. 7-1 is symmetric in both nuclei, $H_{AA} = H_{BB}$, Eq. 7-13 becomes

$$(H_{AA} - E)^2 - (H_{AB} - SE)^2 = 0$$

$$H_{AA} - E = \pm(H_{AB} - SE) \tag{7-14}$$

$$E = \frac{H_{AA} \pm H_{AB}}{1 \pm S}$$

We now write these energies separately as

$$E_+ = \frac{H_{AA} + H_{AB}}{1 + S}$$
$$E_- = \frac{H_{AA} - H_{AB}}{1 - S} \tag{7-15}$$

and substitute them *one at a time* back into Eq. 7-12. This substitution gives the solutions for the coefficients appropriate to each energy. They are

$$C_A = C_B \text{ for } E_+$$
$$C_A = -C_B \text{ for } E_- \tag{7-16}$$

Thus, the wave functions corresponding to the two energies E_+ and E_- are, respectively,

$$\Psi_+ = C_A\chi_A + C_A\chi_B = (2 + 2S)^{-\frac{1}{2}}(\chi_A + \chi_B)$$
$$\Psi_- = C_A\chi_A - C_A\chi_B = (2 - 2S)^{-\frac{1}{2}}(\chi_A - \chi_B) \tag{7-17}$$

where the functions on the right-hand side have been normalized.

EXERCISE 7-2 Show that the proper normalizing factor has been used in Eqs. 7-17.

Before we discuss the significance of Eqs. 7-15 and 7-17, we will generalize what has been done in going from Eq. 7-7 to Eq. 7-17.

1. Use of an LCAO function with n orbitals and the variation principle always leads to a set of n secular equations, each of which contains n coefficients.

2. These secular equations will have a nontrivial solution only if the secular determinant vanishes. This secular determinant will be of dimensions $n \times n$.

3. Solution of the secular determinant leads to a characteristic equation, a polynomial of degree n in E. This characteristic equation has n roots, the lowest of which is an upper bound to the energy of the lowest molecular orbital. The other roots, in increasing order, are upper bounds to the energies of higher energy *orbitals*. (The student should be careful to distinguish between an *orbital* and a *state*. The variation principle will not allow one to make the final statement in step 3 about the excited *states* of a system with more than one electron. For a one-electron system, orbital and state are synonymous, and, in this case only, the higher roots of a linear variation function are upper bounds to the energies of the excited states.)

4. These energies are substituted back into the secular equations one at a time to find the coefficients in the LCAO function that determine the wave function that gives that particular energy. The student should note these general statements because this procedure will be used again in Chapter 8.

We now return to a discussion of the solutions for H_2^+. The integrals H_{AA}, H_{AB}, and S_{AB} can all be evaluated by transforming to elliptical coordinates (see Exercise 7-3). The quantities H_{AA} and H_{AB} both turn out to be negative. The quantity H_{AA} is

$$\int \chi_A\left(-\frac{1}{2}\nabla^2 - \frac{1}{r_A} - \frac{1}{r_B}\right)\chi_A \, d\tau = E_{1s}(\mathrm{H}) - \int \chi_A \frac{1}{r_B} \chi_A \, d\tau \tag{7-18}$$

In Eq. 7-18, it can be seen that the integral H_{AA}, called a Coulomb integral or atomic integral, represents the energy of an electron in a $1s$ orbital of hydrogen plus the attractive energy of nucleus B for this electron. The $1/R_{AB}$ term in the Hamiltonian will be added later. As nuclei A and B are brought closer together, the second term tends to make the energy of H_2^+ more negative and, consequently, increase the stability of the molecule. This energy stabilization is more than counteracted by the nuclear repulsion energy, however, and a plot of $(H_{AA} + 1/R_{AB})$ as a function of R_{AB} gives a curve that has the value $-\frac{1}{2}$ at $R_{AB} = \infty$ and increases monotonically to the value ∞ at $R_{AB} = 0$. The term $(H_{AA} + 1/R_{AB})$ thus gives no contribution to the stability of H_2^+.

The integral H_{AB} is called a resonance integral and is given in Eq. 7-19.

$$\int \chi_A\left(-\frac{1}{2}\nabla^2 - \frac{1}{r_A} - \frac{1}{r_B}\right)\chi_B \, d\tau = E_{1s}(\mathrm{H})S_{AB} - \int \chi_A \frac{1}{r_A} \chi_B \, d\tau \tag{7-19}$$

This integral takes into account the fact that the electron is not restricted to a $1s$ atomic orbital on either atom A or B, but that it can exchange places between the two orbitals. The resonance integral goes to zero at large R_{AB} because both S_{AB} and the second integral on the right in Eq. 7-19 go to zero. H_{AB} becomes more negative as R_{AB} decreases, and the quantity $[H_{AB} - S_{AB}(1/R_{AB})]$ goes through a minimum. It is, therefore, the resonance integral that is responsible for the binding energy in H_2^+ in this model. The significance of this integral can be rationalized in terms of the classical theory of a pair of resonating pendulums [2].

Since both H_{AA} and H_{AB} are negative, E_+ will be the lowest energy, and E_- will be the highest. A plot of the energies E_+ and E_- as a function of internuclear distance is shown in Figure 7-4. The dotted curve in Figure 7-4 represents the exact energy. The student may obtain the plot for E_+ by working Exercise 7-3. The LCAO–MO results for E_- are compared with the exact calculation in Exercise 7-4.

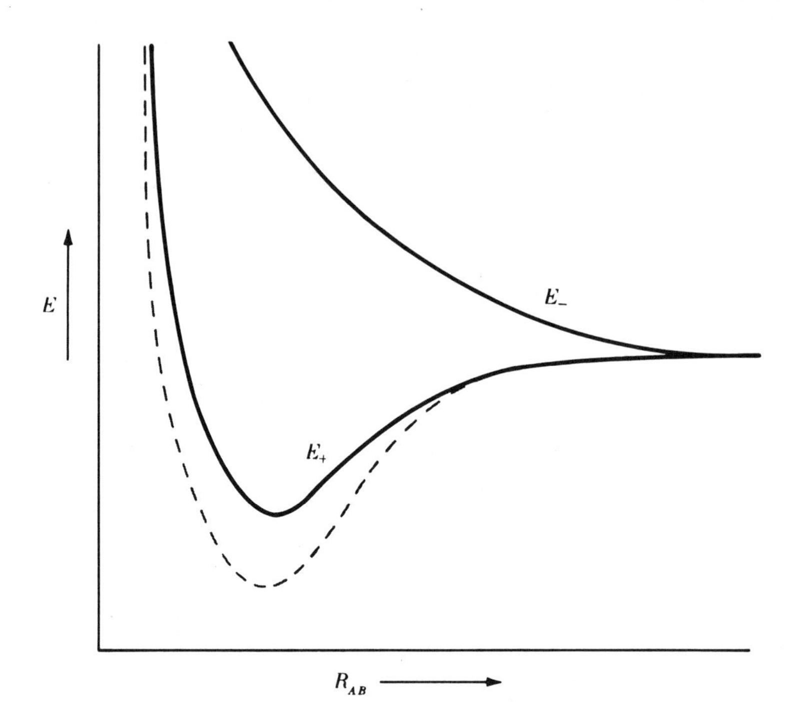

FIGURE 7-4 *Energies of the states* Ψ_+ *and* Ψ_- *for* H_2^+*, as a function of internuclear distance* R_{AB}*. The dotted curve represents the true energy. The energy of the state* Ψ_- *is very close to the energy of this state calculated from the exact solutions of* H_2^+.

EXERCISE 7-3 Evaluate the integrals S_{AB}, H_{AA}, and H_{AB} as a function of the internuclear distance R (use a.u.). Transform to elliptical coordinates and use $1 \leqq \mu \leqq \infty$, $-1 \leqq \nu \leqq +1$ as the limits of integration. On a good piece of graph paper, make a plot of $(H_{AA} + 1/R_{AB})$, $[H_{AB} + S_{AB}(1/R_{AB})]$ and E_+ versus R_{AB}. What is the equilibrium internuclear distance? What is the binding energy? Note which of the two integrals is responsible for the binding.

EXERCISE 7-4 The electronic energy in Hartrees of the $\sigma_u^* 1s$ state for H_2^+ is given as a function of internuclear distance in a.u. below. These energies do not yet contain the nuclear repulsion contribution $1/R_{AB}$. Plot the energy of this state, including the nuclear repulsion term, as a function of internuclear distance and compare this energy with the prediction of the LCAO–MO approximation (data from [1]).

R (a.u.)	$-E$ (H)	R (a.u.)	$-E$ (H)
1.0	0.56481	3.6	0.70041
1.4	0.61208	4.0	0.69555
1.8	0.65223	5.0	0.67729
2.0	0.66754	7.0	0.63913
2.4	0.68858	8.0	0.61066
2.8	0.69911	∞	0.50000
3.2	0.70221		

From a plot of E_+ versus R_{AB}, it can be shown that the LCAO approximation gives a binding energy of 1.76 eV and an equilibrium internuclear distance of 1.32 Å. Comparison of these values with the experimental values of the corresponding quantities (2.791 eV and 1.060 Å) shows that the accuracy of the LCAO function leaves much to be desired. To explore some of the reasons for the poor agreement with experiment, we should compare the electron density computed from the LCAO wave function Ψ_+ with the electron distribution calculated from the exact calculation. The MO electron density is given by

$$\Psi_+^2 = \frac{1}{2 + 2S}(\chi_A^2 + 2\chi_A\chi_B + \chi_B^2) \tag{7-20}$$

and the value of Ψ_+^2 along the molecular axis is compared with the exact electron density in Figure 7-5. Comparison of this figure with Figure 7-3 shows that in Ψ_+ there is a little more electron density halfway between the nuclei than would be obtained from the sum of two noninteracting atomic distributions but not nearly as much as in the exact solution. In addition, the sharp peaking of electron density at the nuclei is absent in the molecular orbital treatment, and for all points 0.75 a.u. beyond the nuclei Ψ_+^2 is greater than the exact electron density.

Is there anything that can be done to improve the LCAO function? One thing that would increase the electron density at the nuclei and pull in the electron density from the wings would be to increase the effective nuclear charge in the 1*s* basis functions used to construct Ψ_+. We can find the optimum effective nuclear charge by introducing an arbitrary parameter η called a *scale factor* into the exponential of the 1*s* wave function. Thus, we write

$$\chi_A = \frac{\eta^{\frac{3}{2}}}{\sqrt{\pi}} e^{-\eta r_A}; \qquad \chi_B = \frac{\eta^{\frac{3}{2}}}{\sqrt{\pi}} e^{\eta r_B} \tag{7-21}$$

and recalculate H_{AA}, H_{AB}, and S. Minimizing E_+ simultaneously with respect to η and R_{AB} gives a binding energy of 2.35 eV at an internuclear distance of 2.02 a.u. (1.07 Å) in much better agreement with experiment than the simple LCAO treatment. The optimum value of η is 1.24, and this whole process of introducing an effective nuclear charge into the basis orbitals of a molecular orbital is called *scaling*.

EXERCISE 7-5 By adding and subtracting η/r_A terms to the Hamiltonian for H_2^+, show that

$$H_{AA} = -\frac{1}{2}\eta^2 + (\eta - 1)\int \chi_A \frac{1}{r_A} \chi_A \, d\tau - \int \chi_A \frac{1}{r_B} \chi_B \, d\tau$$

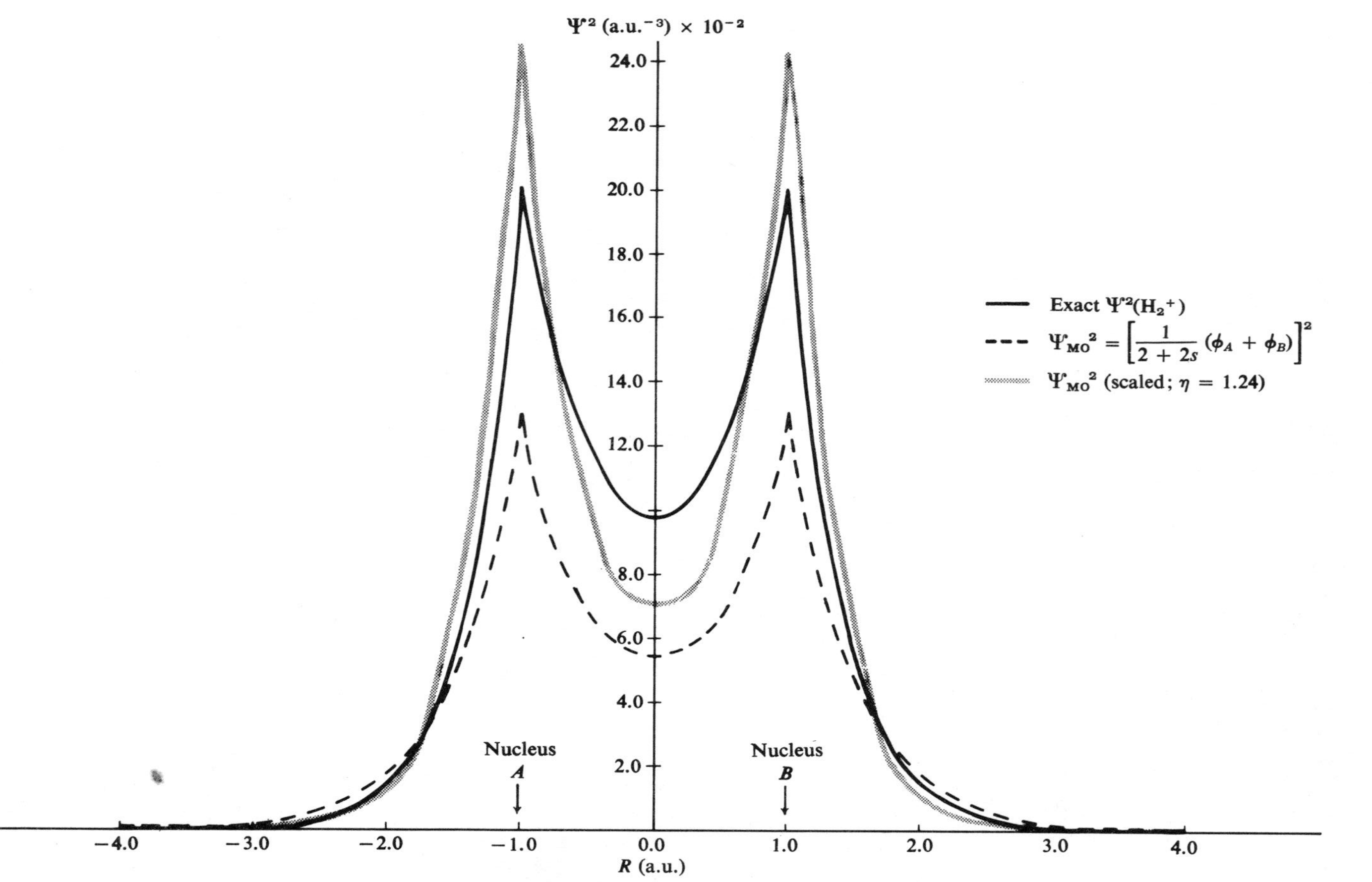

FIGURE 7-5 *Comparison of the probability density* (Ψ^2) *for finding an electron along the* H_2^+ *bond axis. The solid line is from*

Evaluate the integrals and show that

$$H_{AA} = \frac{1}{2}\eta^2 - \eta - \frac{1}{R} + e^{-2\eta R}\left(\eta + \frac{1}{R}\right)$$

EXERCISE 7-6 Evaluate H_{AB} and S_{AB} using the functions in Eq. 7-21 as basis orbitals. Write the expression for E_+ as a function of η and evaluate the optimum value of η at $R = 2.02$ a.u.

A plot of the scaled LCAO function corresponding to Ψ_+ is also seen in Figure 7-5. It can be seen that introducing the scale factor makes the MO wave function peak at the nuclei. In fact, it overcompensates a little. Scaling also improves the agreement in the region between the nuclei but still does not bring the electron density up to the exact value. The scaled MO wave function represents the best that can be done if the basis set consists of only *one* atomic orbital on each hydrogen atom.

Many students wonder about the fact that the effective nuclear charge is greater than one even though the charge on a hydrogen nucleus is exactly one. A partial reply is that in using the LCAO wave function we are trying to *approximate* the solution to H_2^+ with a function that we know is incorrect from the beginning. That is, the exact wave function for H_2^+ is *not* a simple superposition of two hydrogen atom $1s$ orbitals. Thus, parameters that appear in this incorrect approximation need not be identical to exact atomic properties. The effective nuclear charge value being between one and two can be rationalized with the idea of a correlation diagram [9].

It is important for us to summarize at this point some of the advantages of the LCAO method.

1. The theory leads to correct predictions that stable molecules will be formed and gives reasonable values of binding energies and internuclear separations.

2. The theory retains the atoms-in-molecules idea that is intuitively pleasing to the chemist.

3. The theory is computationally manageable. It can be extended to more complex molecules and can be improved to give better agreement with experiment.

4. The theory gives a satisfactory way to treat molecular spectra.

The student should note various ways in which these advantages are manifest as he or she continues in this chapter.

7-3 *THE HYDROGEN MOLECULE*

The hydrogen molecule occupies the same place in the theory of molecular electronic structure that the helium atom occupies in the theory of atomic structure. The hydrogen molecule is extremely important in that all of the features of the two-electron bond are present. An understanding of this molecule gives insight into the nature of the chemical bond in more complex molecules.

The notation used for a discussion of the hydrogen molecule is shown in Figure 7-6. If the Born–Oppenheimer approximation is again made, the Hamiltonian in atomic units becomes

$$\mathscr{H} = -\frac{1}{2}(\nabla_1^2 + \nabla_2^2) - \frac{1}{r_{A1}} - \frac{1}{r_{A2}} - \frac{1}{r_{B1}} - \frac{1}{r_{B2}} + \frac{1}{r_{12}} + \frac{1}{R_{AB}} \tag{7-22}$$

The significance of each of these terms should be familiar to the reader. The eigenvalue equation

$$\mathscr{H}\Psi = E\Psi \tag{7-23}$$

is, like that of helium, impossible to solve in analytical form. We are thus forced to use an approximation method. At the lowest level of approximation, there are two methods to describe the ground state of hydrogen. These are the MO method and the Heitler–London (HL) method. We will discuss the MO method first because it is already familiar from the discussion of H_2^+. (Historically, the HL method was developed first.) After these two simple methods are introduced, various ways to improve the hydrogen mole-

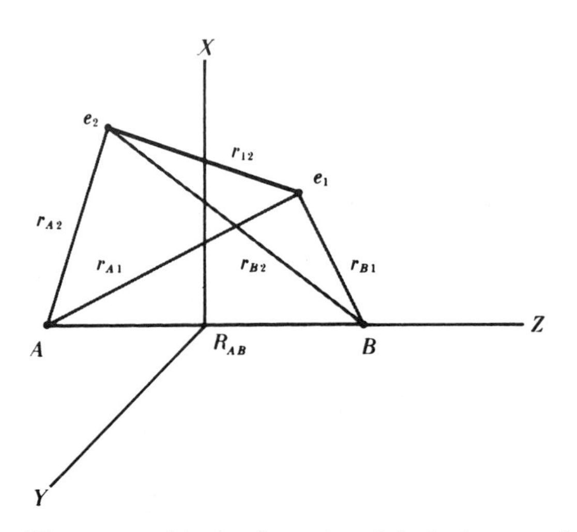

FIGURE 7-6 *Distances used in the discussion of the hydrogen molecule.*

cule wave functions will be discussed. All of the results calculated below should be compared with the following experimental values.

total energy $= -1.174$ H
binding energy $=$ total energy $-$ energy of two isolated hydrogen atoms
$= -1.174 - (-1.000) = -0.174$ H
$= -4.75$ eV
equilibrium internuclear distance $= 0.740$ Å

In the MO method for the hydrogen molecule, we begin investigating solutions to Eq. 7-23 using the approximate Hamiltonian derived from Eq. 7-22 by leaving out the $1/r_{12}$ term. This Hamiltonian can be written as

$$\mathscr{H}_0 = -\frac{1}{2}\nabla_1^2 - \frac{1}{r_{A1}} - \frac{1}{r_{B1}} - \frac{1}{2}\nabla_2^2 - \frac{1}{r_{A2}} - \frac{1}{r_{B2}} + \frac{1}{R_{AB}} \tag{7-24}$$

where the order has been changed to show that the first three terms are a function of only the coordinates and momenta of electron 1, and the second three terms are a function of only the coordinates and momenta of electron 2. This approximate Hamiltonian can be written as

$$\mathscr{H}_0 = \hat{h}_0(1) + \hat{h}_0(2) + \frac{1}{R_{AB}} \tag{7-25}$$

where $\hat{h}_0$ is the electronic Hamiltonian for the H_2^+ ion. Since the $1/R_{AB}$ term contributes only a constant term to the energy within the context of the Born–Oppenheimer approximation, we can make use of the theorem of separation of variables to show that an eigenvalue equation using the Hamiltonian of Eq. 7-25 is satisfied by a product function of the type

$$\Psi = A\omega_i(1)\omega_i(2) \tag{7-26}$$

where ω_i are the solutions to the H_2^+ problem and A is a suitable normalizing constant.

We must now introduce electron spin into the wave function of Eq. 7-26 and adjust the result so that the Pauli principle is obeyed. By applying the same kind of arguments that were used for the case of the helium atom, the student can show that the lowest energy wave function of the type of Eq. 7-26 that satisfies the Pauli principle is

$$\Psi_{\text{MO}} = \omega_1(1)\omega_1(2)\tfrac{1}{2}\sqrt{2}[\alpha(1)\beta(2) - \beta(1)\alpha(2)] \tag{7-27}$$

where the MO, ω_1, can be approximated by the LCAO function

$$\omega_1 = (2 + 2S)^{-\frac{1}{2}}(1s_A + 1s_B) \tag{7-28}$$

EXERCISE 7-7 Show that Ψ_{MO} is an eigenfunction of the operator $\hat{P}(1, 2)$ with eigenvalue -1.

The energy is calculated in a manner similar to that for the helium atom. That is, the expectation value of E is calculated as a function of R_{AB} using Ψ_{MO} and the exact Hamiltonian. Thus,

$$E = (\Psi_{MO}|\,\mathscr{H}\,|\Psi_{MO}) = 2E'(R_{AB}) + \left(\Psi_{MO}\left|\frac{1}{r_{12}}\right|\Psi_{MO}\right) - \frac{1}{R_{AB}} \tag{7-29}$$

where $E'(R_{AB})$ is the expression for the energy of H_2^+ as a function of the internuclear distance R_{AB}. To find the binding energy and the equilibrium internuclear distance, one minimizes E with respect to R_{AB}. If this is done, one obtains

$$E_{MO} = -1.0985 \text{ H}$$

$$\text{binding energy} = E - 2E_H = -1.0985 + 1.000$$

$$= -0.0985 \text{ H} = -2.681 \text{ eV}$$

$$r_e = 0.850 \text{ Å}$$

The agreement with the experimental values is, as expected, not very good.

It is instructive to write the MO function for hydrogen in more detail. Expanded, the space part of Eq. 7-27 becomes

$$\Psi_{MO}(\text{space part}) = \frac{1}{2+2S}[1s_A(1)1s_A(2) + 1s_A(1)1s_B(2) + 1s_B(1)1s_A(2) + 1s_B(1)1s_B(2)] \tag{7-30}$$

It can be seen that the first and fourth terms represent a probability of finding both electrons close to one nucleus. This is equivalent to writing ionic structures for hydrogen of the form $H_A^+H_B^-$ and $H_A^-H_B^+$. One of the great shortcomings of the MO function is that these "ionic terms" enter into the wave function *with the same weight as the covalent terms.* This is contrary to chemical experience because chemists have used hydrogen as a prototype covalent compound. This feature of the MO wave function is also unsatisfactory because it predicts that, upon dissociation, one-half of the hydrogen molecules should dissociate into the ions H^- and H^+. In fact, a hydrogen molecule always dissociates into two hydrogen atoms.

With these considerations in mind, we can write a second wave function for the ground state of the hydrogen molecule that includes only the covalent part of Eq. 7-30. This is the HL wave function for hydrogen. This function is

$$\Psi_{\text{HL}} = N[\phi_A(1)\phi_B(2) + \phi_B(1)\phi_A(2)]\tfrac{1}{2}\sqrt{2}[\alpha(1)\beta(2) - \beta(1)\alpha(2)] \qquad (7\text{-}31)$$

where N is a normalizing constant and where ϕ_A and ϕ_B have replaced the symbols $1s_A$ and $1s_B$. The student should note that the space part of this wave function is symmetric. The spin part must be antisymmetric, therefore, to make the total wave function obey the Pauli principle. To save writing, the labels on the electrons are usually omitted with the understanding that the function for electron 1 is always written first, the function for electron 2 second, and so on. With this convention, Eq. 7-31 becomes

$$\Psi_{\text{HL}} = N[\phi_A\phi_B + \phi_B\phi_A]\tfrac{1}{2}\sqrt{2}[\alpha\beta - \beta\alpha] \qquad (7\text{-}32)$$

Using the exact Hamiltonian to calculate E, one obtains

$$E = -1.1160 \text{ H}$$

$$\text{binding energy} = -0.1160 \text{ H} = -3.140 \text{ eV}$$

$$r_e = 1.67\, a_0 = 0.869 \text{ Å}$$

Note that this calculation gives 70% of the binding energy, which is fairly good.

It can be seen from the above that the HL function gives a better value for the energy than does the MO function. It also predicts the proper behavior at large internuclear distances. It might be tempting to say that the HL method [or valence bond (VB) method as its later modifications were called] is better than the MO method. This conclusion is not justified, however, because both are gross approximations to the actual state of affairs in the molecule. The conclusion that *can* be drawn is that, as an approximation, it is better to leave the ionic terms out of the wave function for hydrogen than to include them and give them equal weight with the covalent terms.

The student should be acquainted with some of the nomenclature arising from VB calculations. Using the wave function in Eq. 7-31, one obtains for the energy

$$E = \frac{\langle(\phi_A\phi_B + \phi_B\phi_A)|\mathscr{H}|(\phi_A\phi_B + \phi_B\phi_A)\rangle}{\langle(\phi_A\phi_B + \phi_B\phi_A)|(\phi_A\phi_B + \phi_B\phi_A)\rangle}$$

$$= \frac{J + K}{1 + S^2} \tag{7-33}$$

where

$$J = \int \phi_A(1)\phi_B(2)\mathscr{H}\phi_A(1)\phi_B(2)\,d\tau_1\,d\tau_2$$

$$K = \int \phi_A(1)\phi_B(2)\mathscr{H}\phi_B(1)\phi_A(2)\,d\tau_1\,d\tau_2$$

and S is the overlap integral. The integral J is called a Coulomb integral (not to be confused with H_{AA}) and gives the energy that a hydrogen molecule would have if its wave function were

$$\Psi = \phi_A(1)\phi_B(2)$$

that is, if the electrons were not allowed to change places. This integral leads to a shallow minimum of about 0.4 eV at an internuclear separation of 1.0 Å. The integral K is called the exchange integral. This nomenclature arises because the electrons have "changed places" in the two halves of the integrand of K. This integral leads to most of the calculated stability for the hydrogen molecule.

EXERCISE 7-8 Write the integrals J and K using the Hamiltonian in Eq. 7-22. By grouping terms and carrying out integrations that give unity or S, show that

$$J = 2H_{AA} + \left\langle \phi_A\phi_B \left| \frac{1}{r_{12}} \right| \phi_A\phi_B \right\rangle + \frac{1}{R_{AB}}$$

$$K = 2H_{AB}S + \left\langle \phi_A\phi_B \left| \frac{1}{r_{12}} \right| \phi_B\phi_A \right\rangle + \frac{1}{R_{AB}} S^2$$

where H_{AA} and H_{AB} were defined in the H_2^+ discussion of Section 7-2.

The generalization of the HL method to more complex molecules is known as the Heitler–London–Slater–Pauling (HLSP) or VB method. In the VB method, the total wave function for a molecule is made up of a product of bond eigenfunctions. If a represents a singly occupied atomic orbital on atom A, and if b represents a singly occupied orbital on atom B, then a bond eigenfunction can be written

$$\Psi = N(ab + ba)(\alpha\beta - \beta\alpha) \tag{7-34}$$

One of these bond eigenfunctions is written for each pair of singly occupied orbitals that can overlap; then a suitably antisymmetrical product is formed to describe the whole molecule.

7-4 *IMPROVEMENTS ON THE WAVE FUNCTION FOR* H_2

In this section, we briefly discuss improvements that can be made in the approximate treatments of H_2. By improvements we mean things that will give a better value of the binding energy. The student should realize that H_2 is being used as a prototype in these discussions and that these same improvements can be made in the wave function for any molecule.

We mentioned previously in comparing the simple MO and VB calculations for the hydrogen molecule that it is better to leave out the ionic terms than to include them and give them equal weight with the covalent terms. A better wave function might arise if the ionic terms were introduced into the wave function multiplied by a variational parameter λ. Such a wave function for H_2 is called the Weinbaum function. Omitting the spin part, this function is

$$\Psi = N[(\phi_A\phi_B + \phi_B\phi_A) + \lambda(\phi_A\phi_A + \phi_B\phi_B)]$$

where λ is the arbitrary variation parameter. Minimizing the energy with respect to λ, one obtains

$$E = -1.1187 \text{ H}$$

$$\text{binding energy} = -0.1187 \text{ H} = -3.229 \text{ eV}$$

$$\lambda = 0.25$$

If one takes the VB wave function as a model for a completely covalent bond, then addition of an optimum amount of ionic terms to the wave function is a quantum mechanical description of the idea of ionic-covalent resonance. Since the Weinbaum wave function gives a better energy than the VB wave function, it is tempting to argue that ionic-covalent resonance is an important feature of chemical bonds, even in homonuclear diatomic molecules. Such a conclusion would be inappropriate, however, because the Weinbaum function introduces another variational parameter and is thus bound to give a better energy because of the variational principle. The introduction of ionic terms into covalent wave functions has many important applications in theoretical chemistry, however. Mulliken's theory of donor-acceptor complexes is one of the obvious examples of such an application [10].

A different approach to the problem of the ground state of a hydrogen molecule was developed by N. Rosen. He argued that a $1s$ atomic orbital is not a good basis orbital to use because the electron distribution around a proton in a hydrogen molecule will not be the same as the distribution in an atom. In a molecule, the electron distribution would be expected to be polarized toward the other nucleus. The way to include this polarization effect is to include a little $2p$ character in the basis orbitals. Rosen therefore used a wave function of the form

$$\Psi_R = N[(1s_A + \kappa 2p_{zA})(1)(1s_B + \kappa 2p_{zB})(2) + \text{symmetrizing terms}]$$

where κ is a variational parameter. Minimizing E with respect to κ, one obtains

$$E = -1.125 \text{ H}$$

$$\text{binding energy} = -0.125 \text{ H} = -3.40 \text{ eV}$$

It can be seen that the Rosen wave function gives the best energy of any function so far discussed.

A third method used to improve molecular wave functions is that of introducing an effective nuclear charge or scale factor into the atomic orbital basis set. We have already seen how this improved both the calculated energy and electron density for H_2^+. This procedure, called scaling the wave function, usually dramatically improves the calculated energy. This improvement comes about because an additional variational parameter has been introduced into the wave function. Scaling has special significance, however, because it insures that the wave function satisfies the virial theorem. The virial theorem relates the expectation value of the kinetic energy $\langle T\rangle$ to the expectation value of the potential energy $\langle V\rangle$. For atoms and molecules where the potential energy terms all depend on distances to the minus one power, the virial theorem states that

$$\langle T\rangle = -\tfrac{1}{2}\langle V\rangle \tag{7-35}$$

Since the total energy of the system equals the sum of the kinetic and potential energies, one can derive the relation

$$\begin{aligned}\langle E\rangle &= \langle T\rangle + \langle V\rangle \\ &= -\tfrac{1}{2}\langle V\rangle + \langle V\rangle = \tfrac{1}{2}\langle V\rangle \end{aligned} \tag{7-36}$$

The physical interpretation of Eq. 7-36 is that the lowering of the total energy of the system upon molecule formation must be due to a *reduction* in the potential energy of the system. Many of the unscaled wave functions for molecules lower the total energy by lowering the kinetic energy of the system, and some theoretical chemists find this objectionable because it violates the virial theorem. For more information on the virial theorem, the student is referred to [3].

For the hydrogen molecule, if a scale factor is introduced into the exponent of the $1s$ atomic orbital, and if the energy is then minimized with respect to this scale factor, the energies of all of the wave functions discussed above are substantially improved. This improvement can be seen in Table 7-1. It should be noted that the scaled MO function gives a better energy than the unscaled Rosen function. This shows that, at least for the hydrogen molecule, wave functions are improved more by scaling than by adding ionic terms or orbital polarization. It is not known if this result holds for more complex molecules, but it is expected to. The scaled Rosen function is quite good in that it gives 85% of the experimental dissociation energy. This energy is still in error by 16 kcal mole^{-1}, a substantial amount when chemical effects are considered, however.

There have been many other calculations on molecular hydrogen. The best is the James–Coolidge wave function as modified by W. Kolos and C. Roothaan [4]. This function is a 50-term variation function, and the interelectron distance is explicitly included in each term. It gives a binding energy of -1.1744 H, in agreement with experiment. Unfortunately (or fortunately, depending on one's viewpoint), all concepts of atomic or molecular orbitals, exchange, Coulomb, and resonance integrals, ionic-covalent resonance, and polarization have disappeared in this function.

An interesting confrontation arose between theoreticians and experimentalists when an improved Kolos–Roothaan–James–Coolidge wave function including relativistic and Born–Oppenheimer corrections gave a calculated

TABLE 7-1 THE EFFECTS OF SCALING ON THE ENERGY OF APPROXIMATE WAVE FUNCTIONS FOR H_2

	$-E$(*unscaled*)		$-E$(*scaled*)		
Wave function	a.u.	eV	a.u.	eV	$-E$(*expt.*)
MO	0.0985	2.681	0.1285	3.495	
HL	0.1160	3.14	0.1391	3.784	
Weinbaum	0.1187	3.229	0.14796	4.024	0.174 a.u. or 4.75 eV
Rosen	0.125	3.400	0.1485	4.039	

energy that was *lower* than the experimental energy as determined spectroscopically. This meant that either the experiments were interpreted incorrectly or there was a fundamental flaw in quantum theory. (The variation principle says that one cannot get a lower energy than the true value.) In this case, the experimentalists found an error in their analysis of the data and the theoretical results were confirmed. This outcome is extremely infrequent for in most cases theory follows experiment, but it was strong confirmation that quantum theory is fundamentally correct and can give accurate information about molecules if the technical computing problems can be solved.

7-5 EXCITED STATES OF H_2^+ AND THE SIMPLE MO TREATMENT OF HOMONUCLEAR DIATOMIC MOLECULES

As was mentioned in the general formulation of the LCAO method, the function Ψ_- can be regarded as an approximation to one of the excited states of H_2^+. It is a dissociative or antibonding state, however. That is, an H_2^+ molecule in the state Ψ_- is *less stable* than a hydrogen atom and a proton at all internuclear distances; consequently, if an H_2^+ molecule could somehow be prepared in state Ψ_-, it would immediately dissociate to a hydrogen atom and a proton.

There are other excited states of H_2^+, however. The molecular orbitals that are approximations to these states can be formed from linear combinations of higher hydrogen-like atomic orbitals. Since these orbitals will be used to discuss the structure of more complex diatomic molecules, we must introduce a nomenclature for them. This nomenclature depends on the folllowing characteristics.

1. The atomic orbitals from which they are formed.
2. The eigenvalue with respect to $\hat{L}_z$, the angular momentum quantum number about the internuclear axis (see Section 7-1).
3. Their property on inversion through the center of the molecule. If the orbital function does not change sign upon inversion, it is called g. If it does change, it is called u. This classification only applies to homonuclear diatomic molecules such as H_2, O_2, and N_2, in which both nuclei have the same atomic number.
4. Their stability with respect to the isolated atoms. If the orbital energy is lower than the sum of the energies of the corresponding orbitals on the isolated atoms, it is called a bonding orbital. If the molecular orbital energy is higher than the sum of the energies of the corresponding atomic orbitals, it is called an antibonding orbital and is designated by a superscript asterisk.

Some examples of this nomenclature are given in Table 7-2.

The student should recall from the discussion of H_2^+ that the combination of a $1s$ orbital on atom A with a $1s$ orbital on atom B gave two molecular orbitals, the $\sigma_g 1s$ and $\sigma_u^* 1s$ of Table 7-2. The general shapes, the nodal

TABLE 7-2 SYMBOLS USED FOR SOME OF THE EXCITED MOLECULAR ORBITALS IN H_2^+ [a]

LCAO *function*	MO *nomenclature*
$1s_A + 1s_B$	$\sigma_g 1s$
$1s_A - 1s_B$	$\sigma_u^* 1s$
$2s_A + 2s_B$	$\sigma_g 2s$
$2s_A - 2s_B$	$\sigma_u^* 2s$
$2p_{0A} - 2p_{0B}$	$\sigma_g 2p$
$2p_{1A} + 2p_{1B}$	$\pi_u 2p$

[a] A second $\pi_u 2p$ orbital can be formed from $2p_{-1}$ orbitals on atoms A and B.

characteristics, and the origin of the g and u designations for these two molecular orbitals are shown in Figure 7-7. In the bonding molecular orbital, when $x \to -x$, $y \to -y$, and $z \to -z$, the wave function still has the same sign. In the antibonding orbital, the wave function has the opposite sign at $p(x, y, z)$ as it does at $p(-x, -y, -z)$.

A general feature of the LCAO method is that a combination of two atomic orbitals on different centers will always give two molecular orbitals, one bonding and one antibonding. Thus, a pair of MO's exactly similar to those shown in Figure 7-7 can be formed from a linear combination of $2s$ atomic orbitals. These molecular orbitals will have a higher energy than the corresponding ones formed from $1s$ atomic orbitals because the principal quantum number of the basic orbitals has been increased and, therefore, the basis orbitals are at a higher energy.

The situation is somewhat more complicated when linear combinations of $2p$ orbitals are taken. Since there are six $2p$ orbitals on the two atomic centers, we would expect to be able to derive six molecular orbitals from them. Two different ways of combining $2p$ orbitals are shown in Figures 7-8 and 7-9. In Figure 7-8, we can see that the two $2p_z$ orbitals can combine to form a pair of σ orbitals, one bonding and one antibonding. The orbitals are of σ type because there is no component of angular momentum about the bond axis. In Figure 7-9, the combination of two $2p_x$ orbitals to form a $\pi_u 2p$ orbital is shown. Another pair of orbitals exactly equivalent to these can be formed from a linear combination of $2p_y$ orbitals. The $\pi_u 2p$ and $\pi_g^* 2p$ molecular orbitals will therefore each be doubly degenerate. These four π orbitals plus the $\sigma_g 2p$ and $\sigma_u^* 2p$ give the required total of six molecular orbitals. Since the $2p$ atomic orbitals have a higher energy than the $2s$ atomic orbitals in atoms with more than one electron, the energies of the molecular orbitals constructed

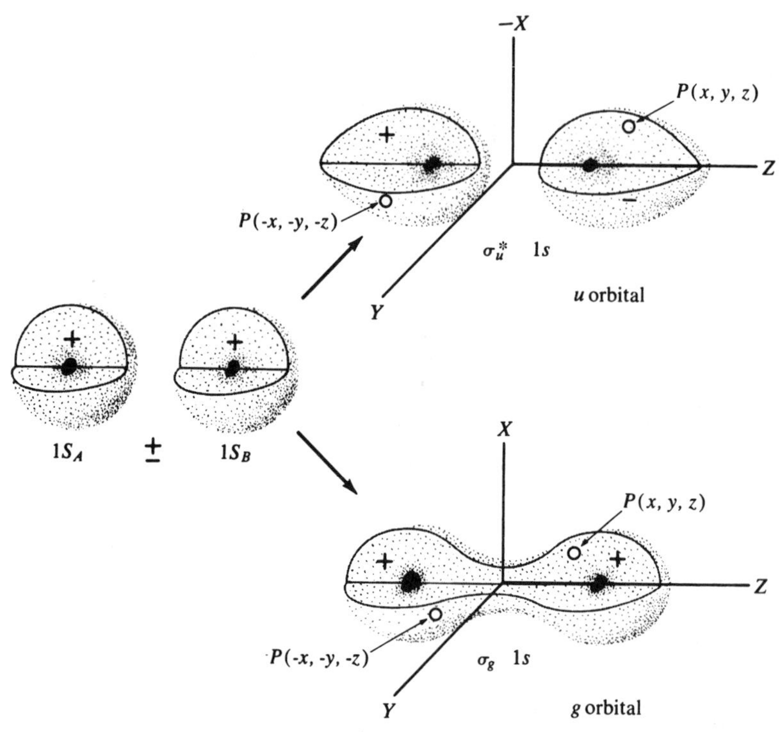

FIGURE 7-7 *Drawings of the $\sigma_g 1s$ and $\sigma_u^* 1s$ orbitals showing their g and u character. Note that the node in the σ_u^* orbital keeps the electrons out of the region between the two nuclei.*

from $2p$ atomic orbitals will be higher than that of the $\sigma_g 2s$ and $\sigma_u^* 2s$ molecular orbitals.

We can summarize these considerations by constructing an energy-level diagram for homonuclear diatomics using the molecular orbitals for H_2^+ as our starting point. We can then, by placing electrons into these orbitals in accordance with the Pauli principle, write the zero order MO wave function of any homonuclear diatomic molecule composed of second-row atoms. The electron configurations generated in this way can be used to discuss the stability and structure of these molecules. Such an energy level diagram is shown in Figure 7-10.

The order of the energies of the MO's formed from the $2p$ atomic orbitals is a function of the internuclear distance and of the energy separation between the $\sigma_g 2p$ and the $\sigma_g 2s$ molecular orbitals. This latter quantity in turn depends on the energy spacing between the atomic $2s$ and $2p$ orbitals. When this spacing is large, the $\sigma_g 2p$ and the $\sigma_g 2s$ do not interact, and the order of the levels is that shown in Figure 7-10. When the spacing between the $\sigma_g 2p$ and

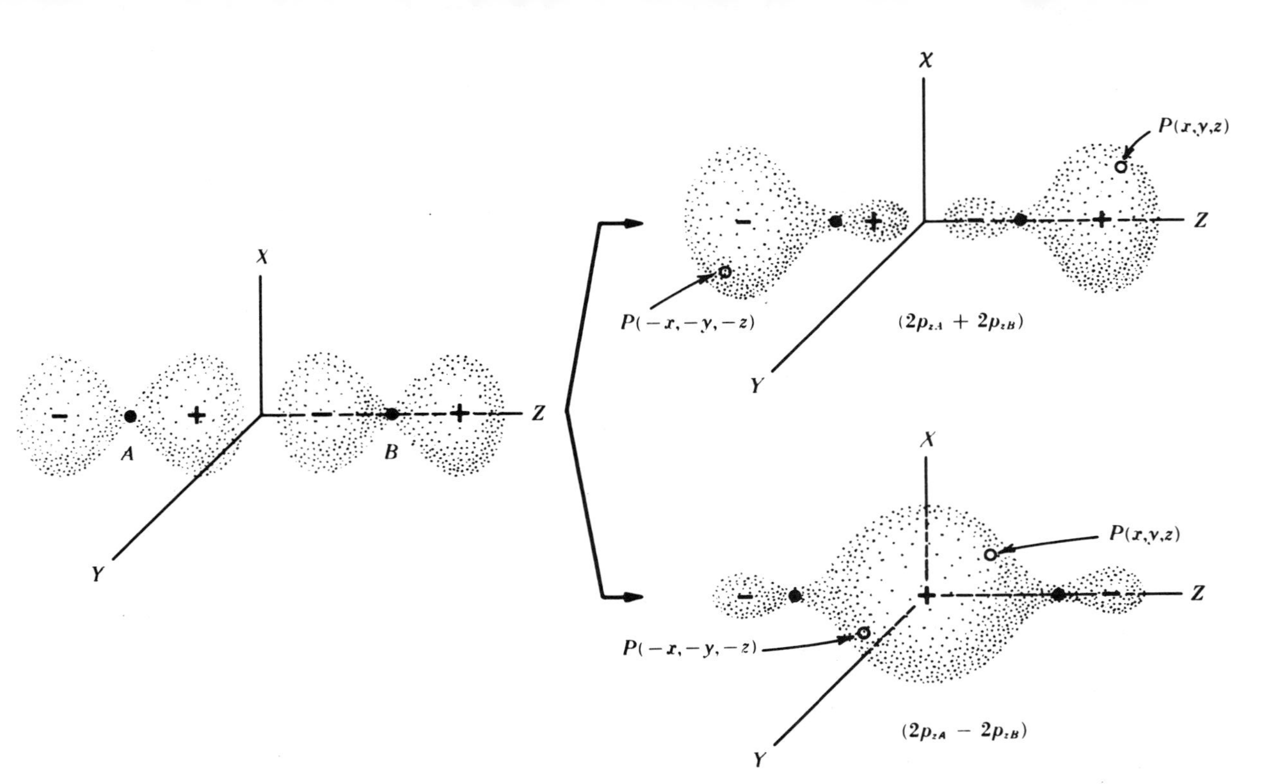

FIGURE 7-8 *The formation of the $\sigma_g 2p$ and $\sigma_u^* 2p$ orbitals from linear combinations of $2p_z$ atomic orbitals. The student should understand the g and u nomenclature for these orbitals.*

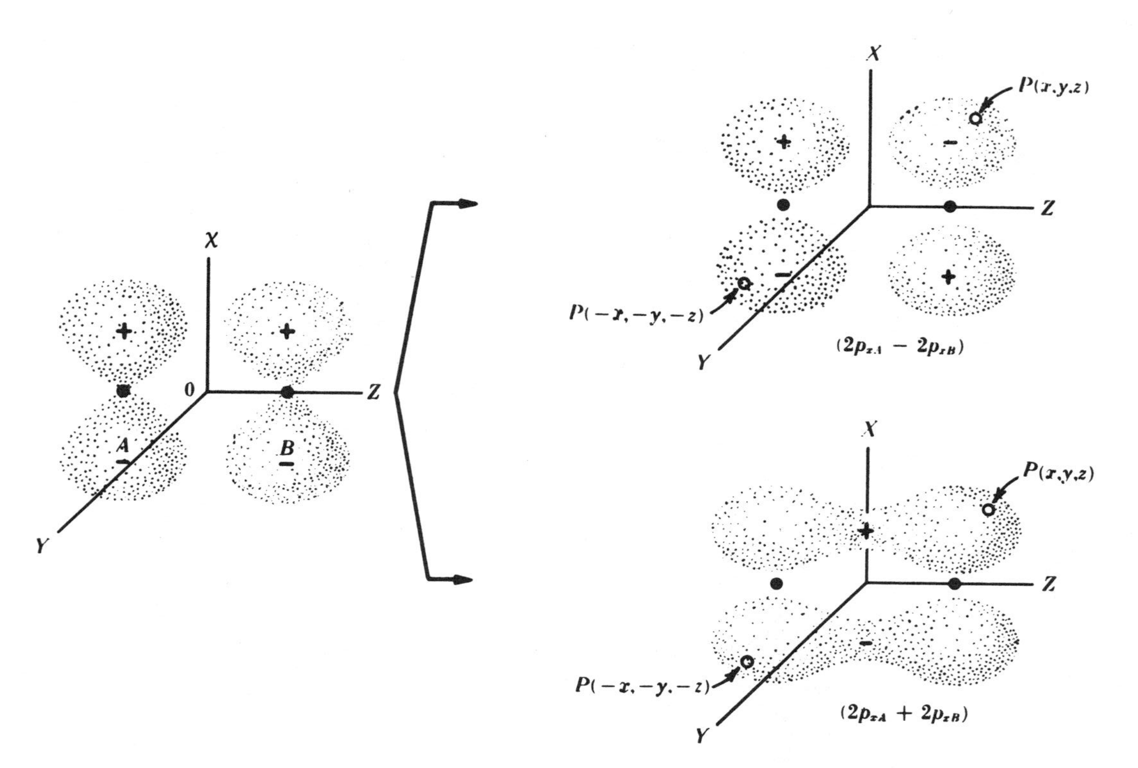

FIGURE 7-9 *The formation of $\pi_u 2p$ and $\pi^*_g 2p$ orbitals from linear combinations of $2p_x$ atomic orbitals. An identical pair of molecular orbitals can be constructed from the $2p_y$ orbitals.*

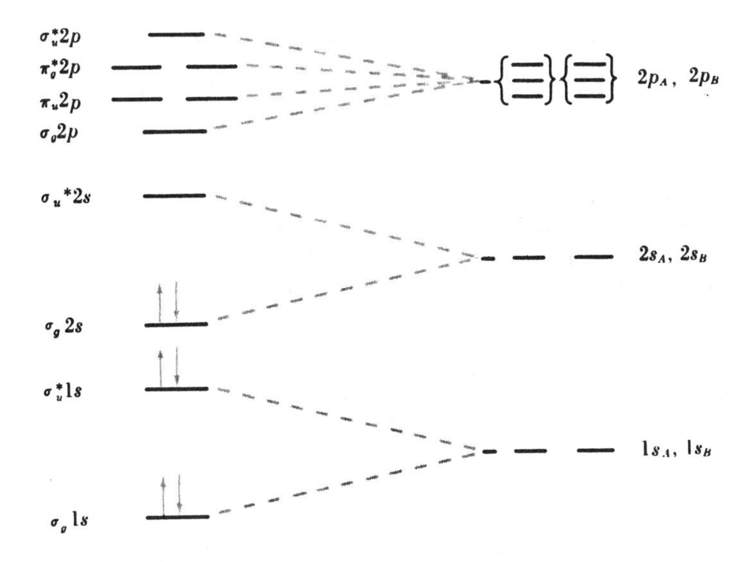

FIGURE 7-10 *The energies and names of the one-electron orbitals used in the simple MO treatment of diatomic molecules. The arrows represent the six electrons of the* Li_2 *molecule. The order shown for the orbitals* $\sigma_g 2p$ *through* $\sigma_u^* 2p$ *only holds for the oxygen and fluorine molecules. For other diatomic molecules the* $\sigma_g 2p$ *orbitals lie higher in energy than the* $\pi_u 2p$ *orbital. Note that all the* π *orbitals are doubly degenerate.*

$\sigma_g 2s$ is smaller, they interact and this interaction lowers the energy of the $\sigma_g 2s$ and raises the energy of the $\sigma_g 2p$. In some cases, the $\sigma_g 2p$ is raised until it is above the $\pi_u 2p$ orbitals. The significance of this effect for individual molecules is discussed later.

We now discuss the structure of homonuclear diatomic molecules by putting the appropriate number of electrons into the orbital scheme in accordance with the Pauli principle. Since the space part of the MO is described by three quantum numbers, only two electrons corresponding to the two possible values of the spin quantum numbers can be placed in each orbital in Figure 7-10. The questions we hope to answer from this simple picture are the following.

1. Is a stable molecule formed?
2. What is the multiplicity of the ground state?
3. What are the relative values of binding energies and bond lengths in a series of stable molecules?

The first molecule that we discuss is He_2. This molecule would have four electrons, and a consideration of Figure 7-10 shows that two electrons would have to go into each of the $\sigma_g 1s$ and $\sigma_u^* 1s$ orbitals. Since both the bonding and antibonding orbitals formed from the hydrogen-like $1s$ atomic orbitals are filled, there will be no net stabilization of an He_2 molecule over a pair of isolated helium atoms. Therefore, in accord with experiment, He_2 is not predicted to be stable.

The Li_2 molecule has six electrons. Besides the four electrons in the $\sigma_g 1s$ and $\sigma_u^* 1s$ orbitals, it will have two electrons in the $\sigma_g 2s$ orbital. This orbital is a bonding orbital, and therefore Li_2 is stable. Since the electrons must be paired, the ground state of Li_2 is a singlet. These arguments are confirmed by experiment. The binding energy and other properties of some diatomic molecules are given in Table 7-3.

Students should be able to convince themselves by an argument similar to the preceding one that Be_2 should not be stable. It is not.

The B_2 molecule has ten electrons. A consideration of Figure 7-10 indicates that the last two electrons would go into the $\sigma_g 2p$ orbital and that B_2 should therefore be stable and have a singlet ground state. This is only partly true. The B_2 molecule is stable, but it has a triplet ground state. This comes about because of the effect discussed earlier. For B_2, the $\pi_u 2p$ orbitals must lie below the $\sigma_g 2p$ or be degenerate with it. For degenerate orbitals, Hund's rule requires that the electrons go into different orbitals with parallel spin in the lowest energy state.

For C_2, the $\pi_u 2p$ orbitals are still below the $\sigma_g 2p$ orbital in energy. This statement is based on the experimental work of Ballik and Ramsay, which shows that the ground state of C_2 is a singlet [5]. The electron configuration of the molecule must therefore be $(\sigma_g 1s)^2(\sigma_u^* 1s)^2(\sigma_g 2s)^2(\sigma_u^* 2s)^2(\pi_u 2p)^2(\pi_u 2p)^2$. The first excited triplet state is only approximately 600 cm^{-1} above the ground state, however. Since the spectrum of C_2 was usually observed in flames, this excited triplet state had a high population, and for a long time this was thought

TABLE 7-3 EXPERIMENTAL DATA FOR DIATOMIC MOLECULES[a]

Molecule	*Binding energy* (eV)	*R*(Å)	*Multiplicity of ground state*
Li_2	1.03	2.672	1
Be_2	not obs.		
B_2	(3.0)	1.589	3
C_2	(5.9)	1.2422	1
N_2	9.576	1.09	1
O_2	5.080	1.207	3
F_2	1.6	1.435	1

[a] Data from G. Herzberg, *Spectra of Diatomic Molecules* (D. Van Nostrand, Inc., Princeton, N.J., 1950), Table 39.

to be the ground state. The C_2 molecule now has four electrons in bonding orbitals that are not balanced by electrons in corresponding antibonding orbitals. It should have a greater binding energy and smaller internuclear distance than B_2. A consideration of the data in Table 7-3 shows this to be the case.

Nitrogen, N_2, has all of the 2*p* bonding orbitals completely filled. It should have a singlet ground state, the largest binding energy, and the shortest internuclear distance of all the homonuclear diatomic molecules. This is borne out by the data in Table 7-3.

The discussion of the structure of O_2 and F_2 is left as an exercise.

EXERCISE 7-9 Making use of the type of arguments used earlier, predict the multiplicity of the ground state of O_2 and F_2. Estimate the relative order of binding energies and internuclear distances for the series N_2, O_2, F_2. Compare your results with the data in Table 7-3. The explanation of the correct multiplicity for O_2 was one of the triumphs of this simple theory.

The student may wonder what the relation is between the MO description of diatomic molecules and the classical chemical structures for these molecules. A bridge can be built between the two models by defining an effective number of bonds related to the difference in the number of bonding and antibonding electrons. We define the effective number of bonds as

$$N_e = \tfrac{1}{2}\,(\text{number of bonding electrons} - \text{number of antibonding electrons})$$

Applying this formula to calculate the effective number of bonds for the nitrogen molecule, one obtains

$$N_e = \tfrac{1}{2}(10 - 4) = 3$$

This result is in accord with the practice of writing a triple bond in N_2. The student may verify that this formula leads to a double bond for oxygen and a single bond for fluorine, in agreement with the usual chemical notation.

7-6 *EXCITED STATES AND ELECTRONIC SPECTRA OF DIATOMIC MOLECULES*

The excited states of a diatomic molecule are formed when an electron in the molecule is excited to a MO which has a higher energy than the orbital occupied in the ground state. The nomenclature for these states is based on the same features as the nomenclature for the one-electron MO's. Similar to the procedure for atomic states, capital Greek letters designate the total angular momentum about the bond axis, and a left superscript designates the multiplicity of the state. In addition, the state is labeled with a *g* or a *u* as a right subscript depending on whether the wave function changes sign upon changing x, y, and z to $-x$, $-y$, and $-z$. This g and u designation only holds for homonuclear diatomic molecules, of course. For example, consider the case of Li_2. In the ground state, the electron configuration of Li_2 is $(\sigma_g 1s)^2 (\sigma_u{}^*1s)^2(\sigma_g 2s)^2$. Since all of the electrons are in σ orbitals, the total angular momentum about the internuclear axis must be zero. Furthermore, the electron spins are all paired, and the multiplicity is 1. The $\sigma_g 2s$ orbital does not change sign upon inversion, so the state will be g. The ground state of the lithium molecule is therefore a $^1\Sigma_g$ state. For Σ states, an additional right superscript is added, characterizing whether the wave function changes sign on reflection through any plane passing through the internuclear axis. A *plus* is used if the function does not change sign, a *minus* is used if it does. The complete symbol for the ground state is thus $^1\Sigma_g{}^+$. To determine whether a Σ state is plus or minus requires an examination of each term of the determinantal wave function for the molecule, and the subject is not pursued further at this point. Some additional electron configurations and their symbols are given in Table 7-4. The student should try to interpret the symbols (except for $+$ and $-$) on the basis of the above discussion.

EXERCISE 7-10 What does the fact that the ground state of $N_2{}^+$ is $^2\Sigma_g$ instead of $^2\Pi_u$ tell about the ordering of the one electron MO's for nitrogen?

We are now in a position to talk about the excited electronic states of the hydrogen molecule and electronic transitions that can occur between these states. Let us first write down some of the possible excited states for H_2. As mentioned above, these are formed by exciting one electron from a $\sigma_g 1s$ MO to a higher energy orbital. Some of the possible states are listed in Table 7-5. In this table, a bar over the orbital indicates β spin, and the lack of a bar indicates α spin.

We first consider the energetics of some of these states. The $^1\Sigma_u$ and $^3\Sigma_u$ states have both one electron in a bonding and one in an antibonding orbital. The $^3\Sigma_u$ state is unstable with respect to a pair of hydrogen atoms, and it is a dissociative state. That is, if a hydrogen molecule could somehow be prepared in this state, it would immediately dissociate into a pair of hydrogen

TABLE 7-4 THE STATE SYMBOLS FOR TWO ELECTRONIC CONFIGURATIONS OF C_2 AND ONE FOR $N_2{}^+$

Molecule	*Electron configuration*	*Symbol*
C_2	$(\sigma_g 2s)^2(\sigma_u^* 2s)^2(\pi_u 2p)^4$	$^1\Sigma_g{}^+$
	$(\sigma_g 2s)(\sigma_u^* 2s)^2(\pi_u 2p)^3(\sigma_g 2p)$	$^3\Pi_u$
$N_2{}^+$	$(\sigma_g 2s)^2(\sigma_u^* 2s)^2(\pi_u 2p)^4(\sigma_g 2p)$	$^2\Sigma_g$

TABLE 7-5 THE GROUND AND SOME EXCITED CONFIGURATIONS OF H_2 AND THEIR STATE SYMBOLS[a]

Electronic configuration	*State*	*Spectroscopic notation*
$(\sigma_g 1s)\overline{(\sigma_g 1s)}$	$^1\Sigma_g$	X
$(\sigma_g 1s)\overline{(\sigma_u^* 1s)}$	$^1\Sigma_u$	
$(\sigma_g 1s)(\sigma_u^* 1s)$	$^3\Sigma_u$	
$(\sigma_g 1s)\overline{(\sigma_g 2s)}$	$^1\Sigma_g$	
$(\sigma_g 1s)\overline{(\sigma_u^* 2s)}$	$^1\Sigma_u$	B
$(\sigma_g 1s)\overline{(\pi_u 2p)}$	$^1\Pi_u$	C
$(\sigma_g 1s)(\pi_u 2p)$	$^3\Pi_u$	c

[a] In the spectroscopic notation, the ground state of a molecule always is designated X. Excited singlet states are designated $A, B, C, \ldots,$ in order of increasing energy. Excited triplet states are likewise designated $a, b, c, \ldots.$ The second three configurations in the table are not given symbols because they are either repulsive or a transition between that configuration and the ground configuration is not allowed.

atoms. The $^1\Sigma_u$ state must dissociate into a proton and a hydride ion, and, with respect to these products, it is stable. The next four states arise from the interaction of one hydrogen atom in its ground state with an *excited* hydrogen atom with its electron in either a $2s$ or a $2p$ orbital. With respect to one normal and one excited hydrogen atom, the $^1\Sigma_u$, $^1\Pi_u$, and $^3\Pi_u$ are all stable. A potential energy curve similar to that found for the ground state will describe the energy of these states as a function of internuclear distance. A schematic picture of the ground state and the degenerate Π states is shown in Figure 7-11. A transition can take place between these two electronic states provided it is allowed by the selection rules for electronic transition.

In Chapter 5, it was pointed out that the intensity of a spectral transition is proportional to the square of the transition moment. For electronic motion in a molecule, the dipole moment operator is

$$\boldsymbol{\mu} = e \sum_i Z_i \mathbf{r}_i \tag{7-37}$$

where $\mathbf{r}_i$ is the position vector of the ith particle and Z_i is the charge on this particle. The sum i is over all electrons and nuclei in the molecule. For electronic transitions, the transition moment is then

$$\mathbf{R}^{kn} = e\left(\Psi_k \middle| \sum_i Z_i \mathbf{r}_i \middle| \Psi_n\right) \tag{7-38}$$

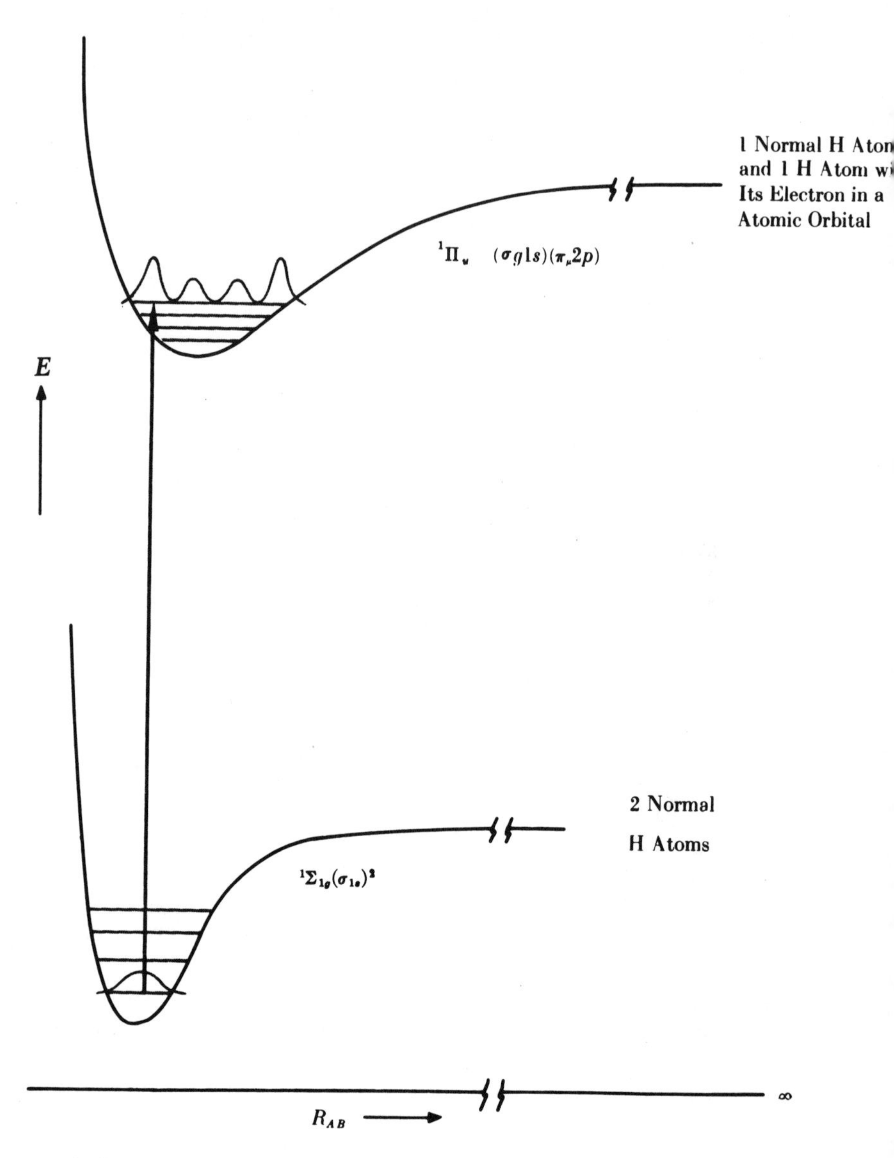

FIGURE 7-11 *The* $^1\Sigma_g$ *ground state and* $^1\Pi_u$ *excited states of* H_2. *Several vibrational sublevels are shown for each state. A vertical transition from the* $v = 0$ *vibrational level of the* $^1\Sigma_g$ *state to the* $v = 3$ *vibrational level of the* $^1\Pi_u$ *state is also shown. Such a transition is called a Franck–Condon transition.*

We do not explicitly evaluate any integrals like Eq. 7-38, but several properties of the integrals can be derived without evaluating them. First, the dipole moment operator does not contain the spin. On account of this, the transition moment will vanish if Ψ_n and Ψ_k have different multiplicities because spin eigenfunctions having different eigenvalues with respect to $\hat{S}_T^2$ are orthogonal (Theorem II of Chapter 3). Transitions between triplet and singlet states are therefore said to be spin forbidden. This is a general result which holds for all molecules. (In some cases single-triplet transitions may become weakly allowed because of spin-orbit interaction. The spectral bands from these transitions are always very weak, however.) Second, the vector $\mathbf{r}_i$ is an odd function of the coordinates of the ith particle. That is, $\mathbf{r} \rightarrow -\mathbf{r}$ as $x, y,$ and $z \rightarrow -x, -y,$ and $-z$, respectively. It is a general theorem that integrals over all space will vanish unless the integrand is an *even* function of all the variables. Applied to the transition moment, this means that the integral in Eq. 7-38 will vanish unless the product of $\Psi_k^*\Psi_n$ is *odd*. Transitions between two g states or two u states in a diatomic molecule will not be allowed, therefore. It is said that these transitions are symmetry forbidden. It is for this reason that the two states shown in Figure 7-11 were chosen. The transition between the $^1\Sigma_g$ and $^1\Pi_u$ states is allowed.

When thinking about electronic spectra, it is important to keep in mind that in its ground and excited electronic states the molecules will also have vibrational and rotational energy. Under moderate resolution, an electronic transition will exhibit vibational fine structure due to the fact that the electronic transition may originate at one of several possible vibrational levels of the ground state and terminate at one of the possible vibrational levels of the excited electronic state. Under high resolution, each of the bands due to the vibrational fine structure may sometimes be seen to be made up of lines due to transitions between different rotational levels. This rotational structure on electronic transitions can only be observed in molecules with small moments of inertia, however.

To analyze the electronic spectrum of a diatomic molecule, we first make the approximation that there is no coupling between electronic and vibrational motion. This is not a new approximation because it can be shown that this follows directly from the Born–Oppenheimer approximation. We can write the total energy for a diatomic molecule as

$$E_t = E_e + E_v \tag{7-39}$$

and the total wave function as

$$\Psi_t = \Psi_e \Psi_v \tag{7-40}$$

In Eqs. 7-39 and 7-40, the rotational energy and wave function have been neglected. In cases where rotational fine structure can be observed in an electronic transition, the rotational energies can be included in Eq. 7-39 by a straightforward application of the methods used in the discussion of vibration-rotation spectroscopy.

We can now write a general expression for the frequencies observed in the electronic spectrum of a diatomic molecule. Using T_e and G to designate electronic and vibrational contributions to the energy of a state, molecules will absorb light with energy

$$\omega_{e,v} = (T_e' - T_e'') + (G' - G'') \tag{7-41}$$

where, again, we use a single prime to designate the upper state and a double prime to designate the lower state. If we call the energy difference in cm^{-1} between the minimums of the potential energy curves of the two electronic states $\bar{\nu}_e$, and substitute the expression for an anharmonic oscillator for G, Eq. 7-41 becomes

$$\begin{aligned}\omega_{e,v} = \bar{\nu}_e &+ \omega_e'(v' + \tfrac{1}{2}) - (\omega_e x_e)'(v' + \tfrac{1}{2})^2 \\ &- \omega_e''(v'' + \tfrac{1}{2}) + (\omega_e x_e)''(v'' + \tfrac{1}{2})^2 \end{aligned} \tag{7-42}$$

In Eq. 7-42, it is important to realize that the ω_e' is the extrapolated fundamental vibration frequency in the *excited electronic state*. It may differ from ω_e'', the extrapolated fundamental vibration frequency in the ground state. Likewise, $(\omega_e x_e)'$ is the anharmonicity constant for the excited state. For diatomic molecules composed of first and second row atoms, only the lowest vibrational state of the ground electronic state ($v'' = 0$) is populated. For the absorption spectrum of these molecules, the electronic band structure is given by

$$\omega_{e,v} = \bar{\nu}_e + \omega_e'(v' + \tfrac{1}{2}) - (\omega_e x_e)'(v' + \tfrac{1}{2})^2 - \tfrac{1}{2}\omega_0'' \tag{7-43}$$

where $\omega_0'' = \omega_e'' - \frac{1}{2}(\omega_e x_e)''$ is twice the zero-point vibrational energy in the ground electronic state.

There is no vibrational selection rule which limits the values which v', the vibrational quantum number in the upper electronic state, can have. Rather, the number of vibrational bands which are observed depends on the Franck–Condon principle and the overlap of vibrational wave functions in the ground and excited electronic states.

We first discuss the Franck–Condon principle. This principle states that an electronic transition takes place so rapidly that the nuclei do not move appreciably during a transition. It should be noted that the line corresponding to an electronic transition in Figure 7-11 is drawn vertically, that is, drawn so that the internuclear distance is the same in the excited state as in the ground state. Such a transition is called a Franck–Condon transition. In general, the equilibrium internuclear distance in the excited electronic state will not be the same as that in the ground electronic state, and the two potential energy curves will be displaced from each other. Because the nuclei do not move appreciably when an electronic transition takes place, the transition must terminate at a vibrational level of the upper electronic state in which the nuclei have a high probability of having the same internuclear distance as they had in the ground state. This is shown in Figure 7-11 by the fact that the electronic transition terminates at the $v' = 3$ state.

The Franck–Condon principle can be rationalized quantum mechanically by considering the overlap of the vibrational wave functions in the ground and excited electronic states. Using the wave function of Eq. 7-40 in the expression for the transition moment, we can write, following [7],

$$\mathbf{R} = e\left(\Psi_e'\Psi_v'\left|\sum_i Z_i\mathbf{r}_i\right|\Psi_e''\Psi_v''\right) \tag{7-44}$$

If we now divide the dipole moment operator into two parts, one depending on the electrons and the other depending on the nuclei, Eq. 7-44 becomes

$$\mathbf{R} = e\left(\int \Psi_e'^*\Psi_v'^*\mathbf{M}_e\Psi_e''\Psi_v''\,d\tau_e\,d\tau_n + \int \Psi_e'^*\Psi_v'^*\mathbf{M}_N\Psi_e''\Psi_v''\,d\tau_e\,d\tau_n\right) \tag{7-45}$$

where $\mathbf{M}_e$ and $\mathbf{M}_N$ are the electronic and nuclear contributions to the dipole moment, and $d\tau_e$ and $d\tau_n$ are the appropriate volume elements for the electronic and nuclear coordinates. The quantity $\mathbf{M}_N$ depends only on the nuclear coordinates. Therefore, the second integral in Eq. 7-45 becomes

$$\int \Psi_e'^*\Psi_v'^*\mathbf{M}_N\Psi_e''\Psi_v''\,d\tau_e\,d\tau_n = \int \Psi_e'^*\Psi_e''\,d\tau_e \int \Psi_v'^*\mathbf{M}_N\Psi_v''\,d\tau_n = 0$$

because the two electronic wave functions are orthogonal (Theorem II, Chapter 3. The quantum mechanical statement of the Franck–Condon principle is that the variation of Ψ_e' and Ψ_e'' with a change in nuclear coordinates is very slow, and therefore the first term in Eq. 7-45 can be written

$$\mathbf{R} = e\left(\int \Psi_e'^*\mathbf{M}_e\Psi_e''\,d\tau_e \int \Psi_v'^*\Psi_v''\,d\tau_n\right) \tag{7-46}$$

where the first integral is the electronic transition moment, and the second integral is the overlap integral between Ψ_v' and Ψ_v''. The transition probability will therefore be proportional to the square of the electronic transition moment and the square of the overlap integral between the appropriate vibrational levels in the ground and excited electronic states. These overlap integrals are called Franck–Condon factors.

7-7 LOCALIZED BONDS, HYBRID ORBITALS, AND DIRECTED VALENCE

A large portion of modern chemistry has been correlated and explained by the concept of the electron pair bond. These electron pair bonds retain much of their identity regardless of what the rest of the molecule containing the bond is like. This retention of bond properties is the basic assumption made in the listing of bond energies, and the successful use of these bond energies to calculate thermodynamic properties justifies this assumption. Further, it was pointed out in Chapter 5 that one was able to carry over infrared frequencies of bonds from molecule to molecule to a fairly high degree of accuracy. This is the basic feature of the organic chemist's use of infrared spectroscopy to identify an unknown compound.

The valence bond method is especially convenient for a description of the electron pair bond. It was pointed out earlier that if we have an orbital ϕ_A containing one electron on atom A and an orbital ϕ_B containing one electron on atom B, then we can write an antisymmetrized bond eigenfunction for an AB bond in the form

$$\Psi_{AB} = N(\phi_A \phi_B + \phi_B \phi_A)(\alpha\beta - \beta\alpha) \tag{7-47}$$

As an application of the use of bond eigenfunctions in polyatomic molecules, we first consider the water molecule. An oxygen atom in the ground state has electronic configuration $1s^2 2s^2 2p^4$, and is in a ^3P state. We can state roughly that there is one electron in each of two $2p$ orbitals with parallel spins. These two electrons can be used to form bonds with other atoms that also have a singly occupied orbital; for example, hydrogen atoms. This simple picture predicts oxygen to be divalent, which it is. The wave function for the water molecule can be thought of as a product of two bond eigenfunctions, each of which has the form

$$\Psi_{\mathrm{OH}} = N[\phi_{2P}(\mathrm{O})\phi_{1S}(\mathrm{H}) + \phi_{1S}(\mathrm{H})\phi_{2P}(\mathrm{O})](\alpha\beta - \beta\alpha) \tag{7-48}$$

where $\phi_{2P}(\mathrm{O})$ is a $2p$ orbital localized on oxygen and $\phi_{1S}(\mathrm{H})$ is a $1s$ orbital localized on hydrogen. The binding energy of a water molecule in this model is predicted to be equal to twice the energy of an OH bond. This is approximately correct.

This simple theory, along with one additional principle, enables one to make predictions about the stereochemistry of molecules. The additional principle needed is the principle of maximum overlap, and was first enunciated by Pauling [8]:

> *Of two orbitals in an atom, the one that can overlap more with an orbital of another atom will form the stronger bond with that atom. Moreover, the bond formed by a given orbital will tend to lie in that direction in which the orbital is concentrated.*

For the case of water, the overlap between the $1s$ orbital of hydrogen and the $2p$ orbital of oxygen will be greatest when the hydrogen atoms are collinear with the axes of the $2p$ orbitals of oxygen. The bond angles in water are predicted to be 90° by this model because 90° is the angle between the axes of a pair of $2p$ orbitals. The actual value of the bond angle in water is 104.5°. The quantitative agreement is not good, but the discrepancy is not serious in view of the naivete of the theory.

For ammonia (NH_3), a similar argument can be developed. The ground state of nitrogen is a ^4S state, and there is one electron in each of the three $2p$ orbitals. We would then predict that nitrogen would be trivalent and should form a stable compound with three hydrogen atoms. Again, we predict that the bond angles in NH_3 would be 90°. The experimental value is 108°. The quantitative agreement is again not good, and a troubling doubt about this simple picture begins to appear because the agreement is getting worse.

For carbon compounds, an additional refinement must be made because the simple picture used above for water and ammonia no longer predicts the correct valence. A carbon atom has a ^3P ground state, corresponding to two unpaired electrons, one in each of two $2p$ orbitals. On the basis of the arguments used above, we would expect carbon to be divalent. It is well known, of course, that carbon is usually tetravalent. Further, it is known that all of the bonds in compounds such as CH_4 and CCl_4 are equivalent. To revise our simple theory in order to explain these facts, we postulate that, in order to get larger overlap, and consequently stronger bonds, atoms can make use of *hybrid orbitals*. For carbon, we can think of these hybrid orbitals as arising as follows. First, a carbon atom in its ground state is excited to the excited state described in Eq. 7-49.

$$C(^3P)2s^2 2p2p \rightarrow C(^5S)2s2p2p2p \tag{7-49}$$

This requires an energy of 33,735.2 cm^{-1} or 96 kcal/mole. Next the $2s$ orbital and the three $2p$ orbitals are combined to form four equivalent orbitals, called sp^3 hybrids, which make angles of 109°28′ with each other. The carbon atom then utilizes the one electron in each of these hybrid orbitals to form four equivalent bonds.

The simplest form that these hybrids can take is

$$\begin{aligned}
\phi_1 &= \tfrac{1}{2}(s + p_x + p_y + p_z) \\
\phi_2 &= \tfrac{1}{2}(s + p_x - p_y - p_z) \\
\phi_3 &= \tfrac{1}{2}(s - p_x + p_y - p_z) \\
\phi_4 &= \tfrac{1}{2}(s - p_x - p_y + p_z)
\end{aligned} \tag{7-50}$$

Some properties of these hybrid orbitals are calculated in the following exercises.

EXERCISE 7-11 Represent the p_x, p_y, and p_z orbitals by unit vectors along an x-y-z axis system and let these vectors point to the center of the faces of a cube. Where, in this cube, will the functions ϕ_1, ϕ_2, ϕ_3, and ϕ_4 point?

EXERCISE 7-12 Show that the four sp^3 hybrid orbitals above are mutually orthogonal.

Each bond in methane can be thought of as formed by the overlap of an sp^3 hybrid orbital on carbon with a $1s$ orbital on hydrogen, and corresponding bond eigenfunctions can be written. The energy gained by forming four bonds instead of two is more than enough to compensate for the excitation of an electron from a $2s$ to a $2p$ orbital.

We are now in a position to explain the deviation of the bond angles in NH_3 and water from 90°. In NH_3, the bond angles are almost 109°, and we can postulate that sp^3 hybrids are used to form bonds in this compound. In H_2O, addition of a little s character to the two $2p$ orbitals would spread the bond angle to 104°. See [2] for additional discussion of this point.

Other types of hybridization have been used to explain other types of bond geometries. A combination of a $2s$ and two $2p$ orbitals can be used to form a set of three sp^2 hybrid orbitals. These hybrids all lie in a plane and make bond angles of 120° with each other. An example of a compound utilizing sp^2 hybrid orbitals is boron trichloride (BCl_3), which is known to be planar, and have bond angles of 120°. The sp^2 hybrids are sometimes called trigonal hybrids.

A combination of a $2s$ oribtal with one $2p_z$ orbital gives a set of two sp or diagonal hybrid orbitals. The formation of these hybrids is shown in Figure 7-12. These hybrid orbitals are collinear. An example of a compound using sp hybrids is mercuric chloride ($HgCl_2$). This is known to be a linear molecule. This molecule further illustrates that hybridization is not restricted to the $n = 2$ atomic orbitals.

It was pointed out at the beginning of this section that the valence bond (VB) theory can be used to perform calculations of molecular properties beginning with a wave function that is a product of bond eigenfunctions. The

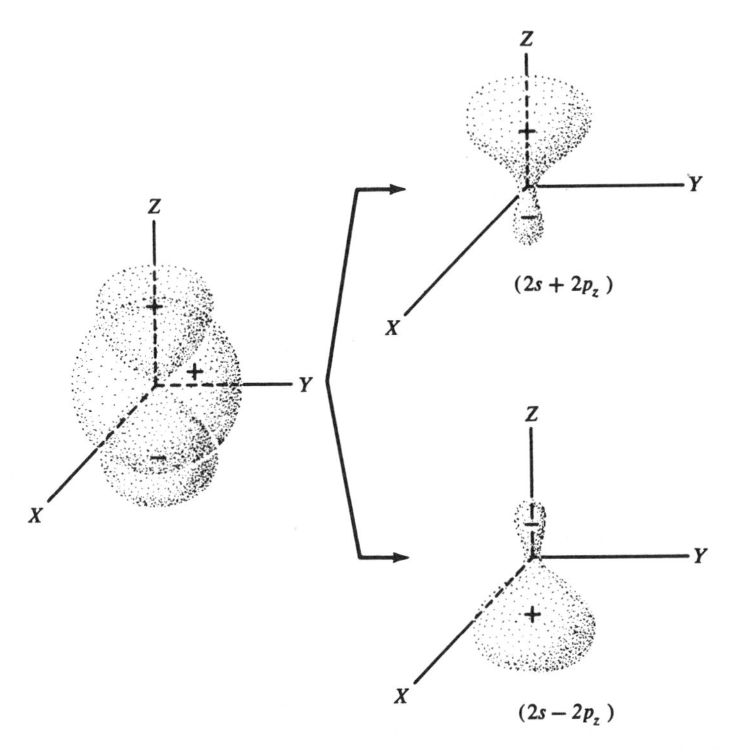

FIGURE 7-12 *Diagram showing the formation of the two sp hybrid orbitals from plus and minus linear combinations of an ns and* np_z *orbital.*

early development of this theory to systems containing more than six electrons was hindered because of computational difficulties caused by the fact that the basis orbitals were not orthogonal. Recently ways have been found to circumvent these computational problems by making suitable approximations, and the VB method is again being used for small molecules. There are a number of advantages of the VB method that make it desirable to continue to investigate its use. The first is that VB wave functions include some electron correlation and, therefore, give a better energy at the first computational level than the MO method. VB wave functions have an added benefit; some effects that depend on electron correlation, such as electron spin densities in certain π-electron radicals (see Chapter 9), can be obtained from them at the lowest computational level. Using the MO method, one must include configuration interaction before these effects appear. Second, the VB treatment yields a qualitatively correct description of dissociation while the MO approximation does not. This makes the VB wave function a more appropriate starting point for calculating potential energy surfaces that involve the breaking of chemical bonds. Finally, the VB orbitals can be readily interpreted in terms of chemical bonding concepts because of their localized nature. Thus, it will be possible to pick out lone-pair, inner-shell, and bonding-pair electrons from a VB wave function. Hybrid orbitals can also be used as basis orbitals and, thus, classical geometric bonding concepts such as those discussed above can be built into the calculation. Students interested in more details about VB calculations should see [9, 10].

7-8 THE σ-π DESCRIPTION OF ETHYLENE AND ACETYLENE

The ethylene molecule is known to be planar, and all of the bond angles are close to 120°. Further, it contains unsaturation represented by a double bond in its formula. We can explain these results by assuming that the structural backbone of the molecule is composed of bonds between sp^2 hybrids on the two carbon atoms and $1s$ orbitals on the hydrogen atoms. This basic framework of the ethylene molecule is called the "σ-bond framework." When an atom utilizes sp^2 hybrids, there is still one p orbital left over. In ethylene, these remaining p orbitals contain one electron each, and they can overlap to form a π bond. The double bond in ethylene can be thought of, therefore, as being composed of a σ and a π bond. It is the π bond which is reactive, and, since the σ-bond framework is composed of sp^2 hybrids, the molecule would be planar and the bond angles would all be 120°.

In acetylene, the two carbon atoms must use sp hybrids in its σ-bond framework because acetylene is known to be a linear molecule. If an atom forms sp hybrids, there are two p orbitals left over on each carbon atom. These can interact to form two π bonds at right angles to each other.

The separation of conjugated organic molecules into a σ-bond framework and a system of π electrons is the starting point for the discussion of the unusual properties of these molecules in Chapter 8.

EXERCISE 7-13 In the light of the above, discuss the structure of allene,

$$H_2C{=}C{=}CH_2$$

in terms of bond geometry and orbital hybridization. The three carbon atoms are collinear.

7-9 SUMMARY

1. Some general features of the quantum mechanical solution to the $H_2{}^+$ problem were discussed. Specifically, the Born–Oppenheimer approximation, which states that the electronic and nuclear motions can be treated independently, was introduced.

2. The variational method utilizing an LCAO function was applied to the calculation of the allowed energies in H_2^+. This method is based on the variational principle, which states that the energy calculated from any appropriate approximate wave function will always be higher than the true energy of the ground state.

3. The MO and VB treatments of the hydrogen molecule were compared. The VB wave function contains no ionic terms, whereas the MO wave function includes the ionic terms and covalent terms with equal weights.

4. Methods for improving the wave functions for molecular hydrogen include adding some ionic character to the VB function, including polarization effects in the basis orbitals, and scaling.

5. A simple MO theory for complex diatomic molecules was developed that was capable of predicting molecular stability and ground state multiplicities.

6. The origin of electronic spectra of diatomic molecules was discussed, and several selection rules were derived from a consideration of the electronic transition moment. The Franck–Condon principle was introduced. This principle states that the intensity of the vibration components of an electronic transition depends on the overlap of vibration wave functions in the ground and excited electronic states.

7. The concept of orbital hybridization was introduced. A simple VB model employing these hybrids was used to discuss the bonding and geometry of several molecules.

8. The student should be familiar with the following terms: Born–Oppenheimer approximation, linear variation function, secular equations, secular determinant, characteristic equation, basis set, Coulomb integral (MO), resonance integral, Coulomb integral (VB), exchange integral, g and u, σ and π orbital designations, scale factor and scaling, Franck–Condon principle, hybrid orbitals, and σ-bond framework.

REFERENCES

1. D. R. Bates, K. Ledsham, and A. L. Stewart, *Phil. Trans. Roy. Soc.* **A246**, 215 (1953).
2. C. A. Coulson, *Valence* (Oxford University Press, London, 1961), pp. 79ff.
3. W. Kauzmann, *Quantum Chemistry* (Academic Press Inc., New York, 1957).
4. W. Kolos and C. C. J. Roothaan, *Rev. Mod. Phys.* **32**, 219 (1960).
5. E. A. Ballik and D. A. Ramsay, *J. Chem. Phys.* **31**, 1128 (1959).
6. A. C. Wahl, *Science* **151**, 961 (1966).
7. G. Herzberg, *Spectra of Diatomic Molecules* (D. Van Nostrand, Inc., Princeton, N.J., 1950), pp. 200ff.

8. L. Pauling, *The Nature of the Chemical Bond* (Cornell University Press, Ithaca, N.Y., 1960), p. 103.
9. F. L. Pilar, *Elementary Quantum Chemistry* (McGraw-Hill, New York, 1968), pp. 519–521, 551ff.
10. See D. M. Chipman, *J. Am. Chem. Soc.* **100**, 2650 (1978) for a recent application and a summary of recent developments.

Chapter 8

THE ELECTRONIC STRUCTURE OF CONJUGATED SYSTEMS

IN CHAPTER 7, the localized bond model was used in the discussion of the structure of water, ammonia, methane, boron trichloride, and other molecules. It was found that this model gave a qualitatively correct picture of the energetics and stereochemistry of the compounds. The model also accounted for additivity of bond energies, the approximate constancy of the force constants, and the valence of an atom. This localized bond model becomes inadequate, however, for the description of any molecule when electrons can be delocalized over more than two atoms. The best known example of a molecule for which the localized bond model breaks down is benzene. Using classical chemical structure theory, benzene must be represented as a "resonance hybrid" between the two limiting structures I and II.

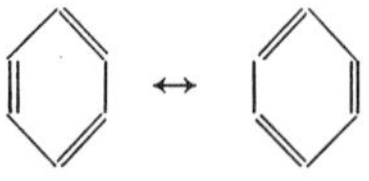

$\leftrightarrow$

I II

That is, the structure of benzene cannot be adequately represented by either structure I or II by itself. It is somewhere between the two. It is the purpose of this chapter to show how these "delocalized" bonds may be treated by MO theory.

8-1 THE LCAO–MO THEORY FOR CONJUGATED HYDROCARBONS

The Hamiltonian for a molecule containing n electrons and η nuclei is, within the context of the Born–Oppenheimer approximation,

$$\mathscr{H} = \sum_{i=1}^{n}\left(-\frac{1}{2}\nabla_i^2 - \sum_{\mu=1}^{\eta}\frac{Z_\mu}{r_{i\mu}}\right) + \sum_{i<j}\sum\frac{1}{r_{ij}} \tag{8-1}$$

In Eq. 8-1, the nuclear repulsion terms have been omitted. For ethylene, the simplest hydrocarbon that has both σ and π electrons, $n = 16$ and $\eta = 6$. Even an approximate calculation of the allowed energies for ethylene would require considerable effort. When dealing with conjugated organic molecules, it is usual to simplify the calculations by first assuming that the π electrons can be treated independently of the σ electrons. In an actual calculation, we will seek a π-electron wave function Ψ_π that will be antisymmetric upon exchange of any two π electrons and adjusted so that the π-electron energy

$$E_\pi = \frac{(\Psi_\pi|\mathscr{H}_\pi|\Psi_\pi)}{(\Psi_\pi|\Psi_\pi)} \tag{8-2}$$

is a minimum. Performing this procedure amounts to postulating that the total wave function for a conjugated hydrocarbon can be written as

$$\Psi = \Psi_\sigma\Psi_\pi \tag{8-3}$$

where Ψ_σ is an antisymmetrized wave function for the σ electrons. To make the wave function in Eq. 8-3 obey the Pauli principle, it will also have to be made antisymmetric upon interchange of σ and π electrons. Since such antisymmetrization will have no effect at the level of approximation used in this chapter, it will be neglected in the remaining discussion.

The Hamiltonian used in Eq. 8-2 is then taken as an effective Hamiltonian. That is, $\mathscr{H}_\pi$ represents the motion of the π electrons in the potential field of the nuclei and an average field of the σ electrons. Therefore, we write

$$\mathscr{H}_\pi = \sum_{i=1}^{n_\pi}\hat{h}_{\text{core}}(i) + \sum_{i<j=1}^{n_\pi}\sum\frac{1}{r_{ij}} \tag{8-4}$$

where now the sums are only over the π electrons. In the π-electron approximation, the ethylene molecule is reduced to a two-electron problem.

In Eq. 8-4, $\hat{h}_{\text{core}}$ may be thought of as including the kinetic energy of the π electrons, the potential energy between the π electrons and the nuclei minus the shielding effect of the σ electrons, and any additional interaction effects between σ and π electrons.

Implicit in this π-electron approximation is the assumption that the σ-bond framework remains the same for all π-electron states.

The student may well wonder if this approximation, which appears rather drastic, is justified. To answer this, we must appeal to the following rationalizations.

1. The distinctive chemistry of the conjugated organic molecules has been correlated by considering only the π-electron systems. This distinctive chemistry appears to be relatively independent of the σ-bond framework.

2. The characteristic spectra of these molecules can be rationalized using the π-electron approximation.

3. A large amount of other experimental data, such as ionization potentials, dipole moments, relative reactivities, and so on, can be at least qualitatively described by the independent π-electron model. We thus, once again, appeal to the fact that a theory works in order to justify its use. This does not mean that making this approximation introduces no difficulties, however. Obtaining exact quantitative agreement with experiment is fraught with many difficulties. For example, the calculated energy separation between excited singlet and triplet states of conjugated organic molecules is always too large. For a more rigorous discussion of the π-electron approximation, see [1, 10].

We now return to the problem of treating the π electrons by the Hamiltonian of Eq. 8-4. Even if $\hat{h}_{\text{core}}$ could be written exactly, eigenvalue equations using Eq. 8-4 would not be exactly soluble because of the electron repulsion terms. We therefore must seek approximate solutions for the π electrons in the same way as for diatomic molecules. The most drastic of these approximation methods is to replace the Hamiltonian in Eq. 8-4 by

$$\hat{\mathscr{H}}_{\pi}^{0} = \sum_{i} \hat{h}_{\text{eff}}(i) \tag{8-5}$$

where $\hat{h}_{\text{eff}}(i)$ is a one-electron operator which incorporates the $1/r_{ij}$ terms into the nuclear potential in some average way. The quantity $\hat{h}_{\text{eff}}(i)$ can be written

$$\hat{h}_{\text{eff}}(i) = \tfrac{1}{2}\nabla^{2}(i) - \sum_{\mu=1}^{\eta} \frac{Z'_{\mu}}{r_{i\mu}} \tag{8-6}$$

where the sum is over all the nuclei in the molecule and where Z'_{μ} is an effective nuclear charge that somehow incorporates the average screening of all of the σ electrons as well as shielding due to the remaining π electrons. Seeing that $\hat{h}_{\text{eff}}(i)$ is a function only of the coordinates and momenta of a single electron, we can use the separation of variables method to reduce the

n_π π-electron problem to n_π *identical* one-electron problems. The final problem to be solved is thus

$$\hat{h}(i)\phi_i = \varepsilon_i \phi_i \tag{8-7}$$

where ϕ_i is a one-electron molecular orbital and ε_i is the corresponding orbital energy. At this level of approximation, the total π-electron energy and wave function will be

$$E_\pi = \sum_{i=1}^{n_\pi} \varepsilon_i \qquad \Psi_\pi = \prod_{i=1}^{n_\pi} \phi_i(i) \qquad \text{(8-8a and b)}$$

Of course, electrons must be placed in the independent electron orbitals in such a way that the Pauli principle is satisfied.

8-2 THE SIMPLE HÜCKEL METHOD

The next step in our approximate calculation of the electronic structure of conjugated hydrocarbons is to assume that the ϕ_i can be written as a linear combination of atomic orbitals. We thus write

$$\phi_i = \sum_{\mu=1}^{\eta} C_{i\mu} \chi_\mu \tag{8-9}$$

where χ_μ is a $2p_z$ orbital localized on atom μ, and where the sum extends over all atoms in the conjugated molecule. We now apply the variation method to find the best values of the energy for the ground and excited one-electron orbitals. Students should review the outline of this procedure in Chapter 7 if they have forgotten it. Minimizing the energy with respect to the $C_{i\mu}$ leads to a set of η linear equations in η unknowns. These, in turn, have a nontrivial solution only if the secular determinant

$$|H_{\mu\nu} - \varepsilon_i S_{\mu\nu}| = 0 \tag{8-10}$$

The features that distinguish the Hückel method from other LCAO methods are the following approximations used to simplify the determinant in Eq. 8-10.

1. All $H_{\mu\mu}$ are equal and are set equal to a Coulomb integral α.
2. All $H_{\mu\nu} = \beta$ if atoms μ and ν are bonded to each other.
   $= 0$ if atoms μ and ν are not bonded to each other.
3. All $S_{\mu\nu} = 1$ if $\mu = \nu$.
   $= 0$ if $\mu \neq \nu$.

Use of this method will now be illustrated in calculations of the π-electronic structure for ethylene and butadiene.

8-3 ETHYLENE

The ethylene calculation is a two-atom and two-electron problem. The nomenclature is shown in Figure 8-1. The LCAO function is

$$\phi_i = C_{i1}\chi_1 + C_{i2}\chi_2 \tag{8-11}$$

The secular equations are

$$\begin{aligned} (H_{11} - S_{11}\varepsilon_i)C_{i1} + (H_{12} - S_{12}\varepsilon_i)C_{i2} &= 0 \\ (H_{21} - S_{21}\varepsilon_i)C_{i1} + (H_{22} - S_{22}\varepsilon_i)C_{i2} &= 0 \end{aligned} \tag{8-12}$$

The secular determinant is

$$\begin{vmatrix} H_{11} - S_{11}\varepsilon_i & H_{12} - S_{12}\varepsilon_i \\ H_{21} - S_{21}\varepsilon_i & H_{22} - S_{22}\varepsilon_i \end{vmatrix} = 0 \tag{8-13}$$

Applying the Hückel approximation, this becomes

$$\begin{vmatrix} \alpha - \varepsilon_i & \beta \\ \beta & \alpha - \varepsilon_i \end{vmatrix} = 0 \tag{8-14}$$

The solutions of Eq. 8-14 are

$$(\alpha - \varepsilon_i)^2 - \beta^2 = 0$$

$$\varepsilon_i = \alpha \pm \beta$$

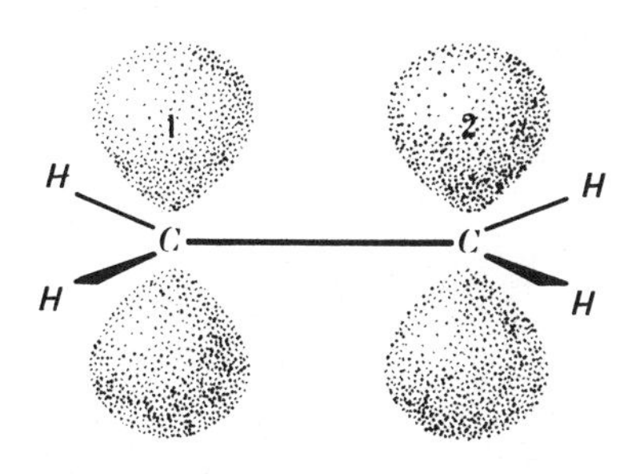

FIGURE 8-1 *Labeling of the $2p_z$ orbitals used in the π-electron calculation for ethylene.*

It should be noted that these solutions are identical in form to the solutions for H_2^+ except that the overlap integrals have been set equal to zero. The two orbital energies and wave functions are thus

$$\varepsilon_1 = \alpha + \beta \qquad \phi_1 = \tfrac{1}{2}\sqrt{2}\,(\chi_1 + \chi_2)$$
$$\varepsilon_2 = \alpha - \beta \qquad \phi_2 = \tfrac{1}{2}\sqrt{2}\,(\chi_1 - \chi_2) \tag{8-15}$$

It should be noted that both α and β are negative quantities (compare H_{AA} and H_{AB} in the H_2^+ case) so that ε_1 is the lowest energy. The total π-electron wave function is obtained by putting the *two* π electrons into ϕ_1 with their spins paired. Thus, for ethylene,

$$\Psi_\pi = \phi_1\phi_1\,\tfrac{1}{2}\sqrt{2}(\alpha\beta - \beta\alpha) \tag{8-16a}$$

$$E_\pi = 2\varepsilon_1 = 2\alpha + 2\beta \tag{8-16b}$$

In the Hückel approximation, the integrals α and β are never evaluated. This avoids the embarrassing problem of specifying the exact form of $\hat{h}_{\text{eff}}$. Rather, the quantities α and β are given empirical values to make the theory fit experiment. A wide range of β values have been given in the literature [2]. One chooses a $-\beta$ value derived to fit the particular experimental data of interest. Several types of data (spectral and polarographic reduction potentials, for example) are fit by a value of 2.37–2.39 eV. The quantity α can be identified with the energy of a p electron on an isolated carbon atom, and is usually taken as the zero of energy. Our π-electron calculation on ethylene can be summarized by the diagram shown in Figure 8-2.

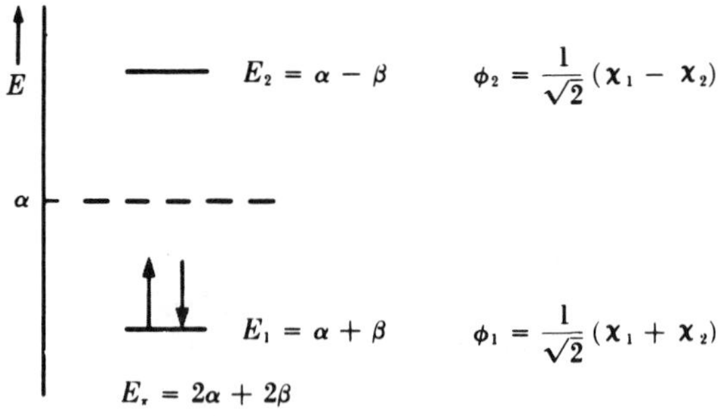

E

$E_2 = \alpha - \beta \qquad \phi_2 = \frac{1}{\sqrt{2}}(\chi_1 - \chi_2)$

α

$E_1 = \alpha + \beta \qquad \phi_1 = \frac{1}{\sqrt{2}}(\chi_1 + \chi_2)$

$E_\pi = 2\alpha + 2\beta$

FIGURE 8-2 *Energy level diagram for the π electrons in ethylene. The two π electrons have been placed in the orbital ϕ_1 with antiparallel spins. The total π-electron energy is equal to the sum of the orbital energy of each electron.*

8-4 BUTADIENE, $CH_2{=}CH{-}CH{=}CH_2$

The calculation of the π-electronic structure for butadiene is somewhat more complex than that for ethylene. Butadiene exists in *cis* and *trans* forms, but in simple Hückel theory there will be no difference in the energy of the two forms because non-nearest neighbor interactions are neglected. (The *cis* and *trans* geometry will be important in determining spectroscopic selection rules, however. See Section 8-6.) For purposes of this initial calculation, therefore, we can regard the molecule as linear. The numbering of the atomic orbitals is shown in Figure 8-3. The MO's will have the form

$$\phi_i = \sum_{\mu=1}^{4} C_{i\mu}\chi_\mu \tag{8-17}$$

Since there are four orbitals in the basis set, there will be a set of four secular equations and a 4×4 secular determinant. Using the Hückel approximations, the student should verify that the secular determinant is

$$\begin{vmatrix} \alpha - \varepsilon_i & \beta & 0 & 0 \\ \beta & \alpha - \varepsilon_i & \beta & 0 \\ 0 & \beta & \alpha - \varepsilon_i & \beta \\ 0 & 0 & \beta & \alpha - \varepsilon_i \end{vmatrix} = 0 \tag{8-18}$$

To simplify the solution to Eq. 8-18, we divide each row of the determinant by β and make the substitution

$$x = \frac{\alpha - \varepsilon_i}{\beta}$$

The secular determinant then becomes

$$\begin{vmatrix} x & 1 & 0 & 0 \\ 1 & x & 1 & 0 \\ 0 & 1 & x & 1 \\ 0 & 0 & 1 & x \end{vmatrix} = 0 \tag{8-19}$$

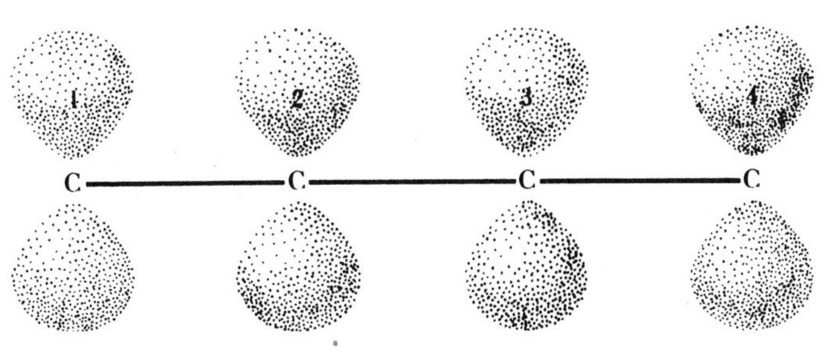

FIGURE 8-3 *Labeling of the atomic orbitals used in the π-electron calculation for butadiene. The hydrogen atoms in butadiene are not shown.*

If Eq. 8-19 is evaluated by the method of minors, one obtains the characteristic equation

$$x^4 - 3x^2 + 1 = 0 \tag{8-20}$$

EXERCISE 8-1 Evaluate the determinant in Eq. 8-19 and show that it leads to the characteristic equation given in Eq. 8-20.

Usually the roots to an equation of this type would have to be found by an iterative method such as the Newton–Raphson method [3]. In this case, however, it turns out that Eq. 8-20 is factorable. Thus

$$x^4 - 3x^2 + 1 = (x^2 + x - 1)(x^2 - x - 1) = 0 \tag{8-21}$$

and therefore

$$\begin{aligned} x^2 + x - 1 &= 0 \\ x^2 - x - 1 &= 0 \end{aligned} \tag{8-22}$$

The four roots of the secular determinant are, therefore,

$$\begin{aligned}
x &= \frac{1 + \sqrt{5}}{2} = +1.618 & \varepsilon &= \alpha - 1.618\beta \\
x &= \frac{1 - \sqrt{5}}{2} = -0.618 & \varepsilon &= \alpha + 0.618\beta \\
x &= \frac{-1 + \sqrt{5}}{2} = +0.618 & \varepsilon &= \alpha - 0.618\beta \\
x &= \frac{-1 - \sqrt{5}}{2} = -1.618 & \varepsilon &= \alpha + 1.618\beta
\end{aligned} \tag{8-23}$$

Using these roots, the energy diagram can be constructed. This is shown in Figure 8-4.

At this point, we introduce a quantity called the delocalization energy (DE). The DE is defined as the difference between the total π-electron energy of the molecule of interest and the π-electron energy of the same number of isolated double bonds. Thus, for butadiene,

$$\begin{aligned}
\text{DE} &= E_\pi(\text{butadiene}) - 2E_\pi(\text{ethylene}) \\
&= 2(\alpha + 1.618\beta) + 2(\alpha + 0.618\beta) - 2(2\alpha + 2\beta) \\
&= 4\alpha + 4.472\beta - 4\alpha - 4\beta \\
&= 0.472\beta
\end{aligned} \tag{8-24}$$

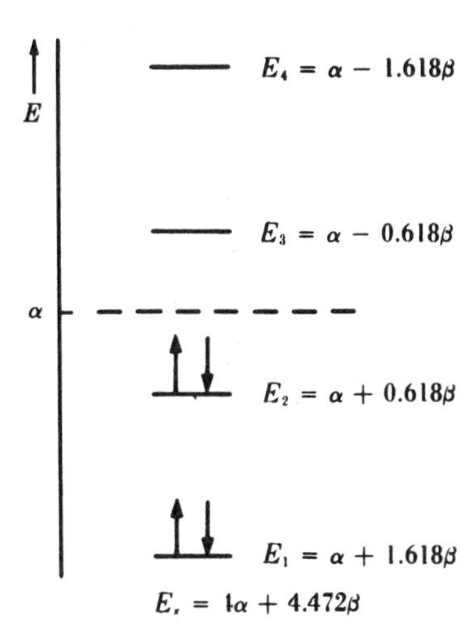

FIGURE 8-4 *Energy level diagram for the π electrons in butadiene. Butadiene has four π electrons, so the two lowest energy orbitals are filled. The total π-electron energy is* $2(\alpha + 1.618\beta) + 2(\alpha + 0.618\beta)$. *The student should note that the orbitals are paired about the zero of energy. This is a general property of conjugated hydrocarbons containing no odd-membered rings.*

The DE is a measure of the extra stabilization that a molecule has due to the fact that electrons can move over the whole molecule. It has often been related to "resonance" energies that can be calculated from experimental data by various means.

For example, Klages [4] has defined the resonance energy (RE) of a compound as follows:

$$\text{RE} = \Delta H^{\text{ref}}_{\text{comb}} - \Delta H^{\text{expt}}_{\text{comb}} \tag{8-25}$$

where $\Delta H^{\text{ref}}_{\text{comb}}$ is the calculated heat of combustion of a reference structure assuming isolated double bonds, and $\Delta H^{\text{exp}}_{\text{comb}}$ is the measured heat of combustion of the compound. (Another scheme has been developed which defines RE's in terms of heats of hydrogenation of unsaturated compounds.) To calculate $\Delta H^{\text{ref}}_{\text{comb}}$, we make use of a table of bond contributions to heats of combustion. These tables are compiled by comparing combustion data for a large number of aliphatic and alicyclic compounds with no conjugated double bonds. Some of the bond contributions to RE's given by Klages are given in Table 8-1. Using the numbers in Table 8-1, we can calculate $\Delta H^{\text{ref}}_{\text{comb}}$ for butadiene. It is

$$\Delta H^{\text{ref}}_{\text{comb}} = 6(54.0) + 49.3 + 2(119.1) = 611.5 \text{ kcal}$$

TABLE 8-1 SOME BOND CONTRIBUTIONS TO HEATS OF COMBUSTION OF A HYPOTHETICAL REFERENCE STRUCTURE FOR A CONJUGATED MOLECULE WHERE THE DOUBLE BONDS ARE ASSUMED TO BE LOCALIZED[a]

Bond	*Contributions to* ΔH_{comb} (kcal)
C—H	54.0
C—C	49.3
C=C (unsubstituted)	121.6
C=C (monosubstituted)	119.1
C=C (*cis*-disubstituted in a six-membered ring)	117.4
C=C (tetrasubstituted)	112.0
correction for a six-membered ring	+1.0

[a] The resonance energy of a compound is defined as the difference between the heat of combustion of the reference structure and the actual heat of combustion.

The experimental value of the heat of combustion is 607.91 kcal [5]. The resonance energy calculated from Eq. 8-25 is therefore 3.6 kcal. Comparing this resonance energy with the value of DE calculated for butadiene gives $\beta = 7.6$ kcal, which is much too low. This result emphasizes that the connection between RE's and DE's calculated from MO theory is a tenuous one. In fact, Dewar and Schmeising have developed an alternate set of bond energies that depends on the type of σ-bond hybridization [6]. Using this set, they can "explain" most of the "RE" without invoking π-electron delocalization.

The whole subject of RE's is a fascinating and controversial one. The main problem is to derive a satisfactory heat of combustion for the "reference" compound. For more information the student should see [2].

The next step in a π-electron calculation on butadiene is the calculation of the coefficients of the four MO's that correspond to the four energies found above. This can be done by substituting the energies into the four secular equations, *one at a time*, and solving for the four sets of coefficients. This procedure sometimes leads to arithmetic errors, however, and a more systematic procedure has been developed for calculating coefficients. This procedure automatically normalizes the MO's also.

The method is based on the fact that the value of a particular coefficient is proportional to the signed minor or cofactor of the appropriate element of a row in the secular determinant (usually the first row). To use this method, one proceeds as follows.

1. Choose a row in the secular determinant, usually the first.
2. Calculate the values of the minors of each element of this row in terms of x and numbers. Give the minors the appropriate sign of $(-1)^{i+j}$, where i is the label of the row, and j is the label of the column for which the minor is evaluated.

The first two steps will be illustrated for butadiene. The four signed minors or cofactors are

$$M_1 = +\begin{vmatrix} x & 1 & 0 \\ 1 & x & 1 \\ 0 & 1 & x \end{vmatrix} = x^3 - 2x \tag{8-26a}$$

$$M_2 = -\begin{vmatrix} 1 & 1 & 0 \\ 0 & x & 1 \\ 0 & 1 & x \end{vmatrix} = -(x^2 - 1) \tag{8-26b}$$

$$M_3 = +\begin{vmatrix} 1 & x & 0 \\ 0 & 1 & 1 \\ 0 & 0 & x \end{vmatrix} = x \tag{8-26c}$$

$$M_4 = -\begin{vmatrix} 1 & x & 1 \\ 0 & 1 & x \\ 0 & 0 & 1 \end{vmatrix} = -1 \tag{8-26d}$$

Once this has been done, the following steps should be carried out.

3. Calculate the values of the minors for the roots, taken one at a time.
4. Square each value obtained in step 3, and sum the squares.
5. Take the square root of the sum of the squares, and divide each value obtained in step 3 by this quantity.
6. The resulting numbers will be the appropriate coefficients. If this procedure is done in tabular form, it is an easy matter to check for errors, and one is less likely to make an error in the first place.

Steps 3–6 for butadiene are illustrated in Table 8-2 for the root $x = -1.618$.

TABLE 8-2 ILLUSTRATIVE CALCULATION OF THE COEFFICIENTS IN THE LOWEST ENERGY MOLECULAR ORBITAL IN BUTADIENE[a]

μ	M_μ	$M_\mu{}^2$	$C_\mu = M_\mu/(\sum m_\mu{}^2)^{\frac{1}{2}}$
1	-1	1.000	-0.371
2	-1.618	2.618	-0.600
3	-1.618	2.618	-0.600
4	-1	1.000	-0.371
		$\sum M_\mu{}^2 = 7.236$	

[a] The final coefficients may all be multiplied by -1 without changing the wave function. All coefficients would then be positive.

This procedure is only learned by practice, which is the reason for the following exercise.

EXERCISE 8-2 Calculate the coefficients for the orbitals which correspond to the other three roots for butadiene.

For small molecules this systematic procedure may not be the quickest. For these cases, the secular equations can be solved as a set of simultaneous equations. Using butadiene as an example, one can set $C_{i1} = 1.0$, use the first secular equation to determine C_{i2} as a function of C_{i3} and C_{i4}, use the second to determine C_{i3} as a function of C_{i4}, use the third to solve for C_{i4} (and then the remainder of the $C_{i\mu}$), and use the fourth equation as a check. Of course, the resulting functions will not be normalized, but that is easily accomplished. Computer programs for performing Hückel calculations are now commonly available.

The wave functions and energies for butadiene are summarized in Table 8-3. The student should verify that the number of nodes in the MO's increases from 0 to 3 as the energy of the orbitals increases.

In many respects, the coefficients of the atomic orbitals are of more interest than the orbital energies because, from the coefficients, additional quantities that are related to the π-electron distribution in the molecule can be defined. We are interested in three such quantities: the electron densities, bond orders, and free valency indices. These are defined as follows.

1. The electron density on atom μ, q_μ. This is the probability of finding a π electron on atom μ. It is defined as

$$q_\mu = \sum_i N_i C_{i\mu}^2 \tag{8-27}$$

where N_i is the number of electrons in the ith *molecular orbital* (N_i can have the values 0, 1, or 2), and where $C_{i\mu}$ is the coefficient of *atomic* orbital μ in the ith molecular orbital. To illustrate, q_1 for butadiene will be calculated.

$$\begin{aligned} q_1 &= 2C_{11}^2 + 2C_{21}^2 + 0C_{31}^2 + 0C_{41}^2 \\ &= 2(0.371)^2 + 2(0.600)^2 = 1.000 \end{aligned}$$

TABLE 8-3 A SUMMARY OF THE ENERGIES AND COEFFICIENTS OF THE FOUR MOLECULAR ORBITALS OF BUTADIENE

i	ε_i	$\phi_i = \sum_{\mu=1}^{\mu} C_{i\mu}\chi_\mu$
1	$\alpha + 1.618\beta$	$0.371\chi_1 + 0.600\chi_2 + 0.600\chi_3 + 0.371\chi_4$
2	$\alpha + 0.618\beta$	$0.600\chi_1 + 0.371\chi_2 - 0.371\chi_3 - 0.600\chi_4$
3	$\alpha - 0.618\beta$	$0.600\chi_1 - 0.371\chi_2 - 0.371\chi_3 + 0.600\chi_4$
4	$\alpha - 1.618\beta$	$0.371\chi_1 - 0.600\chi_2 + 0.600\chi_3 - 0.371\chi_4$

The net charge on the atom, is defined as

$$\xi_\mu = 1 - q_\mu \tag{8-28}$$

From atom 1 in butadiene, the net charge is zero.

EXERCISE 8-3 Suppose that one of the π electrons in butadiene was ionized to give the butadiene positive ion. What would the net charge on atom 1 be in this positive ion?

2. The bond order of bond $\mu\nu$. This quantity is a measure of the amount of multiple bond characters in a bond. Bond orders have been correlated with bond lengths and with vibrational force constants. The bond order is defined as

$$P_{\mu\nu} = \sum_i N_i C_{i_\mu} C_{i_\nu} \tag{8-29}$$

To illustrate, P_{12} for butadiene will be calculated.

$$P_{12} = 2(0.371)(0.600) + 2(0.600)(0.371) + 0 = 0.894$$

The bond order for ethylene is 1.000, so the above result indicates that the 1, 2 bond in butadiene is not quite a double bond.

3. The free valency index of atom μ, F_μ. This quantity is less useful than the other two, but it is sometimes used as a measure of free radical reactivity. It is defined as

$$F_\mu = \sqrt{3} - (\text{sum of bond orders of all bonds to atom } \mu) \tag{8-30}$$

The $\sqrt{3}$ arises because it is the maximum value which the sum of the π-bond orders to a single carbon atom can have. This statement does not apply to molecules where there are triple bonds, of course. To illustrate, F_1 for butadiene is

$$F_1 = 1.732 - 0.894 = 0.838$$

The free valency index is a quantitative expression of Thiele's idea of "residual valence."

EXERCISE 8-4 Calculate the remaining electron densities, bond orders, and free valency indices for butadiene.

8-5 THE USE OF SYMMETRY TO SIMPLIFY QUANTUM MECHANICAL CALCULATIONS

It is obvious, no doubt, to the student that, as the number of atoms in a conjugated molecule increases, the labor and time spent in a simple Hückel calculation becomes large. If the molecule possesses some symmetry, it is possible to reduce this computational labor.

A symmetry operation is any operation performed on a molecule that leaves the molecule indistinguishable from its original configuration. Examples of symmetry operations are reflection in a plane, inversion through a center, rotation about an axis of symmetry, and so on. Since, after a symmetry operation is carried out, the molecule is indistinguishable from its original configuration, the Hamiltonian for a molecule must be invariant upon any symmetry operations. In mathematical language, this means that the Hamiltonian and an operator corresponding to any symmetry operation commute.

Symmetry is used to simplify quantum mechanical problems as follows.

1. Linear combinations of the atomic orbital basis set are found that are eigenfunctions of the appropriate symmetry operators for a molecule.

2. These "symmetry orbitals" are then used as a new basis set of a Hückel calculation.

3. Use can be made of the Theorem IV in Chapter 3 to set all matrix elements of the form $(\Psi_i|\hat{\mathscr{H}}|\Psi_j) = 0$ if Ψ_i and Ψ_j belong to different eigenvalues of a symmetry operator.

This procedure will now be illustrated for butadiene. For butadiene, it will be sufficient to regard the molecule as having a plane of symmetry passing through the middle of the central bond. This plane of symmetry is shown in Figure 8-5. We introduce the mirror plane operator $\hat{\sigma}_m$ that requires one to reflect the orbital following the operator in this plane. A consideration of Figure 8-5 shows that the operator $\hat{\sigma}_m$ has the following effect on the atomic orbitals:

$$\begin{aligned} \hat{\sigma}_m \chi_1 &= \chi_4 \\ \hat{\sigma}_m \chi_2 &= \chi_3 \\ \hat{\sigma}_m \chi_3 &= \chi_2 \\ \hat{\sigma}_m \chi_4 &= \chi_1 \end{aligned} \tag{8-31}$$

Using these results, a set of four eigenfunctions of $\hat{\sigma}_m$ can be constructed by inspection. Since operating on χ_1 and χ_4 with $\hat{\sigma}_m$ gives χ_4 and χ_1, respectively, a function which gives the same function back again upon operation with $\hat{\sigma}_m$ must contain *both* χ_1 and χ_4 with coefficients of equal magnitude. A similar argument holds for χ_2 and χ_3. The four symmetry functions and their

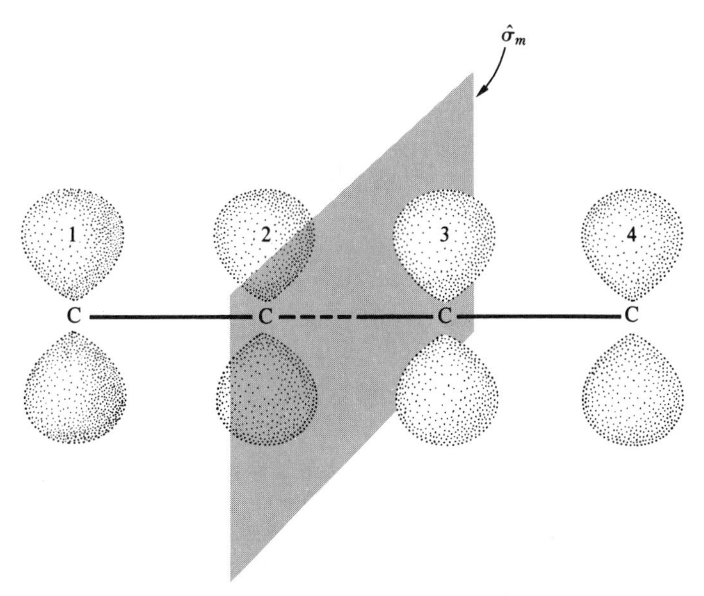

FIGURE 8-5 *The mirror plane in butadiene. The mirror plane operator is only appropriate for cis-butadiene; trans-butadiene has a center of inversion, however, which leads to the same results as the mirror plane.*

eigenvalues λ with respect to $\hat{\sigma}_m$ are, therefore,

$$\begin{aligned} S_1 &= \tfrac{1}{2}\sqrt{2}(\chi_1 + \chi_4) \qquad & \lambda &= +1 \\ S_2 &= \tfrac{1}{2}\sqrt{2}(\chi_2 + \chi_3) \qquad & \lambda &= +1 \\ S_3 &= \tfrac{1}{2}\sqrt{2}(\chi_2 - \chi_3) \qquad & \lambda &= -1 \\ S_4 &= \tfrac{1}{2}\sqrt{2}(x_1 - \chi_4) \qquad & \lambda &= -1 \end{aligned} \tag{8-32}$$

The student should verify the fact that these four functions are eigenfunctions of $\hat{\sigma}_m$ with the eigenvalues given.

Using the symmetry functions as the basis set and applying Theorem IV of Chapter 3, we can immediately write the secular determinant. It is

$$\left|\begin{array}{cc|cc} H'_{11} - E & H'_{12} & 0 & 0 \\ H'_{21} & H'_{22} - E & 0 & 0 \\ \hline 0 & 0 & H'_{33} - E & H'_{34} \\ 0 & 0 & H'_{34} & H'_{44} - E \end{array}\right| = 0 \tag{8-33}$$

where the primes indicate that these matrix elements are integrals over the *symmetry* orbitals. Whenever a determinant can be written so that two or more blocks are connected only by zeros, the determinant is said to be factorable. This means that the determinant above can be written as the product of a pair of 2 × 2 determinants

$$\begin{vmatrix} H'_{11} - E & H'_{12} \\ H'_{21} & H'_{22} - E \end{vmatrix} \begin{vmatrix} H'_{33} - E & H'_{34} \\ H'_{43} & H'_{44} - E \end{vmatrix} = 0 \tag{8-34}$$

and, therefore, that each of the 2 × 2 determinants equals zero.

$$\begin{vmatrix} H'_{11} - E & H'_{21} \\ H'_{21} & H'_{22} - E \end{vmatrix} = 0 \qquad \begin{vmatrix} H'_{33} - E & H'_{34} \\ H'_{43} & H'_{44} - E \end{vmatrix} = 0 \tag{8-35}$$

The important point to notice is that a 4 × 4 secular determinant has been reduced to a pair of 2 × 2 secular determinants, and this greatly simplifies the work of finding the energies and coefficients. Exercise 8-5 illustrates the completion of the problem and the fact that the same results are obtained using the symmetry orbitals as were obtained using the original atomic orbital basis set.

EXERCISE 8-5 Evaluate the matrix elements $H'_{\mu\nu}$, for $\mu, \nu = 1$–4 using the symmetry orbitals as basis functions. Show that the characteristic equations that one obtains are the same as Eq. 8-21. Calculate the coefficients of the two symmetry orbitals S_1 and S_2 using the root $x = -1.618$. Show that the MO that is obtained is the same as ϕ_1.

The use of symmetry becomes more important for large molecules. For example, a Hückel calculation on naphthalene using the atomic orbital basis set would involve solving a 10 × 10 secular determinant and evaluating ten 9 × 9 minors to calculate the coefficients. We can greatly reduce the amount of work needed for the π-electron calculation on naphthalene by making use of the fact that naphthalene has three symmetry operations that leave the molecule indistinguishable from its original configuration. These symmetry operations are two mirror planes $\hat{\sigma}_m$, $\hat{\sigma}'_m$ and a twofold axis of rotational symmetry $\hat{C}_2$. These symmetry elements and the numbering of the atoms in naphthalene are shown in Figure 8-6. The molecular plane is also a symmetry element, but its use does not result in any simplification of the problem since all $2p_z$ orbitals are already antisymmetric with respect to reflection in the molecular plane.

The problem, then, is reduced to constructing a set of simultaneous eigenfunctions of the three symmetry operators $\hat{C}_2$, $\hat{\sigma}_m$, and $\hat{\sigma}'_m$. In this case, it is more difficult to construct the appropriate symmetry orbitals by inspection than it was for butadiene. A systematic procedure for constructing these

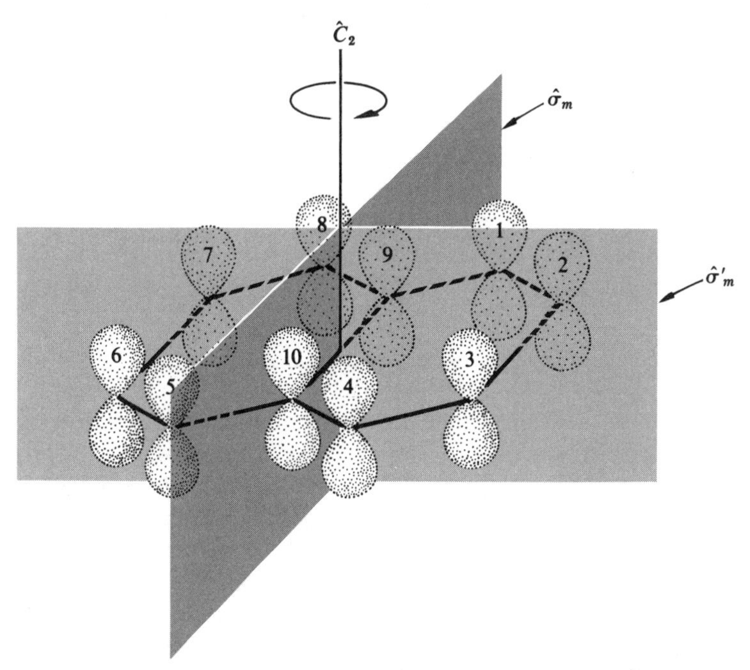

FIGURE 8-6 *The symmetry operations for naphthalene. The $\hat{C}_2$ operator is a rotation of* 180° *about the twofold axis. The $\hat{\sigma}_m$ and $\hat{\sigma}'_m$ operators are reflections in the appropriate planes. The plane of the molecule is also a reflection plane, but all p orbitals must have eigenvalue* -1 *with respect to reflection in this plane.*

orbitals is available, however, and a brief outline of this procedure is now given.

The effects of symmetry on the atomic orbitals of naphthalene can be summarized by specifying the possible eigenvalues of symmetry orbitals with respect to four symmetry operators. The four operators are the three shown in Figure 8-6 plus the identity operator $\hat{E}$ that requires one to leave the molecule alone. The only possible eigenvalue with respect to $\hat{E}$ is $+1$, but we see below that symmetry orbitals can be constructed with eigenvalues ± 1 with respect to the other three operators. The possible combinations of eigenvalues are listed in Table 8-4.

The student will no doubt notice that some combinations are missing from Table 8-4, for example, the set of eigenvalues $+1$, -1, -1, -1. A consideration of Figure 8-6 shows why such a set of eigenvalues would be impossible. The student can see that $\hat{C}_2$ converts χ_1 into χ_5; however, the same result can be obtained by reflecting χ_1 first in $\hat{\sigma}_m$, and then $\hat{\sigma}'_m$. We can thus say mathematically that $\hat{C}_2$ is equal to $\hat{\sigma}_m \hat{\sigma}'_m$. To have eigenvalue -1 with respect to $\hat{C}_2$ then requires that the product of the eigenvalues with respect to

TABLE 8-4 POSSIBLE EIGENVALUES WITH RESPECT TO THE FOUR SYMMETRY OPERATORS FOR THE NAPHTHALENE MOLECULE. EACH SET OF EIGENVALUES IS LABELED TO FACILITATE DISCUSSION

	$\hat{E}$	$\hat{C}_2$	$\hat{\sigma}_m$	$\hat{\sigma}'_m$
A_1	1	1	1	1
A_2	1	1	−1	−1
B_1	1	−1	1	−1
B_2	1	−1	−1	1

$\hat{\sigma}_m$ and $\hat{\sigma}'_m$ be -1, and this is not possible if the eigenvalues with respect to $\hat{\sigma}_m$ and $\hat{\sigma}'_m$ are both -1.

Each of the sets of eigenvalues in Table 8-4 is labeled to facilitate talking about them. By convention, sets containing the eigenvalue $+1$ with respect to $\hat{C}_2$ are labeled with an A, those containing the eigenvalues -1 with respect to $\hat{C}_2$ are labeled with a B. Symmetry orbitals can now be constructed by taking the atomic orbitals one at a time, performing the operations $\hat{E}$, $\hat{C}_2$, $\hat{\sigma}_m$, and $\hat{\sigma}'_m$, and then multiplying the results by the appropriate eigenvalue from Table 8-3. For example, if we take χ_1 and the set of eigenvalues A_1, we can obtain the symmetry orbital

$$\chi_1(A_1) \rightarrow \chi_1 + \chi_5 + \chi_8 + \chi_4 \tag{8-36}$$

since

$$\begin{aligned} \hat{E}\chi_1 &= \chi_1 \\ \hat{C}_2\chi_1 &= \chi_5 \\ \hat{\sigma}_m\chi_1 &= \chi_8 \\ \hat{\sigma}'_m\chi_1 &= \chi_4 \end{aligned} \tag{8-37}$$

Likewise, a symmetry orbital with a B_1 set of eigenvalues formed from atomic orbital χ_2 would be

$$\chi_2(B_1) \rightarrow \chi_2 - \chi_3 - \chi_6 + \chi_7 \tag{8-38}$$

By following this procedure, using different atoms with each set of eigenvalues, the student can construct the set of 10 symmetry orbitals which are given in Table 8-5.

An eigenfunction with a set of eigenvalues given by any row of Table 8-4 is said to belong to the *representation* indicated by the symbol in the left-hand column of the table. Thus the eigenfunction $\chi_2 - \chi_3 - \chi_6 + \chi_7$ is said to belong to the representation B_1.

It should be remembered that all of the symmetry operations commute with each other and with the Hamiltonian. This means that, if we apply Theorem IV of Chapter 3, all integrals of the type $(S_1 |\hat{\mathscr{H}}| S_2)$ will vanish unless S_1 and S_2 have identical eigenvalues with respect to *all four* symmetry operators. Thus all matrix elements between orbitals in different groups in Table 8-5 will vanish. The original 10×10 determinant for the naphthalene calculation using atomic orbitals will then factor into two 3×3 and two 2×2 determinants when symmetry orbitals are used. This factoring results in a considerable saving of effort in the calculation. The completion of the calculation of the π-electron energies and wave functions for naphthalene makes an excellent exercise to test the student's understanding of the material in this chapter.

EXERCISE 8-6 Calculate the π-electron energies and wave functions for naphthalene using the symmetry orbitals in Table 8-5. Also, calculate the electron densities, bond orders, and free valency indices. The orbital energies and coefficients are given in Table 8-6.

The above discussion has been an introduction to the use of group theory to simplify quantum mechanical calculation without using the group theory formalism. The student interested in further study of group theory should see [7].

TABLE 8-5 SYMMETRY ORBITALS FOR NAPHTHALENE[a]

A_1	A_2
$S_1 = \frac{1}{2}(\chi_1 + \chi_4 + \chi_5 + \chi_8)$	$S_4 = \frac{1}{2}(\chi_1 - \chi_4 + \chi_5 - \chi_8)$
$S_2 = \frac{1}{2}(\chi_2 + \chi_3 + \chi_6 + \chi_7)$	$S_5 = \frac{1}{2}(\chi_2 - \chi_3 + \chi_6 - \chi_7)$
$S_3 = \frac{1}{2}\sqrt{2}(\chi_9 + \chi_{10})$	
B_1	B_2
$S_6 = \frac{1}{2}(\chi_1 - \chi_4 - \chi_5 + \chi_8)$	$S_8 = \frac{1}{2}(\chi_1 + \chi_4 - \chi_5 - \chi_8)$
$S_7 = \frac{1}{2}(\chi_2 - \chi_3 - \chi_6 + \chi_7)$	$S_{10} = \frac{1}{2}(\chi_2 + \chi_3 - \chi_6 - \chi_7)$
$S_8 = \frac{1}{2}\sqrt{2}(\chi_9 - \chi_{10})$	

[a] Note that the atomic orbitals χ_9 and χ_{10} only appear in the A_1 and B_1 sets. This is because these orbitals must have eigenvalue $+1$ with respect to $\hat{\sigma}'_m$. The student should verify that the numerical factor in each orbital ensures that it is normalized.

TABLE 8-6 ORBITAL ENERGIES AND COEFFICIENTS OBTAINED FROM A HÜCKEL CALCULATION ON NAPHTHALENE[a]

i	ε_i	C_1, C_4 C_5, C_8	C_2, C_3 C_6, C_7	C_9, C_{10}	Representation
1	$\alpha + 2.303\beta$	0.3006	0.2307	0.4614	A_1
2	$\alpha + 1.618\beta$	0.2629	0.4253	0.0000	B_2
3	$\alpha + 1.303\beta$	0.3996	0.1735	0.3471	B_1
4	$\alpha + 1.000\beta$	0.0000	−0.4083	0.4083	A_1
5	$\alpha + 0.618\beta$	0.4253	0.2629	0.0000	A_2
6	$\alpha - 0.618\beta$	0.4253	−0.2629	0.0000	B_2
7	$\alpha - 1.000\beta$	0.0000	−0.4083	0.4083	B_1
8	$\alpha - 1.303\beta$	0.3996	−0.1735	−0.3471	A_1
9	$\alpha - 1.618\beta$	0.2629	−0.4253	0.0000	A_2
10	$\alpha - 2.303\beta$	0.3006	−0.2307	−0.4614	B_1

[a] The coefficients are arranged in groups that have the same numerical value. The relative signs of the coefficients within each group are determined by the representation to which the molecular orbitals belong. Thus, for ϕ_2, which belongs to B_2, $C_1 = C_4 = -C_5 = -C_8$.

8-6 ELECTRONIC SPECTRA OF CONJUGATED HYDROCARBONS

The basic principles necessary for the understanding of the electronic spectra of π-electron systems are similar to those treated in Chapter 7 in the discussion of the electronic spectra of diatomic molecules. There are several differences in detail, however. One difference comes about because conjugated molecules are usually quite large with large moments of inertia and many vibrational degrees of freedom. This means that rotational fine structure is almost never resolved in the electronic spectra of these molecules. In fact, only the highest frequency vibrations give rise to vibrational fine structure. The electronic spectrum of a conjugated molecule, therefore, usually contains several quite broad bands, some of which may have additional vibrational fine structure.

At first thought it might seem foolish to embark on a calculation of the energies of π-electronic transitions. After all, use of the LCAO method for the hydrogen molecule gave a binding energy that was in error by 43%. How can we expect good agreement for more complex molecules when we are making even more drastic assumptions than were made for molecular hydrogen? The situation turns out to be not so hopeless as the above statements would indicate, however. Although it is true that no π-electron theory yet developed can give exact quantitative agreement with experiment, it is also true that many of the qualitative features of the electronic spectra of conjugated molecules can be rationalized by such a theory.

The *cis*-butadiene molecule will be used as an example in the calculation of π-electron spectral quantities. This molecule is chosen for two reasons: the one-electron energies and wave functions have already been given in Section 8-4, and the molecule shares some of the symmetry operations already discussed for naphthalene.

The symmetry properties of a molecule are extremely important in a discussion of electronic spectra. The student should recall that we were able to derive electronic selection rules for diatomic molecules on the basis of symmetry arguments alone. A similar procedure can be used for the selection rules in symmetric conjugated hydrocarbons except that the arguments are a little more complicated.

The *cis*-butadiene molecule, a suitable coordinate system, and the appropriate symmetry operations are shown in Figure 8-7. By convention, the Z axis is taken along $\hat{C}_2$. The Y axis is perpendicular to the plane of the molecule. There are four possible symmetry operations for butadiene: the identity operation $\hat{E}$, a twofold axis $\hat{C}_2$, and two mirror planes $\hat{\sigma}_m$ and $\hat{\sigma}'_m$. In this case, $\hat{\sigma}_m$ is the plane of the molecule. The possible sets of simultaneous eigenvalues are given in Table 8-7. Such a table is also called a character

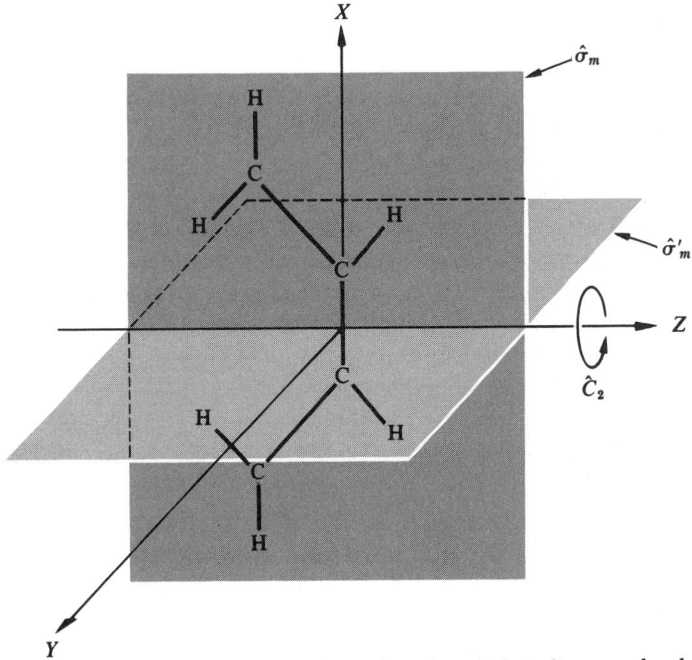

FIGURE 8-7 *The symmetry operations for the cis-butadiene molecule. The molecule lies in the XZ plane. By convention, the Z axis is always taken along* $\hat{C}_2$. *To show how the coordinate x transforms, consider what happens to a unit vector along the X axis when it is operated on by the three symmetry operators.*

TABLE 8-7 POSSIBLE SETS OF EIGENVALUES FOR THE FOUR SYMMETRY OPERATORS IN *cis*-BUTADIENE[a]

	E	C_2	σ_m	σ'_m
$A_1; z$	1	1	1	1
A_2	1	1	-1	-1
$B_1; x$	1	-1	1	-1
$B_2; y$	1	-1	-1	1

[a] The student may show that the coordinates x, y, and z belong to the representation shown by considering the behavior of a unit vector along each axis when it is operated on by the symmetry operators.

table for the C_{2v} group. In the first column of the table, the symmetry labels are given as well as the coordinate, if any, which belongs to that particular representation. The student may show that the coordinates x, y, and z belong to the representation shown by considering the behavior of a unit vector along each axis when it is operated on by the symmetry operators. The remainder of the table is identical to that used in the discussion of naphthalene.

Our first job is to see to which group of eigenvalues or to which representation the four molecular orbitals of butadiene belong. This can be done by operating on each of the MO's in turn by the four symmetry operators and finding the eigenvalue with respect to each. It should be clear immediately that, since all p orbitals are antisymmetric with respect to the molecular plane σ_m, all four molecular orbitals must belong to either A_2 or B_2 representations. By recalling the results of Section 8-5, where the eigenvalue with respect to $\hat{\sigma}'_m$ was used to simplify the calculation, the representations of the four orbitals can immediately be written down. They are

$$\begin{array}{ll} \phi_1, b_2 & \phi_3, b_2 \\ \phi_2, a_2 & \phi_4, a_2 \end{array}$$

where the convention has been introduced that small letters will be used to indicate the symmetry of one-electron orbitals. (Capital letters will indicate the symmetry of a state or configuration.)

Within the one-electron approximation, the ground state wave function for butadiene is

$$\Psi_0 = N\phi_1(1)\bar{\phi}_1(2)\phi_2(3)\bar{\phi}_2(4) \tag{8-39}$$

where, as before, N is a normalizing factor and the bar indicates β spin. The symmetry of this ground state wave function is the product of the symmetries of the one-electron orbitals. The set of eigenvalues characterizing Ψ_0 is thus

$$E = (1)^2(1)^2 = 1 \qquad \sigma_m = (-1)^2(-1)^2 = 1$$

$$C_2 = (-1)^2(1) = 1 \qquad \sigma'_m = (1)^2(-1)^2 = 1$$

and the symmetry of Ψ_0 is therefore A_1.

We next consider some of the excited states of butadiene. These can be thought of as arising from the excitation of one of the electrons in ϕ_1 or ϕ_2 to either ϕ_3 or ϕ_4. Some of these excited state wave functions and their symmetries are given in Table 8-8. The symmetries of these excited states can be obtained in the same way as those for the ground state.

We next use this simple model to calculate the excitation energies and selection rules for π-electron transitions. The excitation energies are the differences between the appropriate orbital energies. Thus

$$E(0 \to 1) = \varepsilon_3 - \varepsilon_2 = -0.618\beta - 0.618\beta = -1.236\beta$$
$$E(0 \to 2) = E(0 \to 3) = \varepsilon_4 - \varepsilon_2 = -2.236\beta$$
$$E(0 \to 4) = \varepsilon_4 - \varepsilon_1 = -3.236\beta$$

It should be noted that, because of the pairing properties of the orbital energies, the $E(0 \to 2)$ and the $E(0 \to 3)$ transitions have the same energy at this level of approximation.

The selection rules are derived by considering the symmetry of the integrand in the transition moment. If the position vector of the ith electron is written out in terms of its components, then

$$\mathbf{R}^{0k} = e(\Psi_k | x\mathbf{i} + y\mathbf{j} + z\mathbf{k} | \Psi_0) \tag{8-40}$$

TABLE 8-8 THE WAVE FUNCTION AND SYMMETRIES OF THE GROUND STATE AND SOME EXCITED CONFIGURATIONS OF BUTADIENE

State	*Wave function*	*Symmetry*
Ψ_0	$\phi_1(1)\bar{\phi}_1(2)\phi_2(3)\bar{\phi}_2(4)$	A_1
Ψ_1	$\phi_1(1)\bar{\phi}_1(2)\phi_2(3)\bar{\phi}_3(4)$	B_1
Ψ_2	$\phi_1(1)\bar{\phi}_1(2)\phi_2(3)\bar{\phi}_4(4)$	A_1
Ψ_3	$\phi_1(1)\bar{\phi}_2(2)\phi_2(3)\bar{\phi}_3(4)$	A_1
Ψ_4	$\phi_1(1)\bar{\phi}_2(2)\phi_2(3)\bar{\phi}_4(4)$	B_1

To evaluate the selection rules, the symmetry of the excited state and of each of the terms x, y, and z must be considered. If the integral $\mathbf{R}^{0k}$ is not to vanish, the integrand must be totally symmetric or belong to the A_1 representation (see Section 7-6). The Ψ_2 and Ψ_3 states are of A_1 symmetry; therefore, only the z component of the transition moment is nonvanishing. This is because only the z coordinate has A_1 symmetry, and the product of the A_1 representation with any other representation will not give an integrand with A_1 symmetry. The states Ψ_1 and Ψ_4 have B_1 symmetry, and the products $\Psi_0\Psi_1$ and $\Psi_0\Psi_4$ therefore have B_1 symmetry. As a consequence, only the x term in the transition moment for these transitions is nonvanishing because a function of B_1 symmetry multiplied by either z or y will not give an integrand which has A_1 symmetry. We therefore find that all of the transitions discussed above will be allowed. We can further predict the spectroscopic behavior of oriented molecules when polarized light is used. If the plane of polarization of a light wave is defined as the plane of oscillation of the electric field vector, then transitions $\Psi_0 \rightarrow \Psi_1$ and $\Psi_0 \rightarrow \Psi_4$ will be observed only when the light beam is X-axis polarized. Likewise, the transitions $\Psi_0 \rightarrow \Psi_2$ and $\Psi_0 \rightarrow \Psi_3$ will be observed only for Z-axis polarized light. For this reason, the transitions $\Psi_0 \rightarrow \Psi_1$ and $\Psi_0 \rightarrow \Psi_4$ are said to be X-axis polarized and the transitions $\Psi_0 \rightarrow \Psi_2$ and $\Psi_0 \rightarrow \Psi_3$ are said to be Z-axis polarized. Comparison of the above calculations with experiment is difficult. The *cis*-butadiene molecule is unstable with respect to the *trans* form and is difficult to study. Because of this, the experimental data are quite sparse. In addition, the lowest energy transition that has been observed is at 2,170 Å. Any other transition will be in vacuum ultraviolet, and experimental studies in this region are rather difficult. The absorption at 2,170 Å corresponds to an energy of 5.71 eV or 551 kJ/mole. If this energy is assigned to the $\Psi_0 \rightarrow \Psi_1$ transition, and if it is compared with the transition energy calculation above from simple Hückel theory, the value of β must be 4.6 eV or 446 kJ/mole. This value is much too high when compared with values of β used in more refined calculations on spectra and other properties. The polarization of this lowest energy absorption band is not yet known, so it is not even absolutely certain that the lowest energy excited state has B_1 symmetry.

The above calculation does not really tell us if a simple Hückel treatment is adequate for a discussion of the electronic spectra of conjugated hydrocarbons. It does, however, illustrate the basic procedure for calculating spectral properties. Furthermore, the treatment given above serves in many cases as the starting point for more refined calculations of π-electron spectra. These refined calculations have been highly successful in calculating the details of the electronic spectra of many conjugated molecules [8, 9]. It now appears, however, that some of the discrepancies between theory and experiment on the spectra of conjugated molecules are due to the inadequacies of the π-electron approximation itself. To make further progress in this field it may be necessary to develop a model which takes account of the effects of the σ electrons on the π-electron distribution.

8-7 SUMMARY

1. The approximations inherent in the π-electron approximation for treating the electronic structure of conjugated molecules were discussed.
2. The general features of the MO method for treating these molecules were introduced. The secular equation arising from this treatment was simplified using the approximations first introduced by Hückel.
3. Examples of simple MO calculations on ethylene and butadiene were given. The discussion of butadiene included a systematic way to calculate the coefficients of the MO's.
4. Resonance energies were defined as the difference between the experimental heat of combustion and the heat of combustion calculated for a reference structure assuming localized double bonds. These RE's were related to the delocalization energies calculated from Hückel theory.
5. The π-electron bond orders, electron densities, and free valence indices were defined.
6. The use of symmetry to simplify quantum mechanical calculations was discussed, and the idea of a symmetry operator was introduced. The use of symmetry was illustrated in calculations on butadiene and naphthalene.
7. The method used to calculate spectral properties of π-electron systems was illustrated by a calculation for *cis*-butadiene. The symmetry of the molecule was used to derive electronic selection rules and the polarization properties of the possible transitions.
8. The student should be familiar with the following terms: delocalization energy, bond order, charge density, free valence index, resonance energy, symmetry operation, identity operator, representation, character table, and polarization of a transition.

REFERENCES

1. R. G. Parr, *Quantum Theory of Molecular Electronic Structure* (W. A. Benjamin, Inc., New York, 1963), Chap. III.
2. A. Streitwieser, Jr., *Molecular Orbital Theory for Organic Chemists* (J. Wiley & Sons, Inc., New York, 1961).
3. H. Margenau and G. M. Murphy, *The Mathematics of Physics and Chemistry* (D. Van Nostrand, Inc., Princeton, N. J., 1943), Vol. 1, pp. 492ff.
4. F. Klages, *Ber.* **82**, 358 (1949).
5. *Selected Values of Chemical Thermodynamic Properties* (National Bureau of Standards, Washington, D.C., 1952), Circ. No. 500.

6. M. J. S. Dewar and H. N. Schmeising, *Tetrahedron* **5**, 166 (1959); **11**, 96 (1960).
7. F. A. Cotton, *Chemical Applications of Group Theory* (Wiley-Interscience, Inc., New York, 1963), Chap. 4.
8. C. Sandorfy, *Electronic Spectra and Quantum Chemistry* (Prentice-Hall, Inc., Englewood Cliffs, N.J., 1964).
9. J. N. Murrell, *The Theory of the Electronic Spectra of Organic Molecules* (J. Wiley & Sons, Inc., New York, 1963).
10. W. Kutzelnigg, G. Del Re, and G. Berthier, "σ and π Electrons in Organic Compounds," *Topics in Current Chemistry* **22** (Springer-Verlag, Berlin, 1971).

Chapter 9

ELECTRON AND NUCLEAR MAGNETIC RESONANCE SPECTROSCOPY

COMPUTATIONS DEALING with the allowed states and energies of an electron or nuclear spin system provide good illustrations of many of the principles that have been discussed previously. Computations on spin problems are convenient because in many cases the quantum mechanics *on the spin system* may be done exactly. This is in contrast to computations on the spatial motion of the electrons, which must be treated by approximate methods for all but the simplest systems. In addition, electron and nuclear magnetic resonance (NMR) spectroscopy give very important information about molecular electronic structure, and students of physical chemistry should have some familiarity with this important field.

It was pointed out in Chapter 6 that an intrinsic spin angular momentum had to be postulated for the electron in order to account for many experimental results. Nuclei, also, have such an intrinsic spin angular momentum, and it is characterized by a nuclear spin quantum number I. Nuclei differ from electrons in that there are a large number of different values which the nuclear spin quantum number can have. The electron spin quantum number, it should be recalled, is restricted to the value $\frac{1}{2}$, whereas nuclear spin quantum numbers range in half-integral units from 0 to 6. Unless chemists wish to study nuclear structure, they must accept the nuclear spin quantum number for any given nucleus as a matter of empirical fact, but there are at least some rules which divide nuclei into classes according to their mass number A and their charge number Z. These rules are the following.

1. If the mass number A is odd, the nuclear spin I is half-integral.
2. If the mass number A and the charge number Z are both even, the spin is zero.
3. If the mass number A is even, but the charge number Z is odd, the spin is integral.

Thus, ^{1}H, ^{19}F, and ^{31}P have nuclear spin $I = \frac{1}{2}$. The nuclei ^{16}O and ^{12}C have $I = 0$, and ^{2}H (deuterium), ^{6}Li, and ^{14}N have $I = 1$. We are only concerned in this chapter with nuclei which have spin $\frac{1}{2}$. This enables us to treat these nuclei and electrons with the same nomenclature.

The properties of nuclear spin angular momentum are essentially the same as those for electron spin angular momentum except that, for most nuclei, the nuclear magnetic moment is parallel to the spin vector and not antiparallel as with electrons (^{15}N, $I = \frac{1}{2}$, is an exception). That is, for nuclei with spin $\frac{1}{2}$, there are two functions α and β, each of which is an eigenfunction of both $\hat{I}^2$ and $\hat{I}_z$, where $\hat{\mathbf{I}}$ is the nuclear spin angular momentum operator and $\hat{I}_z$ is the operator for the z component of nuclear spin. The functions α and β have the property that

$$\begin{aligned} \hat{I}^2\alpha &= \tfrac{1}{2}(\tfrac{1}{2} + 1)\alpha \qquad & \hat{I}_z\alpha &= \tfrac{1}{2}\alpha \\ \hat{I}^2\beta &= \tfrac{1}{2}(\tfrac{1}{2} + 1)\beta \qquad & \hat{I}_z\beta &= -\tfrac{1}{2}\beta \end{aligned} \tag{9-1}$$

For these nuclei, there is a corresponding magnetic moment operator $\hat{\boldsymbol{\mu}}$ equal to

$$\hat{\boldsymbol{\mu}} = g_N \beta_N \hat{\mathbf{I}} \tag{9-2}$$

where g_N is the nuclear g factor that is characteristic of each nucleus, and β_N is the nuclear magneton equal to 5.0505×10^{-24} erg G^{-1}. The student will notice that Eq. 9-2 is identical with Eq. 6-64 for the magnetic moment due to electron spin except that the minus sign is missing. The operators for nuclear spin angular momentum combine and commute in the same way as do those for electron spin angular momentum.

9-1 THE INTERACTION OF AN ISOLATED SPIN $\frac{1}{2}$ PARTICLE WITH AN APPLIED MAGNETIC FIELD

If a spin $\frac{1}{2}$ particle is placed in a static magnetic field, the two states α and β no longer have the same energy. The classical energy of interaction between a magnetic dipole $\hat{\boldsymbol{\mu}}$ and a static field $\mathbf{H}$ is given by

$$E = -\hat{\boldsymbol{\mu}} \cdot \mathbf{H} \tag{9-3}$$

The corresponding magnetic Hamiltonian operator for this interaction is therefore

$$\hat{\mathscr{H}}_e = -(-g_0 \beta \hat{\mathbf{S}}) \cdot \mathbf{H} = g_0 \beta \hat{\mathbf{S}} \cdot \mathbf{H} \text{ (electrons)} \tag{9-4a}$$

$$\hat{\mathscr{H}}_N = -g_N \beta_N \hat{\mathbf{I}} \cdot \mathbf{H} \text{ (nuclei)} \tag{9-4b}$$

The student should recall (Section 1-4) that the dot product of two vectors $\hat{\mathbf{S}}$ and $\mathbf{H}$ is equal to the length of $\mathbf{H}$ times the length of the projection of $\hat{\mathbf{S}}$ on $\mathbf{H}$. Since the direction of $\mathbf{H}$ defines the Z axis, the projection of $\hat{\mathbf{S}}$ on $\mathbf{H}$ is $\hat{S}_z$. Equations 9-4 then become

$$\hat{\mathscr{H}}_e = g_0 \beta H \hat{S}_z \tag{9-5a}$$

$$\hat{\mathscr{H}}_N = -g_N \beta_N H \hat{I}_z \tag{9-5b}$$

The energies of the states α and β are found from the eigenvalue equations

$$\begin{aligned} \hat{\mathscr{H}} \alpha &= E_\alpha \alpha \\ \hat{\mathscr{H}} \beta &= E_\beta \beta \end{aligned} \tag{9-6}$$

The student can easily verify that

$$\begin{aligned} E_\alpha &= \pm \tfrac{1}{2} g_i \beta_i H \\ E_\beta &= \mp \tfrac{1}{2} g_i \beta_i H \end{aligned} \tag{9-7}$$

where the upper sign is for electrons and the lower sign is for nuclei, and where the appropriate values of g and β are used. The quantity H is the magnitude of the magnetic field. Thus the energy of the states α and β diverges as H increases. A plot of the energy of the states α and β for electrons is shown in Figure 9-1. A similar plot holds for spin $\frac{1}{2}$ nuclei except that the labeling of the states is reversed and the appropriate values of g and β must be used.

The phenomena of electron paramagnetic resonance (EPR) and NMR are based upon the fact that transitions are possible between the two states α and β. These transitions occur when an electromagnetic wave of frequency ν

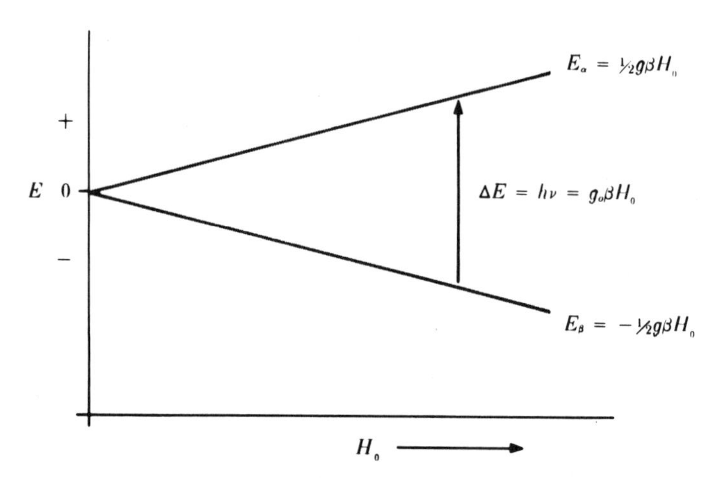

FIGURE 9-1 *A plot of the energy of the states α and β for an isolated electron in an applied magnetic field. A similar plot holds for protons except that the α and β states must be exchanged, and the appropriate values of g and β must be used. Transitions between these two states are studied in magnetic resonance experiments. The transition frequency v is equal to $g\beta H_0/h$.*

is applied to the sample under the condition that

$$h\nu = g_i \beta_i H \tag{9-8}$$

Equation 9-8 is called the resonance condition.

EXERCISE 9-1 Calculate the resonance frequency for a sample containing unpaired electrons in a magnetic field of 3,300 G. Use the Boltzmann distribution law to calculate the population ratio of the α and β states at room remperature (300 K).

9-2 EPR SPECTROSCOPY—HYPERFINE INTERACTIONS

For EPR spectroscopy, the experimental conditions that are usually used to satisfy Eq. 9-8 are a magnetic field strength of about 3,300 G, and a radio frequency of approximately 9,500 MHz. [EPR is sometimes called electron spin resonance (ESR). The terms are synonymous.] For an atom or a molecule to have an EPR spectrum, it must have one or more unpaired electrons. In quantum mechanical language, the wave function for the atom or molecule must have a nonzero eigenvalue with respect to the operator $\hat{S}^2$. In general, there are four main situations in which such atoms or molecules occur. These are the following:

1. Paramagnetic ions in solution and in crystals such as Fe^{3+}, Mn^{2+}, and so on.

2. Organic and inorganic free radicals in solution.
3. Radiation-produced fragments in crystals.
4. Molecules in the triplet state trapped in solids.

Most of the discussion in this chapter will be concerned with case 2.

If the interaction discussed in Section 9-1 were the only one present, EPR spectroscopy would not be very interesting. All atoms or molecules that contained unpaired electrons would give an EPR spectrum consisting of a single absorption at the frequency satisfying the condition described by Eq. 9-8. Fortunately, there are several interactions between unpaired electrons and their environment, and it is these interactions which make EPR fascinating to those interested in the details of molecular electronic structure.

The actual magnetic field which an unpaired electron experiences is the vector sum of the applied field and the *internal* fields of the molecule. Some important types of internal fields are the following:

1. Fields due to other unpaired electrons in the molecule or ion.
2. Fields due to magnetic nuclei in the molecule or ion.
3. Fields due to unpaired electrons in surrounding molecules or ions.

In addition, there may be a contribution to the magnetic moment of the electron from its *orbital* angular momentum. This latter factor affects the frequency of an EPR absorption because it changes the g value from the free spin value.

In what follows, it is assumed that there is no contribution to the magnetic moment of the electron from its orbital angular momentum, and that the only source of internal magnetic fields are magnetic nuclei in the molecule. There are then two types of interaction between the unpaired electrons and these magnetic nuclei.

1. The dipole-dipole interaction. This is analogous to the classical interaction of a pair of magnetic dipoles. The only difference is that the dipole moment of the unpaired electron must be calculated quantum mechanically because an electron is, in fact, distributed over space. This interaction is anisotropic. That is, it depends on the relative orientations of the applied magnetic field and the vector between the unpaired electron and the nucleus of interest. This interaction is an important source of information in single-crystal studies, but it is averaged to zero in liquids because of the rapid tumbling motion of the molecules. The mathematics involved in discussing the dipole-dipole interaction are somewhat complicated, and this interaction will not be considered further here.

2. The isotropic or Fermi-contact interaction. This is a quantum mechanical interaction and has *no classical analog*. It arises when there is a nonzero probability of finding the unpaired electron *at* the magnetic nucleus in question. In the orbital approximation, s orbitals are the only ones that do not vanish at the nucleus. If the wave function for the odd electron has some orbital s character which is centered on a magnetic nucleus, an

isotropic hyperfine interaction will be observed. In fact, this is the only kind of interaction observed for free radicals in solution.

To illustrate the effects of the isotropic hyperfine interaction on an EPR spectrum, consider the case of a system of gaseous hydrogen atoms. In the ground state, the electron in a hydrogen atom has no orbital angular momentum; also, the dipole-dipole interaction vanishes for electrons in s orbitals. The only interaction that must be considered is the isotropic hyperfine interaction between the odd electron and the proton that has spin $I = \frac{1}{2}$. If an EPR experiment is done in very large magnetic fields, both the electron and proton will have their magnetic moments quantized along the direction of the applied field. This gives rise to four different states, each with a slightly different energy. These states are shown schematically in Figure 9-2. A consideration of the selection rules shows that the only transitions that are allowed are those for which the nuclear spin quantum number does not change. Thus transitions between states 3 and 1, and between states 4 and 2, would occur. The EPR spectrum would therefore consist of two lines

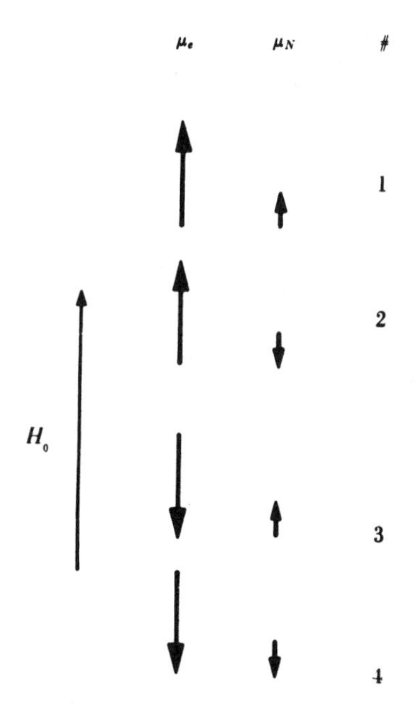

FIGURE 9-2 *Drawing of the possible arrangements of the electron and proton magnetic moment vectors for a hydrogen atom in a strong magnetic field. The numbers of the states correspond to the numbers of the wave functions used later in the quantum mechanical description. The sizes of the electron and nuclear magnetic moment vectors are not to scale. The electron spin vector is actually* ~1,000 *times larger than the proton spin vector.*

corresponding to these two transitions. The spacing between the two lines is the proton hyperfine coupling constant a.

The above has been a qualitative discussion. To show how the energy levels work out in detail, we must calculate them quantum mechanically. The spin part of the Hamiltonian operator for a hydrogen atom in a magnetic field is

$$\hat{\mathscr{H}}_{\text{spin}} = +g_0 \beta \hat{\mathbf{S}} \cdot \mathbf{H} - g_N \beta_N \hat{\mathbf{I}} \cdot \mathbf{H} + ha\hat{\mathbf{S}} \cdot \hat{\mathbf{I}} \qquad (9\text{-}9)$$

In Eq. 9-9, the first two terms represent the interaction of the electron and nuclear spins with the applied field, and the third term represents the electron-nuclear hyperfine interaction, a being the hyperfine coupling constant just defined and h being Planck's constant. Equation 9-9 assumes that all dipole-dipole interactions are zero, and that there is no contribution from electron orbital angular momentum. As mentioned above, both of these assumptions are true for hydrogen atoms. For strong magnetic fields the Hamiltonian 9-9 takes the simple form

$$\hat{\mathscr{H}}_{\text{spin}} = g_0 \beta H \hat{S}_z - g_N \beta_N H \hat{I}_z + ha\hat{S}_z \hat{I}_z \qquad (9\text{-}10)$$

In this case, strong magnetic fields are defined by the condition that $g_0 \beta H$, $g_N \beta_N H \gg a$. For a hydrogen atom in a 10,000-G field, $g_0 \beta H \approx 20$ GHz, $g_N \beta_N H \approx 42$ MHz, and $a \approx 1{,}400$ MHz. Thus the condition $g_0 \beta H \gg a$ holds but not $g_N \beta_N H_0 \gg a$. The above calculation for hydrogen is therefore an approximate one. For the organic radicals discussed later, the approximation $g_N \beta_N H > a$ is fairly good, and the simplified form of the Hamiltonian can be used. For an example of a more complete calculation, the reader is referred to [1]. The spin functions appropriate for this problem are the proper product functions of single electron and nuclear spin functions. These are

$$\Psi_1 = \alpha(e)\alpha(p) \qquad (9\text{-}11a)$$

$$\Psi_2 = \alpha(e)\beta(p) \qquad (9\text{-}11b)$$

$$\Psi_3 = \beta(e)\alpha(p) \qquad (9\text{-}11c)$$

$$\Psi_4 = \beta(e)\beta(p) \qquad (9\text{-}11d)$$

It can be readily seen, by operating on the functions 9-11 with the Hamiltonian 9-10, that all are already appropriate eigenfunctions. The eigenenergies can immediately be written down and are

$$E_1 = \tfrac{1}{2}g_0\beta H - \tfrac{1}{2}g_N\beta_N H + \tfrac{1}{4}ha \tag{9-12a}$$

$$E_2 = \tfrac{1}{2}g_0\beta H + \tfrac{1}{2}g_N\beta_N H - \tfrac{1}{4}ha \tag{9-12b}$$

$$E_3 = -\tfrac{1}{2}g_0\beta H - \tfrac{1}{2}g_N\beta_N H - \tfrac{1}{4}ha \tag{9-12c}$$

$$E_4 = -\tfrac{1}{2}g_0\beta H + \tfrac{1}{2}g_N\beta_N H + \tfrac{1}{4}ha \tag{9-12d}$$

In a typical magnetic resonance experiment, the sample interacts with the oscillating magnetic field of an electromagnetic wave. This oscillating magnetic field is applied perpendicular to the applied field. The appropriate transition moment for calculating intensities is

$$R_{ij} = (\Psi_i|\hat{\mu}_\perp|\Psi_j) = g_0\beta(\Psi_i|\hat{S}_\perp|\Psi_j) \tag{9-13}$$

where $\hat{\mu}_\perp$ is the operator for the component of electronic spin angular momentum perpendicular to the applied field and $\hat{S}_\perp = \hat{S}_x + i\hat{S}_y = \hat{S}_+$. The intensity of the EPR transitions is proportional to

$$g_0{}^2\beta^2\,|(\Psi_i|\,\hat{S}_\perp|\,\Psi_j)|^2$$

Applying Eq. 9-13 to calculate the various R_{ij}, it can be shown, using the methods of Chapter 6, that

$$\begin{aligned} R_{13} &= R_{24} = g_0{}^2\beta^2 \\ R_{12} &= R_{14} = R_{23} = R_{34} = 0 \end{aligned} \tag{9-14}$$

Thus only the transitions $3 \rightarrow 1$ and $4 \rightarrow 2$ are allowed, and these have energies

$$\begin{aligned} E_1 - E_3 &= g_0\beta H + \tfrac{1}{2}ha \\ E_2 - E_4 &= g_0\beta H - \tfrac{1}{2}ha \end{aligned} \tag{9-15}$$

There will therefore be two lines in the EPR spectrum of atomic hydrogen centered at $g_0\beta H$ and separated by ha. The student can see that all of the results of the qualitative treatment given first are confirmed by the quantum mechanical calculation.

EXERCISE 9-2 The student should work through all of the details of Eq. 9-12 through 9-15 by carrying out the appropriate operations.

EXERCISE 9-3 A more exact Hamiltonian for a hydrogen atom is

$$\hat{\mathscr{H}}_{\text{spin}} = g_0 \beta H \hat{S}_z + ha\hat{\mathbf{S}} \cdot \mathbf{I}$$

where the nuclear Zeeman term has been neglected. This Hamiltonian differs from Eq. 9-10 in that it includes the terms

$$ha[S_x I_x + S_y I_y]$$

Treating these terms as a perturbation, derive corrected expressions for the energy levels of a hydrogen atom using perturbation theory. If $a = 1{,}420$ MHz, what error is made in using Eq. 9-12 to compute the spacing in the spectrum taken using $\nu = 9{,}500$ MHz? *Hint*: Matrix elements can be more conveniently evaluated if the perturbing Hamiltonian is written in terms of $\hat{S}_\pm$ and $\hat{I}_\pm$.

Calculation of the EPR spectrum of more complicated radicals is a straightforward, but time consuming, extension of the principles described above. It is more convenient, for these more complex radicals, to use the qualitative picture first discussed.

As a further example, consider the EPR spectrum of the methyl radical $\cdot CH_3$. This radical is planar with bond angles of 120°. The three protons are magnetically equivalent because they are related to one another by a symmetry operation of the molecule. Figure 9-3 shows the possible alignment of the three nuclear moments for a single electron moment alignment. It can be seen that there are *four* possible internal fields due to the three protons, depending on how the proton moments are aligned. The energy differences

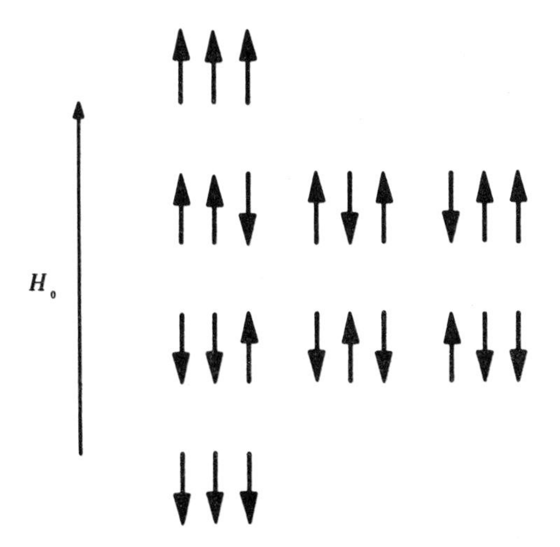

FIGURE 9-3 *Possible alignments of the nuclear moments in the methyl radical. Any of these combinations can occur with any of the two possible electron moment alignments. In a large sample of methyl radicals, the odd electrons will experience four slightly different fields from the four different nuclear moment alignments.*

between nuclear spin levels (see below) are extremely small, and, consequently, all of the configurations shown in Figure 9-3 will have equal populations. The EPR spectrum of a large number of methyl radicals will appear as though each electron spin configuration sees *four* different internal magnetic fields since the nuclear spin configuration in each molecule will be a different one of the four in Figure 9-3. These states will be weighted in the ratio of 1 : 3 : 3 : 1 since there are three times as many ways for the nuclei to give the middle fields in Figure 9-3 as there are for them to give the end fields. The energy level diagram for a methyl radical will therefore look like that of Figure 9-4. It should be pointed out that transitions in Figure 9-4 are represented by arrows of constant length since, in EPR spectrometers, the frequency is held constant and the magnetic field is varied. The EPR spectrum of the methyl radical is therefore made up of four equally spaced lines. The spacing between these lines is independent of H_0 (except in very weak fields) and is, again, equal to the hyperfine coupling constant a.

In general, for a molecule which has N equivalent protons, there will be $N + 1$ hyperfine lines in the EPR spectrum. The relative intensities of these lines will be in the same ratio as the binomial coefficients of the Nth order binomial expansion. If these coefficients are not familiar to the student, they can be calculated by making use of the magic triangle shown in Table 9-1.

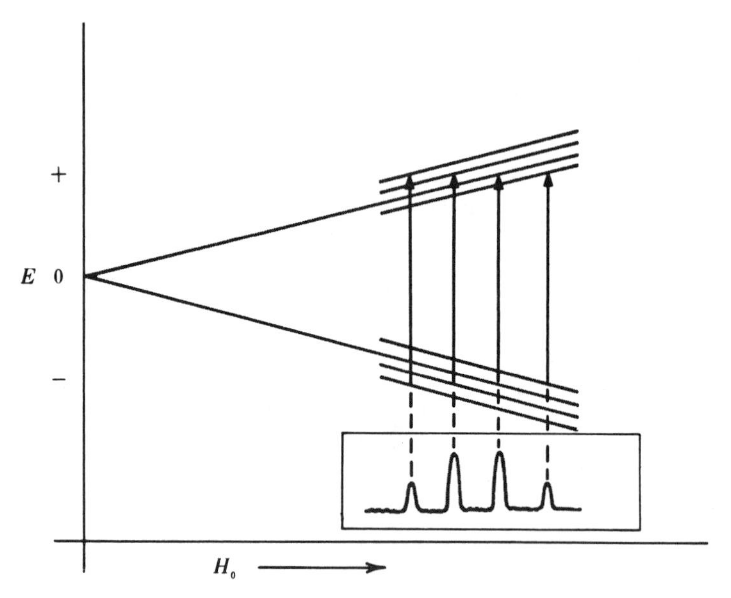

FIGURE 9-4 *The energy level diagram and EPR spectrum of the methyl radical. In EPR experiments, the magnetic field is varied and the frequency is held constant. Because of this, the transitions are represented as arrows with constant length. The lines in the spectrum are separated by the hyperfine coupling constant* $a = 23$ G.

TABLE 9-1 FIRST FIVE LINES OF THE MAGIC TRIANGLE FOR FINDING RELATIVE INTENSITIES OF HYPERFINE LINES IN EPR SPECTRA. WHAT IS THE NEXT LINE?

N	*Relative intensities of* $N+1$ *Lines*								
0					1				
1				1		1			
2			1		2		1		
3		1		3		3		1	
4	1		4		6		4		1

For molecules with nonequivalent protons, the EPR spectrum can be constructed as follows.

1. Divide the protons in the molecule into equivalent groups according to the symmetry of the molecule. Assign each group a coupling constant a_i.
2. Starting with a single line on a piece of graph paper, construct the spectrum that would be expected if only the group of protons with the largest splitting constant were present. (The student must use intuition to decide which group this is if the a_i are not known.)
3. Use each of the lines obtained in step 2 and divide them according to the number of protons in the group with the next largest splitting constant.
4. Continue the process until all of the groups of protons have been used.

The above procedure will be illustrated for the ethyl radical $\cdot CH_2CH_3$. This radical was observed by doing an EPR experiment on liquid ethane while it was being irradiated with a beam of electrons from a van de Graaff generator. If the protons on the radical center are called σ protons and the methyl protons are called π protons, then the splitting constants are [2]

$$a_\sigma = 22.38 \text{ G}$$

$$a_\pi = 26.87 \text{ G}$$

A construction of the theoretical spectrum of the ethyl radical using the above rules is shown in Figure 9-5. There are three π protons and two σ protons, and the initial EPR line will be split first into a 1:3:3:1 quartet with spacing a_π. Each of these lines will then be split into a 1 : 2 : 1 triplet with spacing a_σ to give the final spectrum. This final calculated spectrum should be compared with the experimental spectrum shown in Figure 9-6.

In practice, experimentalists usually have a pretty good idea of what free radical they have, but they have no knowledge of the coupling constants. To analyze an EPR spectrum, a reverse procedure to that described above is

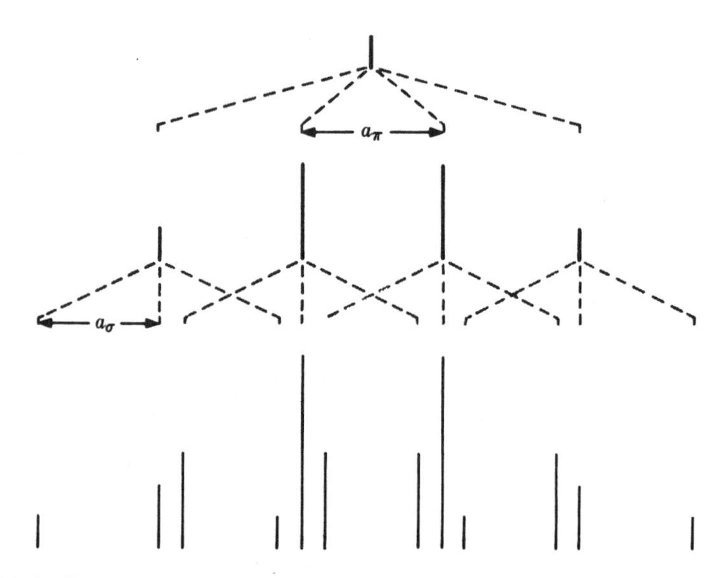

FIGURE 9-5 *Construction of the EPR spectrum of the ethyl radical. A ratio of* $a_\pi : a_\sigma$ *of* 13 : 11, *which is close to the experimental ratio, was used in construction. This spectrum looks enough like the experimental spectrum* (*Figure 9-6*) *that the lines can be assigned and coupling constants measured.*

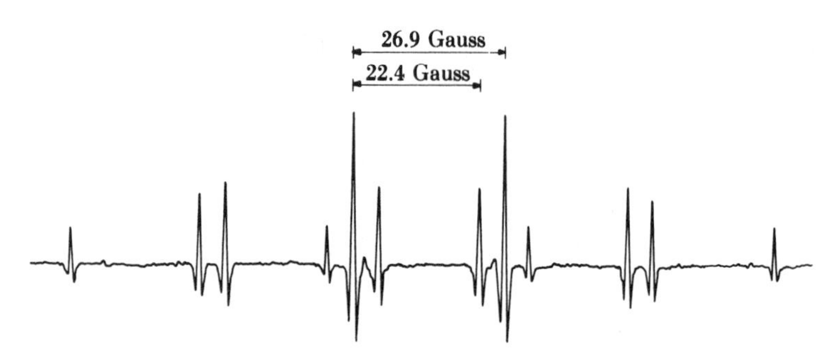

FIGURE 9-6 *Experimental spectrum of the ethyl radical. By comparing this spectrum with the theoretical spectrum in Figure 9-5, accurate values of the coupling constants can be measured. For example, the spacing between the two most intense lines is equal to* a_π; *the spacing between the two end lines is* a_σ. (*Courtesy of R. W. Fessenden and the American Institute of Physics.*)

used. That is, experimentalists construct several trial spectra using different ratios of coupling constants for the radical that they think they have. These coupling constant ratios are guessed at from intuition and past experience. A calculated spectrum that looks like the observed spectrum is (hopefully) finally obtained, and the actual coupling constants are measured from the observed spectrum. A large amount of coupling constant data on conjugated organic radicals is now available for guidance in this process [3].

9-3 COUPLING CONSTANTS AND MOLECULAR ELECTRONIC STRUCTURE

Isotropic hyperfine coupling constants are important in studying the electronic structure of large conjugated organic radicals. It was pointed out above that the isotropic hyperfine coupling constant measures the amount of s character in the odd electron wave function about the nucleus in question. This fact immediately raises an important question. In conjugated aromatic radicals, one observes hyperfine structure from the protons attached to the aromatic ring. Yet the unpaired electron in these radicals is in a π-molecular orbital, and these molecular orbitals vanish in the molecular plane since they are constructed from linear combinations of $2p_z$ atomic orbitals (see Chapter 8). The explanation for this paradox that is now accepted is that there is a small amount of mixing between the σ and π orbitals. That is, the σ-π interactions that were neglected in the discussion of π-electronic structure in Chapter 8 are important when talking about hyperfine interactions. Using this idea, McConnell [4] showed that the hyperfine coupling constant was related to the unpaired spin density on the adjacent carbon by the relation

$$a_\mu = Q\rho_\mu \tag{9-16}$$

where a_μ is the observed coupling constant for proton μ, ρ_μ is the spin density on the carbon atom to which proton μ is attached, and Q is a semiempirical constant equal to 23–27 G. The importance of Eq. 9-16 is that it gives an experimental means of determining the odd electron distribution in conjugated molecules. This experimentally determined electron distribution can then be compared with the results of various theoretical calculations. For example, in a simple Hückel calculation, the spin density at any carbon atom can be very easily calculated. For carbon atom μ, it is equal to $c_{i\mu}^2$, where $c_{i\mu}$ is the coefficient of atom μ *in the molecular orbital i in which the odd electron resides.* Thus for the naphthalene anion radical, for example, the unpaired electron would be in the lowest unoccupied orbital of the neutral molecule, and the odd electron densities can readily be calculated.

Comparison of these calculated spin densities with the experimental values calculated from EPR spectra with the aid of Eq. 9-16 gives important information about the adequacy of various theoretical models.

EXERCISE 9-4 In Figure 9-7, an experimental spectrum of the naphthalene anion radical is shown. Using the methods described above, analyze the spectrum to find the values of the two hyperfine coupling constants a_1 and a_2. Because of the nature of the detection methods used in EPR, the spectrum in Figure 9-7

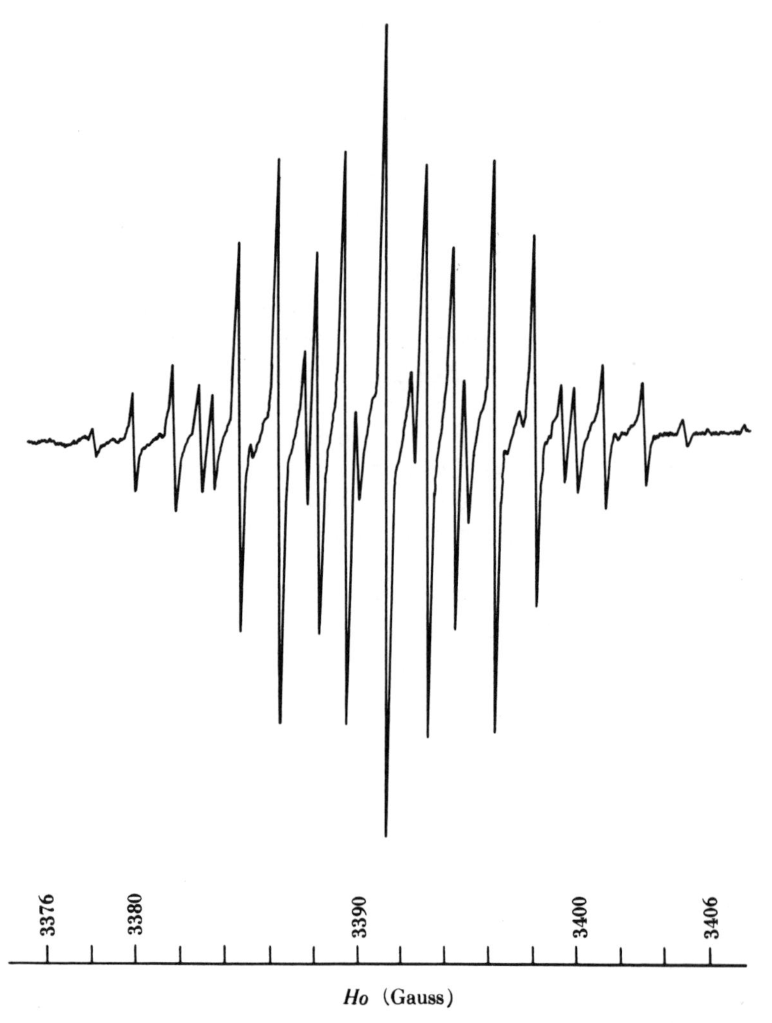

FIGURE 9-7 *Experimental spectrum of the naphthalene anion radical. This is a first derivative spectrum taken at the University of Colorado. The magnetic field strengths are given at the bottom of the figure in gauss. For students interested in calculating the g value of the radical, the cavity frequency was* 9,503 MHz.

is the *first derivative* of the absorption spectrum. The point where the derivative crosses the baseline corresponds to the maximum in an absorption spectrum. The coupling constants should come out close to 5.0 and 1.8 G.

EXERCISE 9-5 Using the results of Exercise 8-6, calculate the spin densities at positions 1 and 2 of naphthalene. Compare these values and their ratio with the corresponding quantities determined from the experimental spectra in Exercise 9-4 and Eq. 9-16. Use $Q = 24$ G.

9-4 NMR SPECTROSCOPY—THE CHEMICAL SHIFT

The resonance condition for spin $\frac{1}{2}$ nuclei is analogous to Eq. 9-8 except that the nuclear magneton and nuclear g factor must be used. Thus, for nuclei,

$$h\nu = g_N \beta_N H \tag{9-17}$$

In many books, NMR is discussed classically rather than quantum mechanically. This is possible because the spacing between nuclear spin levels is much smaller than kT at room temperature. In this classical description, the macroscopic magnetic moment of the sample is thought of as precessing about the applied field with an angular frequency ω_0 called the Larmor precession frequency. The quantity ω_0 is related to the ν of Eq. 9-17 by the relations

$$\omega_0 = 2\pi\nu = \gamma H_0 \tag{9-18}$$

where γ is called the gyromagnetic ratio of the nucleus. By comparing Eqs. 9-17 and 9-18, the student can see that

$$\gamma \equiv \frac{\gamma}{2\pi} = \frac{g_N \beta_N}{h} \qquad \text{or} \qquad \gamma = \frac{g_N \beta_N}{\hbar}$$

EXERCISE 9-6 For protons in a magnetic field of 10,000 G, resonance is observed at a frequency of 42.576 MHz. Calculate g_N and γ for protons.

EXERCISE 9-7 A manufacturer has recently marketed a 100-MHz proton NMR. What magnetic field strength is used in this instrument?

EXERCISE 9-8 For a 60-MHz spectrometer, calculate the relative population of the α and β states of a sample of protons at 25°C.

Since most of the chemical applications of NMR involve proton resonance spectroscopy, the discussion that follows will be restricted to protons.

In NMR, the actual field at a given nucleus is, once again, not merely the applied field, but is the vector sum of the applied field and internal fields. For liquid samples, there are two major contributions to this internal field. The first contribution can be rationalized from the fact that nuclei are magnetically screened due to the diamagnetic circulation of electrons. This source of internal fields gives rise to the phenomenon called the "chemical shift." The second contribution comes from the nonvanishing field at each nucleus due to the magnetic moments of neighboring nuclei. This effect gives rise to "spin-spin splittings." The chemical shift will be discussed in the remainder of this section, and spin-spin splittings will be discussed in Section 9-5.

It is well known that if a molecule with no unpaired electrons is placed in an applied field, the electrons will circulate in such a way as to induce a magnetic field in the molecule opposed to the applied field. This is the same phenomenon as that responsible for the diamagnetic susceptibility of a molecule. Thus, in any molecule, each nucleus will experience an internal field that opposes the applied field due to this circulation of electrons. Furthermore, the strength of this opposing field is proportional to the applied field. We can thus write that the magnetic field at nucleus j is

$$H_j = H_0(1 - \sigma_j) \tag{9-19}$$

where σ_j is called the screening constant of nucleus j. Our transition frequency then becomes

$$\nu = \gamma H_0(1 - \sigma_j) \tag{9-20}$$

For protons, σ_j is of the order of 0–10 ppm, but its absolute value cannot be measured. The reason for this is that, to measure σ_j, one would have to make independent measurements of both frequency and magnetic field to better than 1 ppm. It is possible to measure frequencies to this accuracy, but the most accurate way to measure magnetic fields is to use an NMR probe and measure the frequency necessary to induce resonance. Thus it is not possible to determine H_0 accurately enough to measure σ_j. All chemical shifts are therefore measured with respect to the shift of a proton in a suitable reference compound. Unfortunately, there has been no general agreement on referencing procedures, so that the literature has many chemical shift values using different references. Most modern work, however, uses tetramethylsilane (TMS) as a reference compound.

The chemical shift for a given nucleus δ_j is defined as follows:

$$\delta_j = \frac{H_s - H_r}{H_r} \times 10^6 \text{ ppm} \tag{9-21}$$

where H_s is the value of the magnetic field at the sample absorption peak and H_r is the value of the magnetic field at the reference absorption peak. Fields are used because it is more convenient to run an NMR experiment at constant frequency and vary H_0 slightly than to vary the frequency at constant H_0. To make matters more confusing for the novice, however, field strengths are often expressed in frequency units. The two are equivalent since, for a given nucleus, field is directly proportional to frequency through the gyromagnetic ratio. Substituting Eq. 9-19 into Eq. 9-21, and using the fact that

$\sigma_R \ll 1$, we obtain

$$\begin{aligned}\delta\,(\text{dimensionless}) &= \frac{\gamma H_0(1-\sigma_s) - H_0(1-\sigma_r)}{\gamma H_0(1-\sigma_r)} \times 10^6 \\ &= \frac{\sigma_r - \sigma_s}{1} \\ &= \frac{\nu_s - \nu_r}{\text{spectrometric frequency}} \times 10^6\end{aligned} \tag{9-22}$$

Using this convention, δ is positive if the reference is more screened than the sample, that is, if the reference signal occurs at a *higher* value of H_0 than the sample signal. A chemical shift scale using this convention assigns the protons in TMS the value 0.00. The quantity δ then measures the shift *downfield* from a TMS reference of the protons of interest. A second scale commonly used is the τ scale. In this scale, the protons in TMS are assigned the value 10.000; the chemical shift in τ units is then

$$\tau = 10.000 - \delta$$

There is, at present, some disagreement as to which of these scales is better.

The chemical shift is an important parameter in NMR spectroscopy because, in general, nuclei in different chemical environments will have different screening constants. Thus, in the best cases, there will be one absorption in an NMR spectrum for each group of chemically distinct protons. The relative areas under each of these absorption peaks also give the relative number of protons in each equivalent group. These features have proved valuable in using NMR to identify unknown compounds. For example, suppose we have a compound with known empirical formula C_4H_6. Its NMR spectrum shows two peaks at roughly 4.0 and 7.7 τ with relative heights of 1 and 2, respectively. What is the structure of the compound? There are four possible compounds that can be written that have the correct molecular formula. They are

$CH_3CH_2C{\equiv}CH$
I
ethylacetylene

$CH_2{=}CH{-}CH{=}CH_2$
II
butadiene

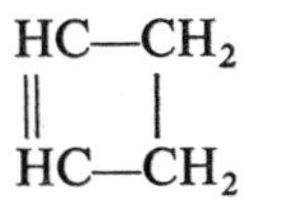

$$\begin{array}{ll} HC & \!\!\!-CH_2 \\ \| & \;\;| \\ HC & \!\!\!-CH_2 \end{array}$$
III
cyclobutene

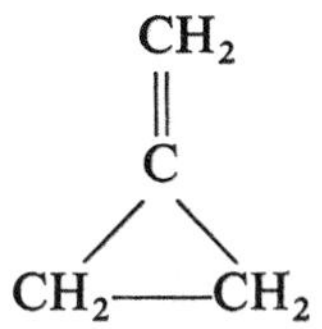

$$\begin{array}{c} CH_2 \\ \| \\ C \\ / \quad \backslash \\ CH_2{-\!\!-}CH_2 \end{array}$$
IV
methylenecyclopropane

The NMR spectrum immediately rules out compound I because it should have three different kinds of protons. Compounds II, III, and IV would all be consistent with the number and relative intensities of the lines in the spectrum, however. The chemical shift difference between the two kinds of protons is quite large, however, and in compound II one would not expect much difference in shielding of the six protons because they are all attached to a double bond. Compound II can, therefore, be ruled out. To make a final decision between compounds III and IV, one must be aware of an empirical fact that protons in a cyclopropane ring would be expected to occur at abnormally high τ values [5]. Knowing this, compound IV can be ruled out, and the unknown substance has been identified as compound III with only its empirical formula and NMR spectrum. This problem and many others can be found in [6].

There is now a great deal of empirical information such as that used in the above discussion about the relation between proton chemical shifts and molecular structure. The interested student should consult [5, 6] for excellent discussions of some of these relationships.

9-5 NMR SPECTROSCOPY—SPIN–SPIN SPLITTINGS

Using the information presented above, we would expect the NMR spectrum of ethyl alcohol (CH_3CH_2OH) to consist of three lines with relative intensities 3 : 2 : 1 corresponding to the three types of protons in the molecule, three methyl protons, two methylene protons, and one hydroxyl proton. This is, in fact, what is observed if the spectrum is observed under *low resolution* (Figure 9-8a). If the spectrum is observed under high resolution, and if the alcohol has not been specially purified, additional splittings appear in the spectrum as shown in Figure 9-8b. These additional splittings have several characteristic features. These are the following.

1. The spacing between lines in the CH_2 quartet is exactly the same as the spacing between lines in the CH_3 triplet.

2. The spacings of the lines in the CH_2 quartet and the CH_3 triplet are independent of the magnetic field strength used in the experiment. The spacing between the CH_2 and CH_3 multiplets (arising from the chemical shift) is directly proportional to the field strengths.

This additional structure arises from the second source of internal magnetic fields mentioned above. Splittings arising from this source are called spin-spin splittings, and the spacing between two lines in the multiplet is called the spin-spin splitting constant J. The existence of spin-spin interactions implies that a given proton has a way of "knowing" how the protons on adjacent atoms are oriented in the applied magnetic field. The interaction cannot be a magnetic dipole-dipole coupling because such a coupling would be averaged to zero in nonrigid molecules by the rapid rotational and vibrational motions

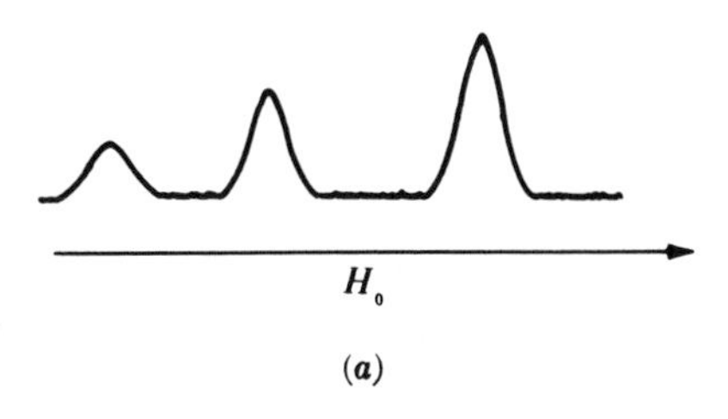

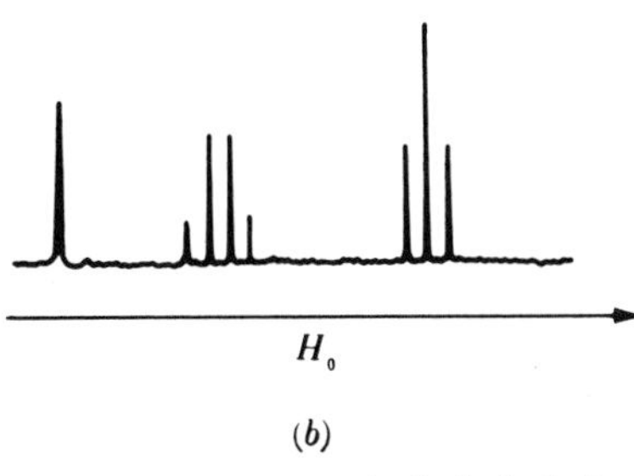

FIGURE 9-8 *Proton resonance spectrum of ethyl alcohol under* (*a*) *low resolution,* (*b*) *high resolution. The relative areas under the three peaks are* 1 : 2 : 3, *and this serves to identify them. The* OH *peak has no spin-spin splittings in* (*b*) *because of chemical exchange. The separations between peaks are not to scale.*

within the molecule. Ramsey and Purcell [8] have suggested a mechanism for this interaction that involves an indirect coupling via the electrons. Consider the case of HF (^{19}F has spin $\frac{1}{2}$). The oriented proton spin would tend to give the electron in the HF bond closest to it a preferred orientation. Because of the Pauli principle, the electron in the HF bond farthest from the proton would have the opposite orientation. This electron, in turn, would give the ^{19}F nuclear spin a preferred orientation. Thus the ^{19}F experiences a small internal field which depends on the orientation of the proton.

The number and relative intensities of the lines within a multiplet can be rationalized by a qualitative picture similar to that used for calculating the number and intensity of the hyperfine components in an EPR spectrum. That is, there is a magnetic interaction of *each* equivalent group of protons with the protons (or other magnetic nuclei) on *adjacent* atoms. For example, if we observe the resonance of the methylene protons in ethyl alcohol, there are four different fields at which resonance can occur, depending on how the methyl protons are aligned in that particular molecule. These four fields arise from the same nuclear configurations that were responsible for the hyperfine splittings in the methyl radical (see Figure 9-3). In a sample with a large number of alcohol molecules, the resonance of the methylene protons on different molecules will occur at any of four slightly different fields. Since there are three times as many ways to get the middle configuration in Figure 9-3 as there are the end configurations, the middle lines in the methylene quartet will be three times as intense as the end lines.

A similar argument can be used to explain why the methyl protons are split into a triplet with relative intensities 1 : 2 : 1. There are three possible configurations of the two methylene protons, so that the protons on CH_3 group will experience three slightly different fields in a sample of ethyl alcohol molecules, with a statistical distribution of 1 : 2 : 1.

For cases where the chemical shift between two types of nuclei is much larger than the spin-spin splitting constant, the following rules enable one to calculate the spin-spin splitting pattern.

1. Divide the magnetic nuclei into groups of equivalent nuclei.
2. No spin-spin splittings are observed between members of the same group of equivalent nuclei.
3. The number and intensity of lines from one equivalent group of protons will depend on the number of *groups* of equivalent nuclei on *adjacent* atoms and on the number of nuclei in each group.
4. A group of N equivalent nuclei on an adjacent atom will give rise to $N + 1$ lines in the spectrum of the group being observed.
5. The intensities of these lines are in the ratios of the binomial coefficients of the expansion $(a + b)^N$, and these ratios can be found using the magic triangle in Table 9-1.
6. For cases where a nucleus is coupled to more than one group of equivalent nuclei with different J's, the spin-spin multiplet must be constructed like a complex hyperfine splitting pattern. That is, one begins with the splittings due to the group with the largest spin coupling constant, and constructs the pattern for coupling with that group alone. Each of the resulting lines is then coupled to the group of protons with next largest coupling constant, and so on.

EXERCISE 9-9 The NMR spectrum of pure 2, 3, 3, 4, 4-pentafluorocyclobutene can be analyzed by assuming that the proton has spin coupling constants of 9.5, 6.9, and 1.5 Hz with the cross ring fluorines, vinyl fluorine, and adjacent fluorines, respectively. Construct the multiplet pattern for this compound.

The student will, no doubt, have noticed that there is nothing in the above prescription for constructing spin-spin multiplets that explains the absence of any spin-spin splittings from or on the OH proton in the ethanol spectrum. For very pure ethanol samples, such spin-spin splittings are observed. That is, the OH resonance is a triplet, and the CH_2 resonance is a *pair* of quartets. This is exactly the behavior predicted on the basis of the above rules. In the absence of special purification procedures, most ethanol contains traces of an acid or base, and these substances catalyze the exchange of the OH proton between different molecules. If this chemical exchange is fast enough, the OH proton will experience only an average magnetic field, and a single line results. Similarly, the effects of spin-spin splittings of the OH proton on the CH_2 group also disappears.

In instances where the difference in chemical shifts between two groups of magnetically distinct nuclei becomes of the same order of magnitude as the spin-coupling constant, the spectrum becomes much more complicated. To interpret such spectra, a complete quantum mechanical analysis must be done. No new principles are involved in such a calculation, but a detailed discussion of the techniques for this analysis is beyond the scope of this book. The interested student is referred to the excellent treatments of this subject in [7, 9].

9-6 LINE WIDTHS, RELAXATION TIMES, AND THE UNCERTAINTY PRINCIPLE

Compared with other kinds of spectroscopy, the spacing between nuclear spin levels is very small. That is, ΔE for proton spin states in a 14,000-G field is much less than kT at room temperature. The two spin states will be almost equally populated with a slight excess of nuclei in the lower state (see Exercise 9-8).

When the spectrometer is swept through the resonance condition, this slight excess of population of the lower state is destroyed unless there are mechanisms by which nuclei in the upper state can "relax" back to the ground state, and thus maintain the Boltzmann distribution. Such relaxation mechanisms are called spin-lattice relaxation (sometimes called longitudinal relaxation), and are characterized by a spin-lattice relaxation time T_1. If T_1 is very long, nuclei cannot relax back to the lower state when excited in a resonance experiment, the Boltzmann excess in the lower state is destroyed, and the absorption signal disappears. This phenomenon is called saturation. If T_1 is very short, the energy of both the upper and lower states will become uncertain because of the uncertainty principle, and line broadening will result.

Spin-lattice relaxation arises from random fluctuations of the internal field of the sample. These fluctuations, in turn, arise from the tumbling motion of other molecules that possess magnetic nuclei and from the rotational and vibrational motion of other magnetic nuclei in the same molecule. There is a component of this randomly fluctuating field at the resonance frequency, and it is this component which causes spin-lattice relaxation.

Besides this fluctuating field, at any instant of time there will be an internal field that has *zero* frequency with respect to a given proton. In general, this zero frequency component will be slightly different for protons of the same type in different molecules. This internal field will give rise to an additional source of line broadening in addition to the uncertainty broadening discussed above. Since the width of an NMR line is measured in Hz, the reciprocal of a line width has units of time (sec). The actual line width of an NMR signal is characterized by an empirical time called the spin-spin relaxation time (or transverse relaxation time) and is designated T_2. In many cases in liquids, $T_1 = T_2$ and the line width is completely due to uncertainty broadening.

The form of the uncertainty principle to be used in this discussion is

$$\overline{\Delta E}\,\overline{\Delta t} \geq \hbar \tag{9-23}$$

This states that the product of the uncertainty in the energy of a state times its lifetime (the maximum uncertainty in time) must be equal or greater than $\hbar$. The quantity $\overline{\Delta E}$ can be written as $h\overline{\Delta \upsilon}$, however, and substituting this into Eq. 9-23 gives

$$\begin{aligned} h\,\overline{\Delta \nu}\,\overline{\Delta t} &\geq \frac{h}{2\pi} \\ \overline{\Delta \nu}\,\overline{\Delta t} &\geq \frac{1}{2\pi} \geq 0.159 \end{aligned} \tag{9-24}$$

This means that the uncertainty in the frequency of a spectral line times the uncertainty in the lifetime of one or both of the states involved must be equal to or larger than 0.159. For NMR spectroscopy, the quantity relating to $\Delta\nu$ is the width of the line at one half the maximum height. The reciprocal of this width in rad sec^{-1} is defined as $2/T_2$. Thus

$$(\Delta\omega)_{\frac{1}{2}\,\max} = 2\pi(\Delta\nu)_{\frac{1}{2}\,\max} \equiv \frac{2}{T_2} \tag{9-25}$$

and T_2 can therefore be measured directly from the NMR spectrum. The spin-lattice relaxation time T_1 is more difficult to measure. The most accurate method uses a pulsed NMR technique [10].

An understanding of relaxation phenomena is becoming more and more important in NMR experiments. For example, in several recent studies, measurements of the longitudinal relaxation time (T_1) for water have been used to study metal ion binding in biological systems. These studies are based on the observation that the presence of certain paramagnetic ions drastically shortens the T_1 for solvent water. Such a T_1 shortening is reasonable because the presence of the unpaired electrons on a rapidly tumbling ion in solution produces strong oscillating fields at the protons in water, and therefore provides an efficient relaxation mechanism. When a large biological molecule is added to the system, the paramagnetic ion may be bound onto it. If this happens, the rotational motion of the ion will be slowed and this will change its effect on the T_1 for water. A careful examination of these changes can give important information on enzyme-metal ion-subtrate interactions [11]. Line width studies can also be used to measure the rates of certain fast reactions [12].

9-7 SUMMARY

1. The energy levels of a system of independent spin $\frac{1}{2}$ particles in a magnetic field were derived. Transitions between these levels give rise to EPR and NMR phenomena.
2. The origin of hyperfine interactions in the solution of EPR spectra of free radicals was discussed. The EPR spectra for gaseous hydrogen atoms, methyl radicals, and ethyl radicals were described.
3. Hyperfine coupling constants from aromatic radicals were used to calculate odd electron densities by McConnell's relationship $a_i = Q\rho_i$. These experimental odd electron densities can be compared with densities calculated by various theoretical methods to test the adequacy of the theory.
4. The chemical shift in NMR spectroscopy was shown to be due to the screening effect of the electrons in a molecule. This screening effect is different for nuclei in different chemical environments, and the chemical shift is therefore a sensitive indicator of molecular electronic structure.
5. Spin-spin coupling gives rise to multiplet patterns in NMR spectroscopy that are formally the same as hyperfine coupling patterns in EPR spectroscopy. The appropriate spin-spin multiplets were calculated for the groups of protons in ethyl alcohol.
6. The two relaxation times of importance in NMR and EPR were introduced and the relationship between line widths, the uncertainty principle, and the magnitudes of the relaxation times were discussed.
7. The student should be familiar with the following terms: dipole-dipole interaction, Fermi-contact interaction, hyperfine coupling constant, spin density, Larmor precession frequency, diamagnetic susceptibility, chemical shift, spin-spin coupling constant, τ scale, δ scale, T_1, T_2, longitudinal relaxation time, transverse relaxation time, and saturation.

REFERENCES

1. A. Carrington and A. D. McLachlan, *Introduction to Magnetic Resonance* (Harper and Row, New York, 1967).
2. R. W. Fessenden and R. H. Schuler, *J. Chem. Phys.* **39**, 2147 (1963).
3. K. W. Bowers, *Advan. Magnet. Reson.* **1**, 317 (1965).
4. H. M. McConnell, *J. Chem. Phys.* **24**, 764 (1956).

5. L. M. Jackman, *Applications of NMR Spectroscopy to Organic Chemistry* (Pergamon Press, Inc., New York, 1959), p. 52.
6. J. D. Roberts, *Nuclear Magnetic Resonance* (McGraw-Hill Book Co., New York, 1959), Appendix C.
7. J. W. Emsley, J. Feeney, and L. H. Sutcliffe, *High Resolution Nuclear Magnetic Resonance Spectroscopy* (Pergamon Press, Inc., New York, 1966).
8. N. F. Ramsey and E. M. Purcell, *Phys. Rev.* **85**, 143 (1952).
9. J. D. Roberts, *An Introduction to Spin-Spin Splittings in NMR Spectroscopy* (W. A. Benjamin, Inc., New York, 1961).
10. H. Y. Carr and E. M. Purcell, *Phys. Rev.* **94**, 630 (1954).
11. A. S. Mildvan and M. Cohn, *Biochem.* **2**, 910 (1963).
12. W. C. Gardiner, Jr., *Rates and Mechanisms of Chemical Reactions* (W. A. Benjamin, Inc., New York, 1969).
13. T. C. Farrar and E. D. Becker, *Pulse and Fourier Transform NMR* (Academic Press, New York, 1971).

BIBLIOGRAPHY

GENERAL REFERENCES ON QUANTUM CHEMISTRY

1. H. Margenau and G. M. Murphy, *The Mathematics of Physics and Chemistry*, 2nd ed. (D. Van Nostrand, Princeton, N.J., 1956). Contains an excellent introduction to advanced mathematics needed for physical chemistry. A reprint of the 1956 edition was done by Krieger in 1976.
2. H. Eyring, J. Walter, and G. E. Kimball, *Quantum Chemistry* (J. Wiley & Sons, Inc., New York, 1949). A classic quantum chemistry text. The material is highly condensed and needs considerable work between the lines.
3. W. Kauzmann, *Quantum Chemistry* (Academic Press, New York, 1957). A more descriptive text than 2. It has a good discussion of classical vibrations, quadrupole effects, and optical activity.
4. J. M. Anderson, *Introduction to Quantum Chemistry* (W. A. Benjamin, Inc., New York, 1969). Integrates the matrix and operator approaches to quantum chemistry.
5. F. L. Pilar, *Elementary Quantum Chemistry* (McGraw-Hill Book Co., New York, 1968). An excellent comprehensive quantum chemistry text that provides not just an introduction to, but also a comprehensive treatment of, the state of the art in 1965–66.
6. M. Karplus and R. N. Porter, *Atoms and Molecules* (W. A. Benjamin, Inc., New York, 1970). An introductory book containing more material than the present text. The quantum mechanics is presented with a chemical flavor.

7. I. N. Levine, *Quantum Chemistry*, Vol. I: *Quantum Mechanics and Molecular Electronic Structure*; Vol. II: *Molecular Spectroscopy* (Allyn and Bacon, Boston, 1970). A comprehensive treatment of all topics in quantum chemistry.
8. P. W. Atkins, *Molecular Quantum Mechanics: An Introduction to Quantum Chemistry*, 2 vol. (Oxford University Press, New York, 1970).
9. C. R. Gatz, *Introduction to Quantum Chemistry* (Charles E. Merrill, Columbus, Ohio, 1971). A book aimed at undergraduate and first year graduate students with limited backgrounds in physics and math.
10. S. R. La Paglia, *Introductory Quantum Chemistry* (Harper and Row, New York, 1971). An introductory text with level and coverage similar to 6. It contains a good discussion of electron density shifts upon chemical bonding.
11. H. F. Schaefer, *The Electronic Structure of Atoms and Molecules: A Survey of Rigorous Quantum Mechanical Results* (Addison-Wesley, Reading, Mass., 1972). A summary of *ab initio* calculations to 1972 and a discussion of their importance for the next decade. An update of this material can be found in *Ann. Rev. Phys. Chem.* **27**, 261 (1976).
12. D. V. George, *Principles of Quantum Chemistry* (Pergamon Press, Elmsford, N.Y., 1973). A book for undergraduates. It is somewhat more detailed than the present text.
13. S. T. Epstein, *The Variational Method in Quantum Chemistry* (Academic Press, New York, 1974).
14. A. R. Denaro, *A Foundation for Quantum Chemistry* (Halsted Press, New York, 1975).
15. H. F. Hameka, *Quantum Theory of the Chemical Bond* (Hafner Press, New York, 1975). A book for a one-semester course at the undergraduate or first year graduate level. The math background is kept low.
16. W. H. Flygare, *Molecular Structure and Dynamics* (Prentice Hall, Inc., Englewood Cliffs, N.J., 1978).
17. J. N. Murrell, S. F. A. Kettle, and J. M. Teddar, *The Chemical Bond* (John Wiley & Sons, New York, 1978).

CLASSICAL MECHANICS

18. H. J. Goldstein, *Classical Mechanics*, 2nd ed. (Addison-Wesley, Reading, Mass., 1980). The "classic" classical mechanics text.

NONMATHEMATICAL QUANTUM MECHANICS AND VALENCE THEORY

19. C. A. Coulson, *Valence*, 2nd ed. (Oxford University Press, London, 1961).
20. L. Pauling, *The Nature of the Chemical Bond*, 3rd ed. (Cornell University Press, Ithaca, N.Y., 1960).

QUANTUM MECHANICS

21. D. R. Bates, Ed., *Quantum Theory, I. Elements* (Academic Press, New York, 1961). The first three chapters give precise but brief discussions of many of the subjects covered in this book.
22. P. A. M. Dirac, *The Principles of Quantum Mechanics*, 4th ed. (Oxford University Press, London, 1958). An advanced and very general treatment of quantum mechanics. The first chapter is interesting reading on the relation between quantum theory and experimental measurements.
23. H. F. Hameka, *Advanced Quantum Chemistry* (Addison-Wesley, Reading, Mass., 1965). Has good discussions of time dependent, optical, and magnetic problems.
24. A. Messiah, *Quantum Mechanics*, 2 vols. (John Wiley & Sons, Inc., New York, 1961). A definitive and advanced treatment of quantum mechanics. Recommended for people going on for advanced work in theory.
25. L. D. Landau and E. M. Lifshitz, *Quantum Mechanics, Non Relativistic Theory*, 3rd ed. (Pergamon Press, Elmsford, N.Y., 1977). An excellent but difficult book for chemists wishing to do advanced work in chemical physics.
26. R. McWeeney and B. T. Sutcliffe, *Methods of Molecular Quantum Mechanics* (Academic Press, New York, 1969).
27. J. Avery, *Quantum Theory of Atoms, Molecules and Photons* (McGraw-Hill, New York, 1972).

ATOMIC AND MOLECULAR SPECTROSCOPY

28. G. Barrow, *Introduction to Molecular Spectroscopy* (McGraw-Hill Book Co., New York, 1962). Good beginning treatment of rotation and vibration spectra.
29. L. J. Bellamy, *The Infrared Spectra of Complex Molecules*. 3rd ed. (J. Wiley & Sons, Inc., New York, 1975).
30. G. Herzberg, *Atomic Spectra and Atomic Structure*, 2nd ed. (Dover Publications, New York, 1944).
31. G. Herzberg, *Spectra of Diatomic Molecules*, 2nd ed. (D. Van Nostrand, Princeton, N.J., 1950). The definitive text in the field. It contains tables of spectroscopic data in the back.
32. G. Herzberg, *Infrared and Raman Spectra of Polyatomic Molecules* (D. Van Nostrand, Princeton, N.J., 1945).
33. G. Herzberg, *Electronic Spectra and Electronic Structure of Polyatomic Molecules* (D. Van Nostrand, Princeton, N.J., 1966).
34. H. H. Jaffe and M. Orchin, *Theory and Applications of Ultraviolet Spectroscopy* (J. Wiley & Sons, Inc., New York, 1962). Good discussion of electronic spectra of organic molecules. Many experimental data are included and rationalized.

35. J. N. Murrell, *The Theory of the Electronic Spectra of Organic Molecules* (J. Wiley & Sons, Inc., New York, 1963). An excellent discussion of the theory of electronic spectra in conjugated organic molecules. Out of print in 1979.

QUANTUM THEORY APPLIED TO ORGANIC CHEMISTRY

36. R. Daudel, R. Lefebvre, and C. Moser, *Quantum Chemistry: Methods and Applications* (Wiley-Interscience, New York, 1960).
37. A. Streitwieser, Jr., *Molecular Orbital Theory for Organic Chemists* (J. Wiley & Sons., Inc., New York, 1961).
38. A. Streitwieser, Jr., and J. J. Brauman, *Supplemental Tables of Molecular Orbital Calculations*, 2 vols. (Pergamon Press, Elmsford, N.Y., 1965).
39. M. J. S. Dewar, *The Molecular Orbital Theory of Organic Chemistry* (McGraw-Hill Book Co., New York, 1969). A more sophisticated treatment of organic quantum chemistry than references 36 and 37.
40. R. Daudel and C. Sandorfy, *Semiempirical Wave Mechanical Calculations on Polyatomic Molecules: A Current Review* (Yale University Press, New Haven, Conn., 1971).
41. A. Streitwieser, Jr., and P. H. Owens, *Orbital and Electron Density Diagrams: An Application of Computer Graphics* (Macmillan Publishing Co., Inc., New York, 1973).
42. H. E. Zimmerman, *Quantum Mechanics for Organic Chemists* (Academic Press, New York, 1975).

MAGNETIC RESONANCE

43. J. A. Pople, W. G. Schneider, and J. J. Bernstein, *High Resolution Nuclear Magnetic Resonance* (McGraw-Hill Book Co., New York, 1959). An old book, but still a good introductory monograph for chemists. It gives a complete coverage of both theory and experiment up to 1959.
44. J. D. Roberts, *An Introduction to Spin-Spin Splittings in NMR Spectroscopy* (W. A. Benjamin, Inc., New York, 1961). This book, though currently out of print, gives detailed instructions for calculating spin-spin splitting patterns for cases when the chemical shifts and spin-coupling constants are of the same order of magnitude.
45. J. W. Emsley, J. Feeney, and L. H. Sutcliffe, *High Resolution Nuclear Magnetic Resonance Spectroscopy*, 2 vols. (Pergamon Press, Elmsford, N.Y., 1966). A comprehensive discussion of all phases of solution NMR as of 1965.
46. A. Carrington and A. D. McLachlan, *Introduction to Magnetic Resonance* (Harper and Row, New York, 1967). An excellent beginning text that covers both EPR and NMR. Reprinted in 1979 by Halsted Press.

47. L. M. Jackman and S. Sternhell, *Applications of NMR Spectroscopy in Organic Chemistry*, 2nd ed. (Pergamon Press, Elmsford, N.Y., 1969).
48. T. C. Farrar and E. D. Becker, *Pulse and Fourier Transform NMR* (Academic Press, New York, 1971). A good introduction to the theory and practice of Fourier transform NMR. The book also contains a good treatment of relaxation phenomena and pulse techniques,
49. L. M. Jackman and F. A. Cotton, Eds., *Dynamic Nuclear Magnetic Resonance Spectroscopy* (Academic Press, New York, 1975).
50. J. N. Herak and K. J. Adamic, Eds., *Magnetic Resonance in Chemistry and Biology* (Marcel Dekker, Inc., New York, 1975).

MISCELLANEOUS

51. F. A. Cotton, *Chemical Applications of Group Theory*, 2nd ed. (J. Wiley & Sons, Inc., New York, 1971). A good introductory text on group theory. It emphasizes chemical applications.
52. H. H. Jaffe and M. Orchin, *Symmetry in Chemistry* (Krieger Publishing Co., New York, 1976).
53. L. Pauling and E. B. Wilson, Jr., *Introduction to Quantum Mechanics* (McGraw-Hill Book Co., New York, 1935). An old book which contains things that you can't find anyplace else, such as an introduction to valence bond calculations.
54. J. C. Slater, *Quantum Theory of Molecules and Solids*, Vol. 1. *Electronic Structure of Molecules* (McGraw-Hill Book Co., New York, 1963). Contains a good discussion of early refined calculations on small molecules.
55. R. M. Hochstrasser, *Molecular Aspects of Symmetry* (W. A. Benjamin, Inc., New York, 1966).

Index

CONVERSION FACTORS FOR USE IN CHANGING ENERGY UNITS[a]

To convert from energy in	*To an energy in* erg molecule^{-1}	eV	cm^{-1}	kcal mole^{-1}	MHz
	multiply by				
erg molecule^{-1}		6.242×10^{11}	5.035×10^{15}	1.440×10^{13}	1.509×10^{20}
eV	1.602×10^{-12}		8,067	23.05	2.418×10^{8}
cm^{-1}	1.986×10^{-16}	1.240×10^{-4}		2.859×10^{-3}	2.998×10^{4}
kcal mole^{-1}	6.944×10^{-14}	4.337×10^{-2}	349.9		1.048×10^{7}
MHz	6.6256×10^{-21}	4.136×10^{-9}	3.336×10^{-5}	9.541×10^{-8}	

[a] To obtain the wavelength equivalent of an energy difference, express the energy in wave numbers and take the reciprocal.

SELECTED PHYSICAL CONSTANTS[a]

Constant	*Symbol*	*Value*	SI *units*	cgs *units*
Speed of light in vacuum	c	2.997925	$\times 10^{8}$ m·s^{-1}	$\times 10^{10}$ cm s^{-1}
Elementary charge	e	1.602189	$\times 10^{-19}$ C	$\times 10^{-20}$ emu
		4.803242		$\times 10^{-10}$ esu
Electron rest mass	m_e	9.109534	$\times 10^{-31}$ kg	$\times 10^{-28}$ g
Proton rest mass	m_p	1.6726485	$\times 10^{-27}$ kg	$\times 10^{-24}$ g
Avogadro's number	N_A	6.022045	$\times 10^{-23}$ mol^{-1}	$\times 10^{-23}$ mol^{-1}
Atomic mass unit	amu (μ)	1.660565	$\times 10^{-27}$ kg	$\times 10^{-24}$ g
Planck constant	h	6.626176	$\times 10^{-34}$ J s	$\times 10^{-27}$ erg s
	$\hbar = h/2\pi$	1.0545887	$\times 10^{-34}$ J s	$\times 10^{-27}$ erg s
Boltzmann constant	k	1.380662	$\times 10^{-23}$ J K^{-1}	$\times 10^{-16}$ erg K^{-1}
Rydberg constant	R_∞	1.097373177	$\times 10^{7}$ m^{-1}	$\times 10^{5}$ cm^{-1}
Bohr radius	a_0	5.2917706	$\times 10^{-11}$ m	$\times 10^{-9}$ cm
Electron g-factor	$g_e/2$	1.00115966		
Bohr magneton	$\beta(\mu_\beta)$	9.274078	$\times 10^{-24}$ J T^{-1}	$\times 10^{-21}$ erg G^{-1}
Nuclear magneton	β_N (μ_N)	5.050824	$\times 10^{-27}$ J T^{-1}	$\times 10^{-24}$ erg G^{-1}
Gyromagnetic	γ_p	2.6751301	$\times 10^{8}$ s^{-1} T^{-1}	$\times 10^{4}$ s^{-1} G^{-1}
ratio of protons in H_2O	$\gamma\!\!\!- = \gamma/2\pi$	4.257711	$\times 10^{1}$ s^{-1} T^{-1}	$\times 10^{3}$ Hz G^{-1}

[a] Data from R. Cohen and B. N. Taylor, *J Phys. Chem. Ref. Data*, **2** (4), 663 (1973). Used by permission of the American Institute of Physics.